Nonlinearity, Chaos, and Complexity

Nonlinearity, Chaos, and Complexity

The Dynamics of Natural and Social Systems

Cristoforo Sergio Bertuglia
Franco Vaio

Polytechnic of Turin, Italy

OXFORD
UNIVERSITY PRESS

Great Clarendon Street, Oxford OX2 6DP

Oxford University Press is a department of the University of Oxford.
It furthers the University's objective of excellence in research, scholarship,
and education by publishing worldwide in

Oxford New York

Auckland Cape Town Dar es Salaam Hong Kong Karachi
Kuala Lumpur Madrid Melbourne Mexico City Nairobi
New Delhi Shanghai Taipei Toronto

With offices in
Argentina Austria Brazil Chile Czech Republic France Greece
Guatemala Hungary Italy Japan Poland Portugal Singapore
South Korea Switzerland Thailand Turkey Ukraine Vietnam

Oxford is a registered trade mark of Oxford University Press
in the UK and in certain other countries

Published in the United States
by Oxford University Press Inc., New York

© Cristoforo Sergio Bertuglia and Franco Vaio 2005

The moral rights of the authors have been asserted
Database right Oxford University Press (maker)

First published in English 2005

First published in Italian by Bollati Boringhieri Editore, Torino, 2003

All rights reserved. No part of this publication may be reproduced,
stored in a retrieval system, or transmitted, in any form or by any means,
without the prior permission in writing of Oxford University Press,
or as expressly permitted by law, or under terms agreed with the appropriate
reprographics rights organization. Enquiries concerning reproduction
outside the scope of the above should be sent to the Rights Department,
Oxford University Press, at the address above

You must not circulate this book in any other binding or cover
and you must impose this same condition on any acquirer

British Library Cataloguing in Publication Data
Data available

Library of Congress Cataloguing in Publication Data
Data available

Typeset by
SPI Publisher Services, Pondicherry, India
Printed in Great Britain
on acid-free paper by Biddles Ltd., King's Lynn

ISBN 0 19 856790 1 (Hbk) 9780198567905
ISBN 0 19 856791 X (Pbk) 9780198567912

10 9 8 7 6 5 4 3 2 1

To Federico, Maria, Alessandro and Maddalena
To Paola, Francesca, Fulvio and Ludovico

Preface

Since the birth of modern scientific thought, there have been numerous attempts to build up mathematical models able to describe various phenomena in the many different sectors of science. However, in a great number of cases that attempt has turned out to be unsuccessful. Nevertheless, the path followed by scientific research has often passed through the construction of mathematical models. In the past, mathematical models were intended as formulae (i.e. the mathematical laws of the so-called 'hard' sciences, such as physics) containing in themselves 'the whole reality' of the phenomenon described. They were considered instruments that could enable us to forecast the future. On the contrary, in contemporary science, the role of mathematical models is far less ambitious. In general they are intended as research instruments which can be used for describing only a subset of all the characteristic aspects of a phenomenon and as an instrument to organize data and to stimulate the development of and in-depth study of ideas.

In recent years, contemporary science has been making determined efforts to tackle themes such as chaos and complexity, which were overlooked by classical science. To deal with these, mathematical modelling has become an increasingly relevant instrument. Chaos, on one hand, means that a model is unable to see far enough into the future, despite being based on deterministic laws that are well understood in their details, at least theoretically, and are able to give correct short-term forecasts. Complexity, on the other hand, does not mean a confused forecasting: it simply means that it is impossible to build up a model which can account for the sudden and (most of all) unexpected 'changes' that sometimes take place during the evolution of a system, even though the evolutionary path between one change and the next can be well described by deterministic laws. The point is that these deterministic laws cannot account for (or 'explain', as is currently said) the fact that sudden unexpected changes of the system state appear in the evolution of the system at certain points and under certain conditions.

These laws turn out to be insufficient when confronted with a reality which is too nuanced and too complicated to be summarized within a limited set of rules. This is true not only from a practical standpoint. In fact, the rationale for this impossibility is intrinsic and rooted in theoretical grounds, because the laws are known, but they do not allow us to 'see' very far into the future.

Discussion of modelling is the central subject of this book. Particular emphasis is given to the aspects of modelling which concern the relationships between the sciences of nature and the sciences of society (i.e. analogies, transfer of concepts, techniques etc.). Modelling is seen as evolving from the first linear forms, typical of the natural sciences, to the richer forms giving birth to richer and richer dynamics, up to the limit where the techniques of mathematical modelling are insufficient or modelling itself is even impossible.

Part 1 deals with the mathematical model as an instrument of investigation. The general meaning of modelling and, more specifically, questions concerning linear modelling are discussed. We outline here the linear origin of modelling, whose meaning, scope and bounds will be discussed in detail. We present some of the most typical examples of linear models in the sciences of nature, namely models of linear oscillations. The parallel between the sciences of nature and the sciences of society is taken up again here by describing a few models (some linear and some nonlinear) built to describe phenomena of society which are similar in form to those used to describe certain natural phenomena. Among these are the linear model of two interacting populations and the Volterra–Lotka model, which is described with some of its variants and applications to biology and to regional science.

Part 2 deals with the theme of chaos. The question of stability of systems is discussed as well as the forms by which chaotic dynamics manifests itself in systems. After a brief mention of the question of detecting chaos in real systems, we illustrate in detail one of the most elementary growth models: the logistic law, both in its continuous and discrete versions. The discrete logistic model, known as the logistic map, is discussed at length, being one of the simplest models at the origin of chaotic dynamics.

Finally, Part 3 deals with the theme of complexity: a property of the systems and of their models which is intermediate between stability and chaos. We discuss here some cases where the modern approach of complexity is proposed as a new and effective means for interpreting phenomena. As the mathematical modelling of those phenomena that we ascribe to the category called 'complexity' is for the moment far from adequate, our discussion goes further, and, in the last chapters, touches on a new point concerning the central role of mathematics in building models, with particular attention to the limitations of mathematics as an instrument, to its meaning and to its foundations. After a short discussion on the meaning of mathematics and on its foundations, the last chapters of Part 3 are dedicated to some reflections concerning future research on the very foundation of mathematics: possibly the way towards the construction of effective models of complex phenomena will pass through the study of new areas of mathematics itself, which are likely to give new general conceptions and new tools and modelling techniques.

This book was first published in Italian in 2003 with the title '*Non linearità, caos, complessità: le dinamiche dei sistemi naturali e sociali*', by Bollati Boringhieri, Turin. The English edition includes some small changes and additions with respect to the Italian edition, which were considered constructive following the numerous reviews published and discussions with Italian and non-Italian academics, the names of which are too numerous to mention here; many of these, however, are scholars with whom the authors have successfully worked over the years, and who are thus often cited in the book.

The authors would particularly like to express their gratitude for the contribution of Professor Roberto Monaco, who has followed the development of this book from the outset, with valuable suggestions.

Finally, a very special thank you goes to Michelle Johnson for her invaluable contribution to preparing the English text, which she undertook with both great professionalism and good grace.

Cristoforo Sergio Bertuglia
Franco Vaio

Contents

| PART 1 | Linear and Nonlinear Processes | 1 |

1	Introduction	3
	What we mean by 'system'	3
	Physicalism: the first attempt to describe social systems using the methods of natural systems	5
2	Modelling	10
	A brief introduction to modelling	10
	Direct problems and inverse problems in modelling	12
	The meaning and the value of models	14
3	The origins of system dynamics: mechanics	19
	The classical interpretation of mechanics	19
	The many-body problem and the limitations of classical mechanics	22
4	Linearity in models	27
5	One of the most basic natural systems: the pendulum	32
	The linear model (Model 1)	32
	The linear model of a pendulum in the presence of friction (Model 2)	35
	Autonomous systems	37
6	Linearity as a first, often insufficient approximation	39
	The linearization of problems	39
	The limitations of linear models	42
7	The nonlinearity of natural processes: the case of the pendulum	46
	The nonlinear pendulum (Model 3 without friction, and Model 3' with friction)	46
	Non-integrability, in general, of nonlinear equations	47
8	Dynamical systems and the phase space	49
	What we mean by dynamical system	49
	The phase space	50
	Oscillatory dynamics represented in the phase space	54

Contents

9	Extension of the concepts and models used in physics to economics	60
	Jevons, Pareto and Fisher: from mathematical physics to mathematical economics	60
	Schumpeter and Samuelson: the economic cycle	62
	Dow and Elliott: periodicity in financial markets	64
10	The chaotic pendulum	67
	The need for models of nonlinear oscillations	67
	The case of a nonlinear forced pendulum with friction (Model 4)	68
11	Linear models in social processes: the case of two interacting populations	71
	Introduction	71
	The linear model of two interacting populations	72
	Some qualitative aspects of linear model dynamics	73
	The solutions of the linear model	76
	Complex conjugate roots of the characteristic equation: the values of the two populations fluctuate	84
12	Nonlinear models in social processes: the model of Volterra–Lotka and some of its variants in ecology	93
	Introduction	93
	The basic model	94
	A non-punctiform attractor: the limit cycle	98
	Carrying capacity	101
	Functional response and numerical response of the predator	103
13	Nonlinear models in social processes: the Volterra–Lotka model applied to urban and regional science	108
	Introduction	108
	Model of joint population–income dynamics	108
	The population–income model applied to US cities and to Madrid	113
	The symmetrical competition model and the formation of niches	118

PART 2 From Nonlinearity to Chaos — 123

14	Introduction	125
15	Dynamical systems and chaos	127
	Some theoretical aspects	127
	Two examples: calculating linear and chaotic dynamics	131

	The deterministic vision and real chaotic systems	135
	The question of the stability of the solar system	137
16	**Strange and chaotic attractors**	**141**
	Some preliminary concepts	141
	Two examples: Lorenz and Rössler attractors	146
	A two-dimensional chaotic map: the baker's map	150
17	**Chaos in real systems and in mathematical models**	**154**
18	**Stability in dynamical systems**	**159**
	The concept of stability	159
	A basic case: the stability of a linear dynamical system	161
	Poincaré and Lyapunov stability criteria	163
	Application of Lyapunov's criterion to Malthus' exponential law of growth	168
	Quantifying a system's instability: the Lyapunov exponents	171
	Exponential growth of the perturbations and the predictability horizon of a model	176
19	**The problem of measuring chaos in real systems**	**179**
	Chaotic dynamics and stochastic dynamics	179
	A method to obtain the dimension of attractors	183
	An observation on determinism in economics	186
20	**Logistic growth as a population development model**	**190**
	Introduction: modelling the growth of a population	190
	Growth in the presence of limited resources: Verhulst equation	191
	The logistic function	194
21	**A nonlinear discrete model: the logistic map**	**199**
	Introduction	199
	The iteration method and the fixed points of a function	201
	The dynamics of the logistic map	204
22	**The logistic map: some results of numerical simulations and an application**	**214**
	The Feigenbaum tree	214
	An example of the application of the logistic map to spatial interaction models	224
23	**Chaos in systems: the main concepts**	**231**

PART 3	Complexity	237

24	Introduction	239
25	Inadequacy of reductionism	240
	Models as portrayals of reality	240
	Reductionism and linearity	241
	A reflection on the role of mathematics in models	243
	A reflection on mathematics as a tool for modelling	246
	The search for regularities in social science phenomena	249
26	Some aspects of the classical vision of science	253
	Determinism	253
	The principle of sufficient reason	257
	The classical vision in social sciences	259
	Characteristics of systems described by classical science	261
27	From determinism to complexity: self-organization, a new understanding of system dynamics	266
	Introduction	266
	The new conceptions of complexity	268
	Self-organization	271
28	What is complexity?	275
	Adaptive complex systems	275
	Basic aspects of complexity	277
	An observation on complexity in social systems	280
	Some attempts at defining a complex system	281
	The complexity of a system and the observer	285
	The complexity of a system and the relations between its parts	286
29	Complexity and evolution	291
	Introduction	291
	The three ways in which complexity grows according to Brian Arthur	291
	The Tierra evolutionistic model	295
	The appearance of life according to Kauffman	297
30	Complexity in economic processes	301
	Complex economic systems	301
	Synergetics	304
	Two examples of complex models in economics	307
	A model of the complex phenomenology of the financial markets	309

31	Some thoughts on the meaning of 'doing mathematics'	315
	The problem of formalizing complexity	315
	Mathematics as a useful tool to highlight and express recurrences	320
	A reflection on the efficacy of mathematics as a tool to describe the world	323
32	Digression into the main interpretations of the foundations of mathematics	329
	Introduction	329
	Platonism	330
	Formalism and 'les Bourbaki'	331
	Constructivism	336
	Experimental mathematics	340
	The paradigm of the cosmic computer in the vision of experimental mathematics	341
	A comparison between Platonism, formalism, and constructivism in mathematics	343
33	The need for a mathematics of (or for) complexity	348
	The problem of formulating mathematical laws for complexity	348
	The description of complexity linked to a better understanding of the concept of mathematical infinity: some reflections	351

References	356
Subject index	375
Name index	380

PART 1: Linear and Nonlinear Processes

Philosophy is written in this grand book – I mean the universe – which stands continually open to our gaze. But the book cannot be understood unless one first learns to comprehend the language and read the characters in which it is written. It is written in the language of mathematics, and its characters are triangles, circles and other geometrical figures without which it is humanly impossible to understand a single word of it; without these one is wandering about in a dark labyrinth.

Galileo Galilei, The Assayer (1623)

'My good Adso' my master said 'during our whole journey I have been teaching you to recognize the evidence through which the world speaks to us like a great book Alanus de Insulis said that

   Omnis mundi creatura
   Quasi liber et pictura
   Nobis est in speculum

And he was thinking of the endless array of symbols with which God, through His creatures,

   speaks to us of the eternal life.

But the universe is even more talkative than Alanus thought, and it speaks not only of the ultimate things (which it does always in an obscure fashion) but also of closer things, and then it speaks quite clearly.'

Umberto Eco, The Name of The Rose (1980)

Introduction

What we mean by 'system'

In our picture of reality, we almost always focus our attention on the description of the evolution over time of objects that we identify with respect to their surrounding environment and which we call systems, mostly in an attempt to predict their future behaviour.

Systems are objects with varying degrees of complexity, although they are always acknowledged as containing different elements that interact with one other. Various kinds of relations may also exist between the system and the external environment, but these are either of a different nature or of a lesser intensity with respect to those that are encountered inside what we identify as 'system' and that contribute to its definition.[1]

When we create models of reality, we identify, or at least we attempt to identify, the elements that constitute it: this makes the description of reality easier, as we perceive it to be. We often acknowledge such elements, such systems, in their turn, as being made up of subsystems, or as being components of larger systems, and so forth. The identification of systems is common to all sciences, whether natural or social: from the mechanical systems of classical physics, whose basis is, or at least appears to be, relatively simple, such as a planetary system, to those of quantum physics, whose basis is more elaborate, such as an atomic nucleus, an atom or a molecule, to systems of biological sciences, such as cells, apparatus, organisms, and again to those of social sciences, such as economic environments, populations, cities and many more. Thus attempts to transfer the concepts and methods of natural sciences to social sciences seek specific similarities between natural and social systems (Cohen, 1993).

---

[1] The same Greek etymology of the word 'staying together' is already in itself an explicatory definition: a system is something made up of elements that 'stay together', so much so that they can be clearly distinguished from all the rest.

The description of a system's dynamics, namely its evolution over time, is particularly interesting. From this point of view, the concept of a system's state is fundamental, to describe which we utilize the measurement of appropriate magnitudes. A system's state means the set of values, at a given point in time, of all the magnitudes that we consider relevant to that system and that are measurable in some way. In some cases, the magnitudes in question are measurable in a strict sense, as occurs in the so-called 'hard' sciences, those that best adapt to a mathematical language; in other cases, however, said magnitudes can only be deduced or observed on the basis of various types of considerations, such as what happens when not just quantitative magnitudes come into play. This is, more specifically, the situation that arises in social sciences, an area in which the application of mathematical language is less immediate than that of the majority of natural sciences.

The specific interest in the dynamics of a given system is mainly justified by the fact that when, on the basis of acquired experience, we succeed in identifying typical reoccurrences in its behaviour, we can then attempt to make predictions about the future behaviour of said system. Obviously we can only do this by assuming that what has occurred in the past will be repeated in the future, an assumption that is often not justified by experience and is sometimes completely arbitrary.

Nevertheless, in some cases this assumption is sound: these are the simplest cases, in which such recurrences are relatively obvious, and we often hear reference to 'laws', generally stated using mathematical formulae, in particular differential equations. Sometimes, in fact, the evolution of the simplest systems presents characteristics that render them easily understandable on the basis of their phenomenological aspects, and also, at least partly, predictable. This is the case, in particular, of numerous systems that are studied in physics.

In describing a system and its dynamics, we use a language made up of symbols and of relations between the latter that you could say we are familiar with (or at least almost always). The language used improves, the more accurate the description that it is able to provide of the observed phenomenology. Mathematics, to be precise, is a language that is often particularly suited to the formulation of laws or, more generally, to descriptions of observations of reality, in particular those relating to physics.[2] Over the course of the last three centuries, a specific formalism has been devel-

---

[2] The idea that the 'book of nature' was written in mathematical language is not a new concept, as it was already proposed by Galileo Galilei at the beginning of the seventeenth century as one of the distinguishing elements of the 'new science' against the vacuity of traditional knowledge and the authority of the Sacred Writings (see the quotes at the beginning of this Part).

oped in mathematics to describe dynamic processes, that of differential equations: equations whose unknown element is actually the function, defined in an appropriate space, that describes the evolutive trajectory of the system within that specific space.

The interest in the dynamics of processes observed in real life has been helped considerably by mathematical techniques developed from the end of the seventeenth century, such as the birth of differential calculus, initially, and subsequently, the notable development of analytical mechanics and mathematical physics, areas which made extensive use differential equations and that from early on, gave a great boost to their study. Later, the mathematical concepts of dynamic analysis were transferred to disciplines other than physics.

In economics, in particular, the first attempt to use differential equations dates back to Léon Walras (1874) and, later, to Paul Anthony Samuelson (1947), who was the first person to introduce the concept of dynamic analysis to this sector/area alongside the more standard one of static analysis (Zhang, 1991).

Physicalism: the first attempt to describe social systems using the methods of natural systems

The application of concepts and methods that originated in the field of physics, particularly in mathematical physics, to disciplines other than physics, is an approach that is generally known as physicalism. It is a well known fact (Weidlich, 1991; Crépel, 1999) that the first attempts to describe social systems in terms of social physics, on the wave of the successes achieved in physics, date back to the seventeenth century and took hold in the nineteenth century. They involved the more or less direct transfer of the ideas, the models and the equations that described the behaviour of physical systems, at least of those that were known and that had been studied up until then, to the field of social sciences. These represent attempts to describe the behaviour of a society by means of something similar to the equation of the state of perfect gases, by means of concepts similar to those of pressure, temperature, density, etc.

The first considerations on the use of mathematics in social sciences date back to Jean-Antoine de Caritat de Condorcet, a firm supporter of the idea that social sciences could be treated mathematically. In his *Essai sur l'application de l'analyse* (1785), Condorcet considered problems of demographics, political economics, insurance, law and more besides; more specifically, he also dealt with the theory of electoral voting, and in this regard, the difficulties of logic that arise when passing from a set of individual opinions to the collective opinion of an assembly.[3] Condorcet proposed a legitimate new programme of application of the science that was called, at that time, political arithmetic and that thus became social mathematics. This occurred after several decades of discussions on the possibility of combining mathematics with social, political and moral sciences, supported by the likes of Daniel Bernoulli, an idea that was viewed with extreme scepticism by scholars such as Jean Le Rond d'Alambert, who disputed the use of calculus techniques, particularly the calculus of probability, and criticized the too superficial and direct application of algebra.

Pierre Crépel (1999) observed how the desire to introduce measurements and to apply calculus to the sphere of social sciences was evident in the numerous encyclopaedias of that century, as well as in newspapers and letters among others. However, when it came to actually applying the mathematics that was enjoying such enormous success in physics, to constructing a coherent theory of social mathematics and transforming these general ambitions into concrete facts, such attempts, made mostly by a few French and Italian scholars, achieved only a limited success, and none of them were comparable to those conducted by Condorcet several decades earlier. According to Crépel, this illustrates one of the characteristic traits of the Age of Enlightenment: great scientific projects were being conceived, but neither the means of calculation of the era, nor the availability of data, enabled them to be put into effect. The ideas were often brilliant, but their realization was almost always inadequate.

The situation changed in the early nineteenth century. Scholars engaged in political activities, such as Pierre-Simon de Laplace, Jean Baptiste Fourier and Gaspard Monge, applied mathematical methods that they had personally developed and that were the result of their scientific beliefs, to social sciences. Even though romanticism erected a barrier between that which regarded man and the sphere of application of mathematics in the first half of the century, mathematics, particularly the calculus of probability, was often used in the study of social problems. The scholars of that period were

---

[3] Condorcet's paradox is well known as regards the situation that can arise with majority voting when there are more than two candidates to choose from, i.e. when each voter expresses his vote in the form of a ranking of preferences between the alternatives proposed. For example, imagine a situation in which there are three voters, A, B and C, each of which must rank three candidates x, y and z: A might vote x, y, z, while B might vote y, z, x, and C might vote z, x, y. In this case, it would be impossible to draw up a final ranking that takes the majority into account, because: x beats y 2 to 1, y beats z 2 to 1 and z beats x 2 to 1 (Blair and Pollack, 1999).

struck by the regularity of statistical data with regard to the social sphere and used very strict parallels with the natural world.

The Belgian mathematician and astronomer, Adolphe-Lambert Quetelet (1835), was a figure of some importance at that time. He developed and applied the abstract concept of the 'average man', namely a man for whom each characteristic, taken individually, was calculated as the average of identical characteristics of real men (Devlin, 1998). The concept of the average man was staunchly disputed by famous scholars, even though these same scholars belonged to the school of thought that wanted to apply the methods and concepts of physics to social sciences, such as the economist Antoine Augustin Cournot, the first to apply infinitesimal analysis to economics (Cohen, 1993) and the sociologist Émile Durkheim, the author of a so-called 'law of gravitation of the moral world', proposed in parallel to Newton's law of universal gravitation. The objection raised was that the study of the average man was senseless, because the calculation of averages of said measurements did not have any concrete meaning, insofar as the resulting average man did not correspond to any real case and was simply outrageous.[4] Quetelet was to be the teacher of François Verhulst, who, adopting a physicalist approach, proposed a famous model of population development, the logistic function, that he obtained by starting from a strict analogy with the equation of the motion of a body in the presence of friction (see chapter 20).

A physicalist approach like those cited, in reality, works only if there is a strict structural isomorphism in the interactions between the elements that make up one type of system (material points, elementary particles, atoms, molecules, planets, galaxies, etc.) and those that make up the other type of system (individuals, families, businesses, social groups, populations, etc.). Such isomorphism, in general, does not appear to exist, therefore the analogy can only be phenomenological and totally superficial: in most cases it almost appears to be strained interpretation dictated by the desire to find associations between totally different situations at all costs.

Apart from theoretical issues regarding the different foundations of the disciplines, the development of a description of social systems along the lines of natural ones encounters various kinds of insurmountable obstacles. Firstly, we come up against the difficulty of identifying all of the significant variables within the social systems followed by the difficulty of actually

---

[4] To clarify the objection raised, we can observe that in a set of right-angled triangles that are not all similar, the arithmetic average of the measurements of the longest sides, the shortest sides and the hypotenuses do not satisfy the Pythagorean theorem. In other words, the 'average' triangle, i.e. the triangle whose sides are represented by the averages calculated, is not actually right-angled.

measuring the majority of these variables; lastly, and more importantly, social systems are not something that can be experimented with in a laboratory. This does not mean, however, that the search for similarities between the social and natural sciences has always been useless or even detrimental. In this regard, the physicist Ettore Majorana, in a pioneering moderately popular article published in 1924, edited by Giovanni Gentile Jr., son of the philosopher of the same name and friend of Majorana, four years after the mysterious disappearance of the author, acknowledged the fundamental analogy between the statistical laws applied in physics and those of social sciences. The transfer of a proposition formulated in one disciplinary field to another disciplinary field, in which it was able to make progress, is by no means a new concept. However, in the new disciplinary field it then had to find the appropriate justification, firmly rooted in the new disciplinary field (Bertuglia and Rabino, 1990). In recent years, for example, this has occurred with the theory of spatial interaction: William Reilly (1931) transferred Newton's law of universal gravitation to the field of location–transport interrelations to acquire details of consumer goods ('the law of retail gravitation'); later, Alan Wilson (1967, 1970, 1974) found justification for it within the specific disciplinary field, thus illustrating the fertility of Reilly's[5] intuition (we will return to the theory of spatial interaction in Chapter 22).

The mathematical models used in a discipline, or at least some of their elements, as well as the resulting concepts generated by reflections that the

---

[5] Reilly's model assumes: (1) that a consumer is 'attracted' to a particular market centre by a 'gravitational force' that is inversely proportional to the square of the distance that he has to travel to reach said centre; (2) that the consumer is willing to travel distances that increase with the value of the goods he is looking for and with the variety of the offer of a market (the alternatives of choice), i.e. with the size of the specific market, but only up to a certain limit (no consumer, for example, will travel one thousand kilometres just to be able to choose between 20 different types of milk). The balance that is established between the constraint due to the distance and the appeal exercised by the size of the market centre, according to Reilly, permits the establishment of market areas and their borders to be defined. Even in more recent years than those in which the gravitational theory of spatial interaction was developed, there have been frequent attempts to propose or 'construct' varying degrees of similarity between various natural sciences, physics in particular, and various social sciences: see for example Dendrinos (1991b), and Isard, (1999). Nevertheless it should be noted that there have also been frequent cases in which the attempt to transfer and apply certain concepts of physics to social sciences has been made in an improper, erroneous or arbitrary way, or is even intentionally misleading, simply to give a scientific air to the theories proposed, using erroneous or irrelevant reasoning; this is often due to a substantial incomprehension of the physical–mathematical concepts of the person applying the transfers and/or analogies. In these cases, obviously, we are not dealing with physicalism, but as the title of a book by Alan Sokal and Jean Bricmont (1997) says, bona fide 'impostures intellectuelles'.

same models stimulate, may be taken as a cue for developments in disciplines other than the original one. On the other hand, the mathematical model itself is an abstraction constructed from real situations and, as such, it should not come as a surprise that abstractions conducted in different fields possess areas of overlap. We are still dealing with abstractions, not of direct transfers from one disciplinary field to another.

Mechanics is a typical, perhaps the most typical, certainly the oldest, example of science that deals with the description of the evolution of a system over time, achieving, at least in several cases, extremely important results. It represents one of science's greatest achievements in the last few centuries and a cornerstone of the civilization that has developed in the Western world in the last three centuries. The impressive development of mechanics since the eighteenth century has undeniably had fundamental effects on many other cultural fields: for example it has had an enormous influence on philosophy and the general vision of the world. It is therefore natural to refer to mechanics and to its schemes of reasoning, if not actually to its methods of modelling and calculus, in particular to those developed in the classic era.

In Chapter 2 we will look at the fundamental concept of a model: the idea that underlies the entire discussion of this book. In Chapter 3 we will then enter into the merit of what we mean in mathematics by linearity, the concept at the origins of the linear models that have enjoyed immense success, particularly in eighteenth and nineteenth century science. Subsequently, (Chapters 4–10), we will briefly illustrate several simple methods and models that originate in classic mechanics, highlighting in particular, in Chapter 6, some of the intrinsic limitations of linear modelling. Finally, in Chapters 11–13, we will deal with several important models developed in the sphere of social sciences that describe the evolution of interacting populations. More specifically, in Chapter 11 we will take an in-depth look at the properties of a simple linear model, while, in Chapters 12 and 13 we will illustrate the Volterra–Lotka nonlinear model, today recognized as a classic, which attempts to describe the dynamics of two populations that interact in a closed environment, some of its developments and its numerous applications.

2 Modelling

A brief introduction to modelling

During our entire lives we learn to construct mental models of the world that surrounds us. Catching a falling object by anticipating where it is going to fall before it gets there, understanding the thoughts of an interlocutor before he speaks on the basis of a few signs, deciphering the intentions of an opponent before he acts; these are just a few examples of how we read a situation and its various aspects, something that we all do constantly, constructing models based on our store of previous experiences as well as our attempts or mistakes.

As we said in Chapter 1, in science, the study of systems that evolve over time, i.e. of systems whose state changes over time, is particularly interesting. The swinging of a pendulum, the radioactive decay of a substance, the movement of the planets in the solar system, the growth of a plant, the trend in the number of individuals of a particular population in a defined territory, the flow of road traffic along a street or that of telephone communications in a channel and the trend of a stock exchange index, are just some examples of what we mean by a system that evolves over time. In each of the above examples, as in many others, we have a set of states at various points in time and we seek a law of motion that reproduces the succession of these states over time. A large part of scientific work, from the very origins of science onwards, has concentrated on establishing the 'right' law of motion to describe the evolution of a system over time, based on observations of a limited set of successive states (Casti, 2000). When these activities lead to the formulation of recurrences (laws) in a mathematical language, then we have a mathematical model.

Obviously, the more complex the system we want to describe, the more difficulty we have in grasping its working mechanisms and its dynamic aspects. The experiments and measurements conducted on both natural and social systems provide results that have to be organized and analysed in order to be used constructively and to permit any subjacent mechanisms to be identified. This is why in natural and social sciences, mathematical models are often used as tools to describe the evolution of their systems. A mathematical model thus becomes a useful instrument to give the systematic observation of data a qualitative leap: from the collection and organization of the data observed, to the analysis, interpretation and prediction of the system's future behaviour, using a mathematical language (Bellomo and Preziosi, 1995).

In the past, as we said in Chapter 1.1, mathematical modelling was essentially linked to physics, a sphere in which the first models were developed, and this discipline provided all the basic models of mechanics in the eighteenth and nineteenth centuries. Today, mathematical modelling is a discipline that plays a fundamental role in many sciences, both natural and social.

The events that we observe in reality and that constitute the subject of dynamic modelling are generally complicated, made up of numerous elements and are faceted into numerous different aspects, in which an event is linked to other events. It is extremely difficult, therefore, if not impossible, to gain an in-depth understanding of them.

Just as we use tools to extend the capacity of our senses, in the same way we can use dynamic simulation to extend the ability of our minds to conceive working models of reality (Hannon and Ruth, 1997ab). In fact, one of the main differences between so-called instinctive modelling and so-called scientific modelling is actually the extensive, if not exclusive, use of mathematical tools in the latter. Naturally, the use of electronic computers is fundamental to these activities. The latter represent tools that enable us to vastly extend our capabilities, although we should always be aware of their potential and of their limitations.

Generally, models consider certain magnitudes that are dependent on time, called state variables, and describe their evolution over time using either systems of differential equations, in which time varies with continuity, or systems of finite difference equations, in which time varies at discrete intervals. In this regard, we refer to dynamic models, that are constructed on the basis of certain hypotheses and that, in general, are integrated numerically. With regard to models in continuous time, it should be noted that they can almost always only be integrated numerically, implicitly transforming them into models in discrete time, partly distorting their conception in continuous time and, sometimes, even considerably altering their dynamics. We will come back to this point in Chapters 20–22 where we will describe the model representing the growth of a population called 'logistic': we will see how the dynamics in the version of the model in continuous time (the 'logistic function') is considerably different from that envisaged by the version of the same model in discrete time (the 'logistic map').

Dynamic modelling is therefore a process that enables us to extend our knowledge of reality, and the computer is its main tool because it allows the user to trace the evolution of the dynamic system represented by the model, integrating the equations numerically. The main aims of dynamic modelling are, therefore, to describe the 'flow' of a situation and to predict its future evolution.

When constructing a model, we necessarily have to overlook several aspects that are considered secondary and focus on a single part of a larger picture, on a specific set of characteristics of the system that we want to describe. The models are all necessarily abstractions of the situation. Therefore, their importance is not based on how accurately they reproduce a situation that is too complicated to be described in depth, but rather on providing us with results for the assumptions that we make and, in the final analysis, also in enabling us to deal with the implications of our own opinions (Israel, 1986).

Direct problems and inverse problems in modelling

In the construction and use of models, two opposing perspectives can be identified. On one hand, there is the so-called direct problem, namely the problem, based on a given model, of reproducing the properties observed empirically using calculus. On the other hand, we have the inverse problem, namely the actual construction of the model based on observations, or rather the search for an abstract representation of a set of causes and relationships, the consequences of which lead to a reproduction of the observations of the phenomenology. Often, where the direct problem is relatively simple, the inverse problem is generally more complicated: the latter, in fact, is essentially the search and the formalization of causes and mechanisms that have led to a known result. From this point of view, we could say that all problems related to the interpretation of observed data are inverse problems. The main difficulty that arises with inverse problems, therefore, is actually evaluating the compatibility or otherwise of the model with the experimental data: we are dealing with a fundamental issue that warrants careful examination and that can lead to the modification of the model itself (Tikhonov and Goncharsky, 1987).

Whilst on the subject, it could be interesting to interpret the concept of modelling in accordance with that proposed by the mathematician Jacques Hadamard (1932). Let us assume that u is a vector defined by the set of the values of the magnitudes observed in a space U and that the model is made up of a vector z in a certain space Z. Let us also assume that space U and space Z are both so-called metric spaces, i.e. that it is possible to define and calculate distances within them; in our case this means that we assume that we can say 'how far' a set of data (a vector) is from another set of data (another vector). We then define an operator A that relates z to u, i.e.

$$Az = u$$

Usually, the values that make up vector z, which is the fruit of our suppositions, ideas, attempts, etc., cannot be measured and, in addition, we are only in possession of the approximate values of the components of u, which is the data obtained from measurements. The direct problem is the passage from z to u through operator A, while the inverse problem regards identifying z from u, through the operator A^{-1}, the inverse of A, namely identifying the relationships that are at the source of the observations that make up u.

There are numerous cases in which we are faced with inverse problems; for example, in the sphere of applied natural sciences, the measurement of the anomalies of gravitational and magnetic fields on land surfaces (the u variables), with the aim of making suppositions (the z variables) as to what is under the surface (Glasko, Mudretsova and Strakhov, 1987), or the processing of diagnostic data in medicine through the use of mathematical techniques to formulate models and to identify pathologies (Arsenin, 1987). Extending the field somewhat, the inverse problem arises in social sciences every time that, for example, on the basis of demographic and/or economic data (empirical data u), we want to abstract a model, namely a theoretical scheme that takes this into account (whose parameters are the z variables). This is the case, one example of many, of the entire series of spatial interaction models, in which we assume the existence of functional relations between geographical areas (the u variables), which manifest themselves in the form of regularities in the flow structure (the z variables) (Campisi, 1991; Rabino, 1991).

Without going into the details, which are of a rather technical nature, we shall limit ourselves to pointing out that the main mathematical difficulty that arises, when we try to resolve the inverse problem, is that the inverse operator A^{-1} could present discontinuities and/or not be defined in the entire space U; in this case, the problem is known as an 'ill-posed problem'.[6]

---

[6] In this regard, one also talks about a well-posed problem if the problem is characterized by the existence, uniqueness and continual dependence of the solution z on u data. The concept of a well formulated problem is different: a problem is well formulated if the

More often than not inverse problems are ill posed: i.e. they do not have unique dependent solutions with data continuity. This means that different solutions have to be identified, each valid in a particular subset of the available data; in other words, there are different abstractions that can be made based on the analysis of the data; a mathematical model, in fact, is just an abstract approximation of a concrete system that necessarily implies a number of variables lower than that of the object that it wants to represent.

Again according to Hadamard (1932; see Goncharsky, 1987), to be well-posed, a problem must satisfy several conditions: (1) the solution has to exist, at least for the time interval considered; (2) the solution must be unique; (3) the solution must be stable with respect to small disturbances, in other words it has to depend continuously on the initial data, so that, if a small error is committed in the description of the current state, the limited effect of such can be estimated in its future evolution.

As we have already said, in modelling we make simplifications and approximations and, inevitably, errors are committed in deciding what to approximate and how to make the simplifications. For this reason, it is useful to have an estimation of the difference between the solutions of similar models and this is why we have mentioned metric spaces, in which distances can be defined in accordance with the appropriate criteria.[7]

The meaning and the value of models

Well-posed problems, as we have said, lead us to develop a model, which in turn poses new questions: in fact, we observe what we call 'real events', we extract a version to which we add our point of view, i.e. the model, and we draw conclusions that allow us to compare our explanation of the events with reality. If we are good at constructing models, we therefore get

equations are associated with a number of initial and/or boundary conditions that enables it to be solved (Bellomo and Preziosi, 1995). We will not go into the technical details, for which the reader should refer to the cited reference; we will simply observe that a well formulated problem doesn't necessarily imply the existence of a solution in a given time interval. Various problems of optimal control, linear algebra, the sum of Fourier series with approximate coefficients are, in reality, ill-posed problems.

[7] Starting in the 1960s, mathematical techniques have been developed that have proved useful in the search for approximate solutions to ill-posed problems, in particular the work of the Russian mathematician Andrei Nikolaevich Tikhonov, who was awarded the Lenin prize in 1966 for his work in this field. We will not be going into the details of this rather technical subject; we recommend interested readers to consult the vast amount of specific literature available, for example Tikhonov and Arsenin (1979), Tikhonov and Goncharsky (eds.) (1987), Bellomo and Preziosi (1995) and the reference cited herein.

increasingly closer to an effective representation of reality, which we can use as a tool to identify the consequences of the assumptions that form the basis of the model in the future. All models, in fact, are necessarily the result of partial visions of reality; we could say that they constitute projections of it made starting from different points of view; some of these projections are too simple, others too complicated. According to John Casti (1987), good models can be defined as the simplest ones that are able to 'explain' the majority of the data, but not all, leaving room for the model itself, and the theory on which it is based, to improve further.

One of the fundamental steps for the construction of a model is the identification of the state variables: the magnitudes that 'bring with them' our knowledge of the system, during the step-by-step evolution of the model over time. Spatial coordinates and speed are examples of state variables, as we will see in our discussion of the pendulum, as well as the number of individuals in a population, as we will see in models of interacting populations, and also temperature, pressure and volume in models of the dynamics of gas, macroeconomic variables in models of economic systems, etc. Control parameters then have to be defined, i.e. those magnitudes that act on the state variables, such as external forces in mechanical systems, rates of development in economic systems and populations flows in demographic systems, etc.

The different components of a model often interact through retroactive cycles, also called feedback cycles. Basically, they are processes according to which component *A* of the model generates changes in another component *B*, which, in turn, generates a further change in component *A*, which, in return, acts on *B*, and so on. Feedback cycles can be negative if a change in component *A* results in a change in component *B* that contrasts with that of *A* that generated it, and positive in the opposite case. In general, a negative cycle generates a situation of stability or stasis (depending on the situation and on how it is interpreted). This could be the case, for example, of an economic system that distributes little wealth to workers that produce little and badly because they are demotivated due to the little wealth received in exchange for their work, which, in turn, is due to the fact that the workers, as a consequence of their demotivation, produce little wealth.[8] Positive feedback, on the other hand, generally gives rise to situations of increasingly rapid growth, namely situations of instability. A typical example of positive feedback is the price–wage spiral that has established itself in certain

---

[8] An ironic example of negative feedback is provided by a renowned aphorism of several years ago according to which in a system described as being real socialism, workers 'pretend to work for a state that pretends to pay them'.

socio-economic and political circumstances and that acts like one of the mechanisms that generates inflation. An increase in the cost of labour (or of raw materials or of taxes) that affects production translates into an increase in prices, because businesses, to safeguard their profits, tend to offload cost increases on prices. This leads to a general increase in prices that reduces real wages. This drives workers to demand wage increases and so on, in a situation in which inflation ends up feeding itself.[9]

It is clear that the more dynamic, many-faceted, or even vague and conflictual the system we are attempting to describe is, the larger the distance between the model and reality. For example, while situations of stationarity can be modelled in a relatively simple way, other more complicated situations, in which internal relations abound that render the system unstable through positive feedback cycles, are difficult to model. As we will see in Chapter 5, in the case of oscillations of a small amplitude of a pendulum around a state of equilibrium, it is easy to obtain the equation of motion on the basis of linear approximations, obtaining sound results. In the same way one can apply linear approximation, obtaining results that are also plausible, to describe situations of relative stability in social systems, such as those of scarce demographic variation, low inflation, limited economic inequality, reduced competition for resources, low social conflict, etc.

Generally, however, in the case of social systems, we find ourselves faced with situations that are extremely difficult to describe, both because instability is practically the norm, and because very often the situations are different from each other, which makes the determination of regularities or laws, in natural science terms, very hard, if not impossible. In any event, using the appropriate simplificative hypotheses, models can be constructed for social systems similar to those used in natural science, in particular in those used in classical physics.

Just as in natural sciences, also and above all, in social sciences, the meaning attributed to models is not, therefore, a strictly predictive one, typical of the vision of classical sciences indicated with the term 'determinism' (see

---

[9] Naturally, we are over-simplifying; recognizing the cost–price spiral as the unique cause of inflation is a reductive and fundamentally wrong viewpoint. There are a variety of theories on the causes of inflation that are much more complex than our example; they can be grouped mainly into monetary theories, typical of neoclassical schools of thought, according to which inflation originates from the excessive quantity of money in circulation, and non-monetary theories, typical of Keynesian tradition, according to which we should not look to the monetary sphere for the causes of inflation, but in so-called real phenomena, present on the demand side (for example a surplus of demand in situations of full employment) and on the supply side (for example, the price–wage spiral, fiscal inflation, imported inflation, etc.).

Chapter 3). Models play other, multiple roles. On one hand, the observation of the evolution over time of a model enables us to experiment, by examining copies of real situations, approximated according to certain assumptions, in the laboratory; models are, in other words, tools used to formalize hypotheses and to examine their consequences (so-called scenario analysis). On the other hand, models enable the effects of fluctuations, or the effects of the approximations introduced to be highlighted. Basically a model is a tool that is useful to organize and stimulate ideas and knowledge on a given aspect of reality and to make it, so as to speak, 'come alive'.

The hypotheses at the basis of any model, as can be inferred from the above, are never a strict, complete and objective reflection of the real world, neither the natural, nor the social one. Inevitably, these hypotheses reflect our incomplete knowledge of reality (or better, the inaccessibility of full knowledge), the opinions (and the prejudices) of their authors and, lastly, the choices as regards the approximations that have to necessarily be taken. In addition, numerical integrations make the model evolve from its initial conditions made up of necessarily imprecise and, at times, incomplete data. This means that the evolutive trajectory of the model, almost always an unstable trajectory (we will discuss the concept of the stability of an orbit in Chapters 17–19), is, in reality, only an approximate indication of what would happen if the base hypotheses were true, if the approximations introduced didn't have any effect and if the data entered were complete and exact.

In Chapters 5–10 we will consider a series of examples on modelling in physics, the field of science that right from its origins has been best disposed to mathematization. We will be tackling several aspects of the problem of describing the oscillations of a system, a topic that has been studied at length in classical physics, whose importance is such that it surpasses the borders of the territory denoted as belonging to physics. In many areas of science, in fact, oscillations are observed (or sometimes postulated) in systems in which certain magnitudes demonstrate dynamics whose periodicity is more or less pronounced; we adopt the methods used in physics to describe them, in line with the discussions we touched upon in Chapter 1 regarding the transfer of the methods and the concepts of physics to other disciplinary fields. First of all, we will formulate some extremely simple hypotheses: those that lead to the so-called linear pendulum; then we will refine the initial hypotheses, introducing new aspects and new details to the problems in question and thus making their mathematical treatment more complicated. Lastly, we will examine descriptions of the oscillations of systems that are more complicated than a pendulum, such as biological and economic ones.

Before entering into the merits of the discussion of examples of modelling, we will examine some preliminary areas. First of all (Chapter 3), we will take a brief look at several historic aspects of the problem of modelling the dynamics of systems, taking a cue from classical mechanics, the first area in which models began be constructed; then (Chapter 4), we will examine several mathematical concepts regarding linearity, which we will repeatedly be using hereinafter.

3 The origins of system dynamics: mechanics

The classical interpretation of mechanics

Classical mechanics started to emerge at the end of the seventeenth century, drawing its origins from the study of the movements of the planets, within the grandiose and revolutionary framework created in Newton's *Philosophiae naturalis principia mathematica* (1687). The resulting vision of the natural processes is strictly deterministic: the knowledge of the present state of a system enables calculation of the system's state in any future (or past) time without any ambiguity. Classical mechanics enjoyed considerable success right from the start, and for a long period of time it was considered to be of extreme importance, so much so that scientists (or better experts in natural philosophy) gave it their wholehearted support, reinforcing the conviction that Newton's classical mechanics was able to provide complete descriptions of reality, and above all, its future evolution.

For a long time, classical mechanics was seen as an indisputable and extremely effective model of natural reality: it was attributed the role of a key to interpretation, a framework in which reality, or at least many of its various aspects, could be organized and interpreted on the basis of observed data. In the almost ontological vision developed from Newton's time onwards, which in Laplace found its most complete expression (we will come back to this point in chapter 26), classical mechanics possesses several fundamental characteristics that qualify it in a distinct way: (1) it is *reductionist*, i.e. it breaks down the systems that it analyses into simple elements and it sets itself the objective of describing the behaviour of the system solely on the basis of the behaviour of the elements it is comprised of; (2) the models that it provides are *reversible* in time and in space; (3) it is, as we have already said, *deterministic*, i.e. the calculation of dynamics uses equations (equations of motion) that recognize unique solutions that permanently depend on the initial state and on the values of the parameters that describe the systems' state (Bohm, 1957).

The enormous success achieved by the work of Newton[10] led to numerous repeated attempts to transfer the most important elements of Newton's theory to fields other than mechanics, in particular the law of universal gravitation which was founded on the idea of the existence of an attractive force between one body and another, whose intensity decreases by the inverse square of the distance between them. 'The inverse square law', whether taken literally, or considered more generically as a simple metaphor, when seen outside the rigid determinism that characterizes Newton's schema, became a full-fledged scientific paradigm (Cohen, 1993). This took place immediately following the publication of Newton's *Principia*, starting with Archbishop George Berkeley, who, although a harsh critic of the mathematical methods used by Newton (see p. 317, note 140), was the first to attempt to construct a social analogy of Newton's law (Boyer, 1968; Cohen, 1993). Later, various studies by well-known philosophers, economists and sociologists such as David Hume, John Craig, Léon Walras and Émile Durkheim were made, right up to the construction, in more recent years, of gravitational models of spatial interaction (Sen and Smith, 1995).[11]

Over the years, various elements have come to the attention of experts that have put the ambitious claim that seventeenth–eighteenth century mechanics is an effective tool for the complete description of the world, back into perspective. Numerous criticisms of this vision have emerged, above all in the last century and a half, with specific reference to the ontological claim implied in the cited characteristics, which present themselves as the bona fide principles at the basis of classical mechanics. These criticisms have often been expressed in terms of alternative theories that have led to mechanics being reassessed under perspectives other than the original one, less rigid than the latter, but also, necessarily, less optimistic.

Several stages in the historic evolution of the discipline can be identified, through which we pass from the trusting certainties typical of the mechanistic vision to the progressive acknowledgment of the intrinsic difficulties. The following is a very brief outline of the appearance of specific conceptions in the history of scientific thought:

---

[10] Gullen (1995) contains a popular description of the figure and the work of Newton, which we highlight due to the particular originality of the portrayal.

[11] To be precise, we should distinguish, as Cohen (1993) does, between applications of the Newtonian paradigm to social science, made through the formulation of analogies, i.e. functional similarities, and homologies, i.e. the identity of shapes or structures, or through the formulation of metaphors.

(1) *determinism*, namely classical mechanics in the strict sense of the word (Newton, Leibniz, Lagrange, Euler, Laplace: from 1687, the year in which Newton's *Principia* was published, to the first few decades of the nineteenth century);
(2) *statistical indetermination* (Clausius, Lord Kelvin and Boltzmann, in the second half of the nineteenth century);
(3) *quantum indetermination* (Bohr, Heisenberg, Schrödinger, Dirac and many others, starting from the first few decades of the twentieth century);
(4) *deterministic chaos* (firstly Poincaré, at the end of the nineteenth century, then it was abandoned, and resumed by Lorenz, Smale, Yorke, Prigogine and others in the last few decades of the twentieth century).

Points (1)–(3) represent events that followed each other, real historic phases of scientific thought. Point (3), which emerged in an era following that of the first two discoveries on chaos by Henri Poincaré, is still today the subject of scientific research (just think of the important developments in the study of the physics of elementary particles, which started after the Second World War and is still fully underway) and part of its development, from the 1960s onwards, has occurred in parallel with the renewed interest in the concept of chaos, point (4).

We will not be directly discussing the first three stages here, as they are more strictly linked to physics. We will be focusing on deterministic chaos, and on one of the elements that is part of a more general phenomenology, characteristic of many natural and social systems, the name of which is *complexity*. The science of complexity, which takes a global view of the world, regards the intricate and unpredictable interrelations between systems and the elements that they are made up of and focuses on the system, the aspects that regard its organization and its entirety, sets itself the objective of understanding a situation in a holistic sense, surpassing the strictly reductionist point of view adopted by classical mechanics, whose roots are in the Cartesian philosophy, and placing it in a more general framework (Israel, 1999).

It is widely accepted that the reductionist and analytical approach that presents itself as the foundation of classical mechanics, in the years between the end of the seventeenth century to the end of the nineteenth century, produced extremely important theoretical and applicative results; more specifically, it led to the development of mathematical techniques that we would be justified in calling prodigious, and supported the diffusion of newly formulated scientific thought, which had not existed in previous centuries. However, this approach, although extremely effective in some environments, is unable to produce results in other environments; in actual fact its application to some problems of mechanics itself is totally ineffective.

The many-body problem and the limitations of classical mechanics

The first important problem that reductionism encounters and where it fails is the historic so-called many-body problem (or the problem of n bodies), the description of the dynamics of a system made up of more than two point masses in reciprocal gravitational interaction (for a general introduction to the problem see, for example, Birkhoff, 1921, Arnold, 1987). In brief, the problem can be described thus. Let us imagine a system of n point bodies of any mass, with $n > 2$, subject to an attractive force, each of them towards each of the others, of decreasing intensity proportional to the inverse of the reciprocal distance squared, in accordance with Newton's law of universal gravitation. If the initial speeds are all nil, the bodies fall towards the system's centre of mass, which represents the simplest solution; but if the initial speeds are not zero, what will the positions and speeds of the n bodies be at a given time? The problem has obvious origins in astronomy and refers to the description of the dynamics of a planetary system; it can be transformed into the question: given sufficient observational data on the current celestial configuration, and given Newton's gravitational mathematical model, what will the sky look like in say one billion years?

From Euler's era onwards, the problem, even then limited to three bodies only, was tackled numerous times without success: the formulation of the mathematical model is simple, but its integration poses almost insurmountable difficulties, so much so that the search for a general solution is considered one of the most difficult problems in mathematics. Towards the end of the eighteenth century, Louis Lagrange managed to simplify the problem, translating it into a search for solutions in the form of infinite series and transferring the problem of the existence or otherwise of solutions to the problem of the convergence of such series. In this regard, the *particular* solution that Lagrange demonstrated the existence of is renowned, in the case of three bodies of equal mass: there is a closed stable orbit, in the form of a circumference, that each of the three bodies, placed at the tips of an equilateral triangle inscribed in the circumference itself, travels along (Bottazzini, 1999).

In addition to the trivial solution of the bodies that all fall towards the system's centre of mass and Lagrange's solution, numerous other *particular* solutions have been found to the problem of n bodies, but to date no general solution has been identified (Barrow-Green, 1997).

Recently, many classes of stationary solutions have been calculated numerically, consisting of closed stable orbits common to all the n bodies of the system, assumed to be of equal mass, for different values of n (Simó, 2000). Figure 3.1 shows several examples of these orbits calculated for a system of five bodies. We must point out, however, that this does not mean that the problem of n bodies has been solved: they are simply specific

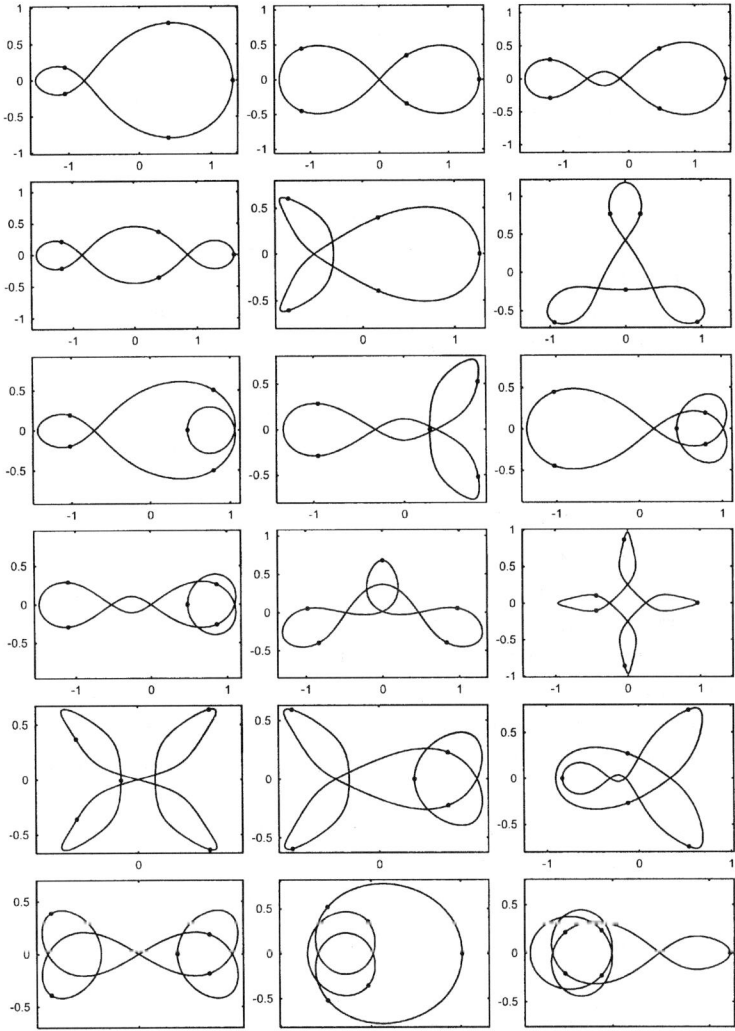

Figure 3.1 Examples of closed stable orbits in a system of five bodies in gravitational interaction.
Source: Simó (2000)

solutions, referred to clearly defined conditions. To date we still do not have a general solution that can be applied to any gravitational system: the reductionist approach, which has always been used in the formulation of the mathematical model of a system, fails.

If one wants to extend this consideration of the description of system dynamics to non-mechanical systems, such as, for example, social systems, where the term 'social' means biological, economic, regional, etc., even more so than in the problem of n bodies, a uniquely reductionist approach is even more ineffective and has to be reassessed and placed in a wider context.

Everyday life is full of examples of situations in which the reductionist approach is not effective. Think, for example, of the description of the dynamics of the Earth's atmosphere, an aspect that is commonly described as weather forecasting: even though conducted with very sophisticated techniques, no forecast manages to be reliable when it is pushed over the period of a few days. At the basis of the impossibility of making reliable predictions for systems such as the atmosphere, there is a phenomenon known today as the butterfly effect. This deals with the progressive limitless magnification of the slightest imprecision (error) present in the measurement of the initial data (the incomplete knowledge of the current state of each molecule of air), which, although in principle negligible, will increasingly expand during the course of the model's evolution, until it renders any prediction on future states (atmospheric weather conditions when the forecast refers to more than a few days ahead) completely insignificant, as these states appear completely different from the calculated ones.

The intrinsic unpredictability of a system's evolution contrasts sharply with the deterministic vision of Laplace (1814), a direct consequence of Newton's revolutionary theories and, in Laplace's era, still almost universally accepted. In Laplacian determinism, in fact, it was acknowledged that with the full knowledge of the initial states and the equations of motion it is always possible to univocally predict future states with arbitrary precision (see chapter 26).

Poincaré[12] had already sensed the existence of a phenomenon that amplified the disturbances in the model of n bodies in 1889, but the matter was

---

[12] Poincaré's first great success originated from his previous unsuccessful attempts regarding the problem of n bodies. This occurred when, in 1889, he won the prize that two years before, Oscar II, King of Sweden and Norway, had promised to award on the occasion of his sixtieth birthday, the 21st January 1889, to the person that solved this age-old problem. Poincaré did not solve the general problem, but he still received the prize, in consideration of his fundamental general study of the differential equations of dynamics and his in-depth analysis of the problem of three bodies in gravitational interaction (this is the specific, most interesting case, which refers to the Sun–Earth–Moon system). In the work he presented,

resumed only in more recent years by Enrico Fermi, in a publication that did not receive much attention (Fermi, Pasta and Ulam, 1955).[13] Only some years later, completely independent of Fermi, Edward Norton Lorenz, in a famous article (1963), which this time attracted the attention of scholars, even though not immediately (Gleick, 1987), focused considerable attention on the phenomenon of the progressive magnification of disturbances that he had by chance observed in the integration of a mathematical model of the dynamics of the Earth's atmosphere. Times had changed in the few years since the article by Fermi, Pasta and Ulam: electronic computers were becoming increasingly widespread, making very efficient means of calculation available to scholars like Lorenz, and allowing a totally new approach to mathematical research.

A new perspective, called holistic, was thus confirmed in the study of systems. According to this perspective, each system should be considered in its entirety: the analytical approach of reductionism is not always effective; reductionism therefore saw its role as absolute protagonist that it had played since the end of the seventeenth century lose ground. This also imposed a review of Leibniz's renowned principle of sufficient reason, a tenet that had firmly established itself in Western philosophy.

We will be looking at Laplacian determinism and Leibniz's principle of sufficient reason in more depth in chapter 26. First, however, in the remaining chapters of the first part and in the second part we propose to take a closer look at situations whose dynamics, even if such have originated from deterministic laws, are unpredictable. In Chapters 5–10 we will examine one of the first greatest themes that classical mechanics tackled successfully, namely the description in analytical terms of the oscillating movement of a pendulum, from the most simple abstract models, dynamic linear models, to those that envisage the introduction of corrective elements that make them less abstract, more realistic and increasingly remote from linearity. Later we will see several examples of how certain elements of models of oscillations,

---

Poincaré developed a new approach to the problem, establishing a new basis and revealing the first weakness in Laplace's rigid determinism. Of Poincaré's numerous publications, the work entitled '*Les méthodes nouvelles de la mécanique céleste*', published in three volumes between 1892 and 1899, is particularly important: it summarizes a good part of Poincaré's fundamental contributions to mathematics – including that conducted on the matter of n bodies – that qualify him as one of the greatest mathematicians of all time (Bottazzini, 1999).

[13] In the biography he wrote about Enrico Fermi, his master, colleague and friend, the physicist Emilio Segrè describes the studies made by the latter as 'an ingenious application of computers to a theoretical example of solving a nonlinear problem' (Segrè, 1971, p. 186). The contribution cited in the text, recalls Segrè, was one of Fermi's last works, written a few months before he died, in 1954, and was published posthumous the following year.

both linear and nonlinear, can be usefully transferred to the field of social science. In Chapter 4, however, before entering into the merits of the cited models, we will continue to examine what is meant by the linearity of a model, with the assistance of some technical details.

4 Linearity in models

A vast quantity of different system dynamics can be found in both natural and social sciences. However, they all have several common characteristics which enable them to be described in fairly simple and, above all, very general mathematical terms. We are referring to the dynamic processes typical of systems close to a configuration of stable equilibrium. Such systems tend towards equilibrium by means of a process that, to a good approximation, can be described by a so-called linear model. In other words, we can approximate the description of the evolution of these systems towards equilibrium by means of a calculable evolutive trajectory, if we know the initial conditions, using a law of motion expressed in the form of an extremely simple differential equation: a linear differential equation.

Linear differential equations are equations in which the unknown function and its derivatives appear *only* summed together or multiplied by continuous functions of the independent variable. They are therefore the following type of equation:

$$x^{(n)} + a_{n-1}(t)y^{(n-1)} + a_{n-2}(t)x^{(n-2)} + \ldots + a_1(t)x' + a_0(t)x = b(t) \quad (4.1)$$

where the $a_i(t)$ and $b(t)$ are continuous functions of the sole independent variable t (they might also be constants, which considerably simplifies the resolution of the problem) and the $x^{(i)}$ are the i-th derivatives of the unknown function $x(t)$.[14] The first member of (4.1) is a linear combination of

---

[14] Note that the presence of nonlinear functions in a model does not necessarily mean that the equation is nonlinear. Consider for example:

$$\dot{x} = \sin \omega t \text{ and } \dot{x} = \sin x$$

The first is a linear equation (non-autonomous, as we will see in chapter 5, because time *also* appears as an independent variable in the second member function, as well as being present in the derivative of the $x(t)$); on the contrary, the second equation (autonomous, because time does not appear explicitly) is nonlinear, as the sole unknown function $x(t)$ is the argument of the nonlinear function $\sin x$.

x and its subsequent derivatives, according to coefficients that are *only* functions of t (independent variable) and not of x (dependent variable, unknown function of t).[15]

The case of the homogeneous equation is particularly important, in which we have $b(t) = 0$. Homogeneous linear differential equations, in fact, have a very specific property, which makes the search for their general solution relatively simple: if in some way we are able to determine any two solutions, called $x_1(t)$ and $x_2(t)$, then we have that the sum of $x_1(t)$ and $x_2(t)$, even if multiplied by arbitrary coefficients c_1 and c_2, is still a solution to the given equation. In other words, any linear combination whatsoever $c_1 x_1(t) + c_2 x_2(t)$ of two functions that satisfy the equation is still a solution of the same: this property is known as the principle of superposition. Without entering into a discussion on the theory of the solutions to a linear equation and without being too rigorous, we can limit ourselves to observing that, by virtue of this principle, the general solution that describes the dynamics of a process according to a linear model (an equation) can be seen as a function that can be broken down into parts, all of which evolve individually over time. At each point in time, any linear combination of these parts provides a solution to the equation. The principle of superposition, therefore, expresses and summarizes the reductionist point of view (see chapter 3), which, therefore, closely links it to the concept of linearity.

A physical process that clearly exemplifies the above is that of constructive or destructive interference between low amplitude waves. For example, it can easily be observed that disturbances of a low amplitude spreading from different points on the surface of a liquid overlap simply by summing their amplitudes. It is easy to recognize a phenomenon of this type by observing the overlapping at a point of two waves, generated, for example by two stones thrown into two different points on the surface of the water. The equation that describes, in a first approximation, the motion of the point of the surface is a linear equation; the phenomenon of the interference between

---

[15] Equation (4.1) can also be condensed in the form $L(x) = b(t)$, having defined the linear differential operator $L(x)$ as:

$$L(x) = x^{(n)} + a_{n-1}(t)x^{(n-1)} + a_{n-2}(t)x^{(n-2)} + \ldots + a_1(t)x' + a_0(t)x$$

A fundamental property of $L(x)$ is expressed by the following relation, in which c_1 and c_2 are two constants:

$$L(c_1 x_1(t) + c_2 x_2(t)) = c_1 L(x_1(t)) + c_2 L(x_2(t)) \tag{4.1'}$$

by virtue of which we can say that the operator $L(x)$ is a linear operator, which justifies calling equations like (4.1) linear equations.

the two waves is mathematically reflected in the fact that even though the two waves have, so as to speak, independent lives, if considered together they represent a new disturbance that develops in accordance with the same differential equation that governed the single waves. If, however, the waves on the surface of the water are of high amplitude, then the linearity of the description ceases to be an acceptable approximation. This happens, for example, with a wave on the surface of the sea that is moving towards the shore, which, when it becomes too high with relation to the distance between the surface and bottom, becomes a roller and breaks.

A second basic example of linear dynamic evolution can be seen in the growth of a sum of capital invested at a constant rate of interest where, to make things simpler from a mathematical point of view, we assume that the growth of the total amount occurs by means of a continuous process, namely that the interest is capitalized instantaneously.

A process of this type, where the interest continuously and instantaneously generates further interest, is commonly called compound interest. Simple considerations of a general nature lead us to define the differential equation that governs the system's dynamics as a function of time t:

$$\dot{x}(t) = k\,x(t) \tag{4.2}$$

where $x(t)$ represents the total amount and k is the growth rate.

Equation (4.2), solved, provides a simple exponential law:

$$x(t) = x(0)e^{kt} \tag{4.3}$$

where $x(0)$ indicates the initial capital.

This example regards the growth of a sum of capital, but extending the concept a little, the situation described could also be interpreted in other ways: the model, for example, could represent the growth of a population whose initial value is $x(0)$, and which increases at a constant growth rate k, in the presence of unlimited environmental resources, i.e. without any element that intervenes to slow down growth. Equation (4.3) expresses the continuous growth of a magnitude (capital, population or other) according to an exponential function whose limit, over time, is infinity.[16]

---

[16] The version of (4.3) in discrete form is a simple geometric progression and is the tendency to grow at a constant rate which, according to Thomas Malthus (1798), a population demonstrates in the presence of resources that grow directly in proportion to time (therefore not at a constant rate). We will return to Malthus' law in chapter 20, where, furthermore, we will also discuss the variation introduced by Pierre-François Verhulst to

In (4.3) the time dependency only affects the exponential, which contains the growth rate k; the value of the total amount at a certain point in time, moreover, is directly proportional to the initial value of the capital $x(0)$. Consequently, two separate amounts of capital invested at the same rate produce, after a certain time, two amounts of interest that, summed, are equal to the interest produced by an initial amount of capital equal to the sum of the two separate amounts of capital: it follows the same pattern as the linearity illustrated in (4.1′). This conclusion, in reality, could be possible for small sums of money; for large amounts of capital, on the other hand, other aspects may intervene that alter the linear mechanism described. It is unlikely, for instance, that an investor who has a large amount of capital to invest, would content himself with the same rate of interest obtained by those with small sums of capital, for which, presumably, lots of small amounts of capital invested separately would together obtain a lower amount of interest than that obtained by a single large amount of capital equal to the sum of the former. Alternatively, after a certain period of time, variations in market conditions may intervene that give rise to variations in interest rates or other aspects still.

The dynamics of a linear model are obviously not a mathematical abstraction made on the basis of the observation of the evolution of a system in certain specific conditions: linear abstraction *can* be acceptable (but not *necessarily*) in cases in which the magnitudes in question undergo 'small' variations, like the small disturbances to the surface of the water or like the effects that the low growth rates have on a small amount of capital over a short period of time. Similarly, the dynamics of any system that is brought towards a state of stable equilibrium by a return force can be considered linear, if we can assume that the intensity of the force is directly proportional to the distance of the system in question from the state of equilibrium. This, for example, is the case of a horizontal beam that supports a load at its centre: for small loads, the extent of the deformation that the beam undergoes is proportional to the load, whereas for heavy loads this is no longer true, to the extent that for too large a load, the beam will break. This is also the case of a particular set of oscillating systems such as, for example, the previously cited movement of a point on a surface of water, or that of a pendulum, when it makes small oscillations.

If the system's distance from equilibrium is not '*small*', then it is not realistic to assume that the return force is proportional to the distance, just as it is not realistic to assume that the growth of an amount of capital does

---

represent a more realistic case in which the limited nature of resources hinders the unlimited growth of the population.

not depend on its initial value, or that a population continues to grow according to Malthus' model even when it reaches high values, i.e. very high growth rates or very long periods of time. In cases such as these, we cannot talk of linearity: the linear model provides an unacceptable approximation of the dynamics in question.

We will return to the concept of linearity in Chapter 25, where we will extend our discussion on its meaning, its value and its limitations. In Chapter 5 we will apply the linear hypothesis to the description of the oscillation of a pendulum. Linearity is the simplest hypothesis to apply to oscillations, it was the first to be formulated and is the most studied, so much so that it has almost become a metaphor applicable to other areas of science, as we mentioned in Chapter 3. Occasionally, in fact, we also speak of periodic oscillations in economic and social systems, which often we attempt to describe as if they were real physical systems that are oscillating.[17] The discussion will occasionally touch on some technical elements, which are useful to clarify how we can construct and refine a model of a phenomenon that, although basic, is of fundamental importance in terms of the frequency with which it is used, through analogies, to interpret other more complicated phenomena.

---

[17] In this regard, and particularly in economics, it would be more correct to talk about the cyclicity of change, rather than periodicity in its strictest sense. It could be argued, in reality, if the alleged oscillations in historic series of data of economic systems were effectively comparable to those of a mechanical system, or if, when attempting to describe systems with extremely complicated evolutions over time and with hardly any repeat patterns, one wanted to establish an unjustified homology with the laws of mechanics. We already mentioned in Chapter 1 the risks that the direct application of the theories and the schemas of physics to social systems may involve in terms of the correct and effective interpretation of the phenomena of social sciences.

5 One of the most basic natural systems: the pendulum

The linear model (Model 1)

Let us start by illustrating the model of an oscillating pendulum, basing ourselves on the simplest assumption: a pendulum that makes small oscillations in obviously completely theoretical conditions, in which there is no form of friction.

Let us imagine that we strike the pendulum, initially still in a position of stable equilibrium, with a small force: it distances itself from equilibrium, but, as it does so, it starts to undergo the action of a restoring force, which increases with the size of the open angle with respect to the vertical (Figure 5.1). The speed of the pendulum decreases until it cancels itself out, when the restoring force becomes equal to the force applied initially that moved the pendulum from equilibrium.

The linear hypothesis is in the assumption that the restoring force and the size of the angle are directly proportional at any given moment during oscillation. This means, for example, that if we hit the pendulum with double the force, then the maximum amplitude of the oscillation, as long as it does not become 'too' large, doubles; in this way, we can assume that the oscillation consists of the sum of two oscillations caused by two equal initial forces, applied separately, that are summed together.

It should be borne in mind that the linear hypothesis, namely this direct proportionality between the action, the force initially applied, and the response, the amplitude of the oscillation, or between the restoring force and the oscillation angle, is more accurate the smaller the angle by which the pendulum distances itself from the vertical. Therefore, we consider the approximation acceptable only for 'small' angles. To be more precise, the linear hypothesis is valid within a limit in which the maximum amplitude of the oscillation tends to zero; if on the other hand the angles and the

The linear model 33

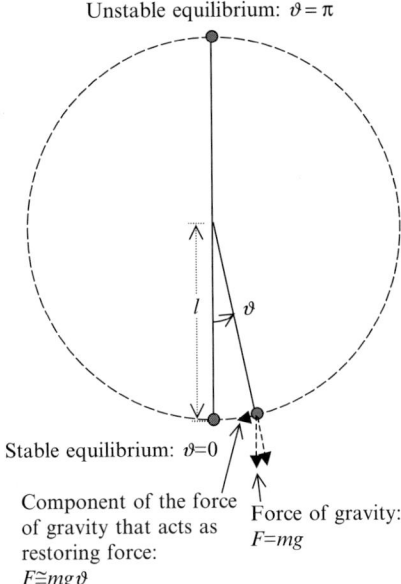

Figure 5.1 Diagram of a pendulum that makes small oscillations in the absence of friction.

restoring forces are both large, other aspects of the oscillatory phenomenon come into play which make the response disproportional to the action and which we will discuss in Chapter 7.

Assuming that the system is linear, i.e. that the action and effect are directly proportional, the equation that describes the dynamics of the pendulum can be extracted from Newton's second law of dynamics, $F = ma$, the first real example of effective mathematical modelling of a dynamic process in the history of science,[18] that we now rewrite as:

$$F = m\ddot{x}(t) \tag{5.1}$$

in which we use the symbol $\ddot{x}(t)$ to indicate the second derivative of $x(t)$ with respect to time, i.e. the acceleration. Equation (5.1) is an actual example of a second order differential equation, which can be either linear, if the force F is a linear function of x and/or of its first derivative or it does

---

[18] We are obviously setting aside the observations of Galileo Galilei on the motion of a pendulum and on the fall of bodies, which, however, were not formulated in terms of differential equations (introduced much later by Newton) and therefore did not lead to a real dynamical system, such as (5.1).

5: The pendulum

not depend on x, or nonlinear, if the force F is a nonlinear function of x and/or of its first derivative. In the linear case that we are about to discuss (the simplest model of oscillation), the pendulum is subjected to a restoring force that only depends on its position and that therefore, in a given position, is constant over time. The mathematical model of a linear oscillation of this type is known in mathematical physics as a 'harmonic oscillator'.

We therefore have an object with a mass m, whose barycentre distances l from a centre of rotation to which it is connected by a rigid rod of a negligible mass, as shown in Figure 5.1 $\vartheta(t)$ indicates the angle, a function of time, that the rod of the pendulum forms with respect to the vertical (position of stable equilibrium) and g indicates the acceleration of gravity. As we have said, the oscillations are assumed to be of a small amplitude, where by 'small' we mean that they are such that the arc described, whose maximum amplitude is ϑ_1, can be considered to have a negligible length with respect to the length of the pendulum l. This implies, and it is here that the linear hypothesis takes shape on a technical level, that one approximates $\sin(\vartheta) \cong \vartheta$. The law of dynamics (5.1), therefore, can be transformed as follows:

$$ml\ddot{\vartheta}(t) = -mg\vartheta(t) \qquad (5.2)$$

The first member of (5.2) is none other than the term $m\ddot{x}(t)$ from Newton's equation (5.1): $l\ddot{\vartheta}(t)$ is obtained by deriving the length $l\vartheta(t)$ of the arc travelled by the mass m during oscillation twice with respect to time, expressing the angle $\vartheta(t)$ in radians (see Figure 5.1). The second member of (5.2), on the other hand, is the component of the force of gravity that acts in a perpendicular direction to the rod of the pendulum: it is the restoring force that pushes the pendulum towards the position of equilibrium and, therefore, is directed in the opposite direction to that of the growth of the angle ϑ, which is why it has a minus sign. The linear hypothesis lies really in the fact that, for small angles, the restoring force is assumed to be proportional to the amplitude of the angle.[19]

Equation (5.2) is a homogeneous linear differential equation with constant coefficients, whose integration leads to a combination of exponentials in a complex field, which translates into simple sinusoidal oscillatory dynamics. Simplifying and assuming $\omega^2 = g/l$, we obtain:

---

[19] In practice, if we limit ourselves to considering an angle of 1 degree as the maximum amplitude of the oscillation, assuming that the intensity of the restoring force is directly proportional to the angle of oscillation, the error that we introduce is around 1 per cent.

$$\vartheta(t) = \vartheta_0 \cos(\omega t) \qquad (5.3)$$

where the constant of integration ϑ_0 indicates the initial amplitude of the oscillation, in the case that the pendulum is initially moved away from equilibrium and then left to oscillate. The solution of (5.2) is a periodic function $\vartheta(t)$, whose period T is constant, as can be obtained from (5.3) with some simple operations:

$$T = \frac{2\pi}{\omega} = 2\pi \sqrt{\frac{l}{g}}$$

Therefore it represents a dynamics which is stable over time.

We will now attempt to improve our model of a pendulum, making the description of the phenomenon more plausible, on the basis of different, more realistic initial hypotheses, while remaining within linearity.

The linear model of a pendulum in the presence of friction (Model 2)

Let us ignore for the moment the hypothesis according to which there is no form of friction present in the motion of the pendulum and let us introduce an element into (5.2) that describes the effect of friction. In this case, taking into consideration that friction manifests itself as a force that opposes motion, the intensity of which, at low speeds, is proportional to the speed of the latter, we can correct equation (5.2) by inserting a term that is added to the restoring force. The speed with which the pendulum moves can be obtained by differentiating the expression $s(t) = l\vartheta(t)$ with respect to time, obtaining $\dot{s}(t) = l\dot{\vartheta}(t)$.

Therefore, (5.2), which is still Newton's renowned law, is now corrected by the addition of a term proportional to $l\dot{\vartheta}(t)$:

$$ml\ddot{\vartheta}(t) = -kl\dot{\vartheta}(t) - mg\vartheta(t) \qquad (5.4)$$

The minus sign, in front of the first term of the second member, representing the force of friction, is justified by the fact that such force manifests itself by acting in the opposite direction to the peripheral speed $l\dot{\vartheta}(t)$, in the same way as the restoring force has a minus sign before it.

Equation (5.4), like (5.2), is a second order linear differential equation with constant coefficients. The solution of (5.4) depends on the values of the coefficients: for certain sets of values, the pendulum makes damped oscillations, whose amplitude decreases exponentially over time; for other sets of values, it can bring itself to the vertical position, the state of equilibrium, at a speed that decreases exponentially over time, without oscillating. Nil amplitude, i.e. when the pendulum is still, is a position of equilibrium that is called asymptotically stable (we will give an initial definition of the concept of asymptotic stability in Chapter 8; several aspects of the concept of asymptotic stability will then be examined in greater detail in Chapters 17–19).

We observe at this point, without going into the mathematical details, that the case of the forced pendulum also merits consideration: the motion of a pendulum subjected not only to the force of gravity and the force of friction, but also to another force which is a function of time. There are a great number of examples of oscillating systems that are similar to that of a forced pendulum: all of those in which a system is subjected to the action of a restoring force that makes it make periodic oscillations, but also undergoes the action of another external variable force. Consider the example of a forked branch of a tree, whose two parts (forks) are able to oscillate; if you make one oscillate and you observe the overall motion of the two parts: the stem of the fork, oscillating, acts on the second stem (fork), making it oscillate; it, in its turn, acts in return on the first fork. Basically you will observe that the motion of oscillation passes alternately but without interruption from one fork to another. This represents a system made up of two coupled pendulums.

We can also consider, as a second example of forced oscillation, the phenomenon of resonance: what we observe, for example in an oscillating system that receives a fixed intensity impulse at constant intervals, integer multiples of the period of oscillation.

Let us return to the model of the pendulum. The equation of the motion of the forced pendulum can generally be written in the following form:

$$ml\ddot{\vartheta}(t) = -kl\dot{\vartheta}(t) - mg\vartheta(t) + F(t) \qquad (5.5)$$

In (5.5), the force $F(t)$ directly acts on the oscillating mass, summing itself algebraically to the return component of the pendulum due to the force of gravity and to friction. In Chapter 10, when we touch upon the chaotic pendulum (Model 4), we will write yet another form of the equation for a particular forced pendulum, in which the force $F(t)$ doesn't act directly on the oscillating mass, as in (5.5), but on the oscillation pivot.

Autonomous systems

With a system that makes forced oscillations of the type illustrated by (5.5), we meet our first example of a particular type of system, called non-autonomous. A system is non-autonomous if it is described by equations that contain a term in which a direct and explicit dependence on time manifests itself, like that caused by an external force $F(t)$ in (5.5), whereas the mechanism that determines the evolution of an autonomous system does not directly depend on time.

Let us clarify what we mean by explicit dependence on time with an example. The concept is important and far-reaching and does not only apply to a pendulum but to any system. It is true nevertheless that in a pendulum that makes small oscillations, the restoring force is not constant, but depends directly on the position of the pendulum itself and it is the latter that depends on time. If we held the pendulum still, out of equilibrium, with our hand, we would feel a constant restoring force (the weight component perpendicular to the rod). However, in the case of a non-autonomous system, like a forced pendulum, we would feel a variable restoring force, as if the pendulum was a live 'pulsating' being, due to the effect of the restoring force that now depends (commonly referred to as 'explicitly') on time. It is evident that the description of the motion of an autonomous system is, in general, much simpler than that of a non-autonomous system, because the equations used to describe it are generally simpler.

In an autonomous system, therefore, time does not appear directly among the variables in the equations that describe the dynamics of the system, which not only makes its analytical treatment much simpler, it also gives the system certain properties that characterize it in a particular way. If the forces in play do not depend explicitly on time, then every time that the system assumes a given configuration, namely a given speed at a given point in space, the dynamics that ensue immediately afterwards are always the same. Basically, the subsequent evolution at a given moment in time is determined with no uncertainty, as the forces, for that given configuration, are known and constant.[20] For example, a certain pendulum is subjected to

---

[20] By using a slightly stricter language and anticipating the contents of Chapter 8 regarding the phase space, we can rewrite the sentence in the text, saying that at a given point in the phase space the conditions are stationary. This means that, when passing at different times through the same point, the orbit always passes there 'in the same way', i.e. the trajectory has the same geometric characteristics, i.e. at that point it has the same tangent. Stationary orbits basically only exist in the case of an autonomous system, as it is only in this case that the conditions for the validity of the theorem of existence and uniqueness of the solutions of a

the same restoring force, every time that the angle that it forms with respect to the vertical position, by oscillating, assumes a given value.

Returning to equation (5.5), we observe that the solution obviously depends on the form of $F(t)$. Apart from specific cases in which the expression of $F(t)$ is particularly simple, such as the case in which $F(t)$ has a sinusoidal trend, as in the cited example of the two coupled pendulums, and the period of oscillation of the pendulum has a rational relationship to that of $F(t)$, in general the evolution of the pendulum can no longer be ascribable to regular periodic oscillations and cannot be calculated in exact terms, but only through numerical integrations.

It is evident that a description of the dynamics of a system as simple as that of a pendulum, that has some claim of resembling an observed situation, cannot be based on the simplified hypotheses presented here, unless as a mathematical abstraction. If we wanted to be more realistic, we would at least need to take into account the fact that the amplitude of the oscillations, although low, is not negligible. The linear model described, both in the form of (5.2) and in the form of (5.4) basically needs to be corrected, by introducing oscillation amplitudes that are not 'small'. Considering amplitudes that are not 'small', however, forces us to abandon the condition of linearity: a problem that warrants further examination. In Chapter 7, we will see how to alter the model of the pendulum, making it nonlinear; but first in Chapter 6, we will continue our exploration of the meaning and the limitations of linearity.

---

differential equation are satisfied. The equations that describe the oscillating systems that we are discussing, as we have said, are variants that originate from Newton's equation $F = ma$; therefore, if the system is autonomous, the forces in play only depend on the position (technically, they are referred to as conservative forces) and the system can be described in terms of energy conservation.

6 Linearity as a first, often insufficient approximation

The linearization of problems

The perception that we have of reality and, in particular, of the dynamic processes that we observe therein, is such that rarely, if at all, do we witness a direct and strict proportionality between cause and effect (in more specific contexts, other expressions could be used, such as proportionality between input and output, between stimulus and response, between action and reaction, and more besides). In fact, at times actions of relatively small intensities produce consequences that appear to be disproportionately large with respect to the size of the action, or, vice versa, actions that we consider to have a large intensity have unexpectedly little effect with respect to our expectations. Situations of this nature are very frequent and typical to our perception of reality.

With reference to the examples cited in Chapter 5, we have repeatedly insisted on the use of the adjective 'small' (small oscillations, small forces, small amounts of capital, etc.). In the case of the pendulum, as we have said, this expression should be understood as 'the smaller the amplitude of the arc traced in the oscillation is with respect to the length of the pendulum, the more precise are the predictions of equation (5.2) [or (5.4) or (5.5)] with respect to the motion actually observed'.

In more general terms, the attribute 'small' depends on the more or less arbitrary choice, even if based on common sense, of a threshold above which we no longer want to retain the linear approximation that we make acceptable. Essentially, with linear approximation we are simply ignoring everything that we consider negligible, describing the evolutionary process, when we wish to remain within the scope of a direct proportionality between action and effect (which is portrayed in the linearity of the differential equation that describes the process). Strictly speaking, the condition that we create with linear approximation is not realistic, certainly, but it is the simplest to perceive and facilitates calculations: if reality was, in actual fact, linear, then we would be able to formulate predictions on the future evolution of a system solely on the basis of the differential equations that model its (linear) dynamics.

The simplest description of the phenomena in question actually entails taking into account only the preponderant terms of the magnitudes in play, ignoring those of lesser importance, which are considered to be corrective terms. In actual fact, this method, called linearization, has enabled us to recognize many invariant elements and a great deal of regularity in natural phenomena. Above all in classic physics, linearization has enabled us to deal with various phenomena with a high degree of approximation, even if other numerous, fundamental problems remain unsolved. The idea is simply that, as the terms that were ignored by linearizing the equations were small, the difference between the solutions of the linearized equation and those of the nonlinear equation assumed 'true', but unknown, ought to be small as well; however, this is not always the case.

During the classical era of science there were no, and to date still are no, analytical techniques able to handle the problem of nonlinearity in general terms. Several methods of numeric integration were well-known; first of all Euler's method, that, in principle, enables a particular integral of an equation to be constructed, step by step; these methods, however, applied to nonlinear equations, often gave rise to situations in which the calculations became too lengthy to be made by hand. This not only made the process of linearization of phenomena natural, but it imposed it, at times, as the only direction in which to proceed. This process was pushed to such a degree that, often, one sought to immediately impose a linear equation, even before having completely understood whether the phenomenon that one wanted to describe displayed the characteristics required to make a description in linear terms acceptable.[21]

It is worth observing, even at this early stage, that linear approximation becomes increasingly unacceptable, the further away we get from a condition of stable equilibrium. In general, in the history of science, we have always preferred to tackle the study of the dynamics of systems when the latter are in a condition that is close to stable equilibrium because there the forces in play are small and therefore the linear approximation can be considered acceptable, which means that the evolution of the systems is generally fairly easy to predict. It is abundantly clear that the study of the dynamics of systems, when the latter are in a condition in which their treatment is relatively easy, leads to more satisfying results than when we are dealing with the description of conditions that are difficult to treat and

---

[21] So much so that in the jargon, the expression 'to linearize a problem' has become synonymous with 'to simplify a problem'.

where formulating predictions becomes virtually impossible. Nevertheless, as has already been shown by Poincaré and, more recently by a number of other scholars (Prigogine, 1962, 1980, 1993; Prigogine and Stengers, 1979, 1988; Nicolis and Prigogine, 1987), a system that is far from equilibrium, and therefore in a condition in which its evolution is characterized by considerable nonlinearity, presents aspects that make its evolution of particular interest due to the unpredictable new situations that may arise.

The world of classical science has shown a great deal of interest in linear differential equations for a very simple reason: apart from some exceptions, these are the only equations of an order above the first that can be solved analytically. The simplicity of linearization and the success that it has at times enjoyed have imposed, so to say, the perspective from which scientists observed reality, encouraging scientific investigation to concentrate on linearity in its descriptions of dynamic processes. On one hand, this led to the idea that the elements that can be treated with techniques of linear mathematics prevail over nonlinear ones, and on the other hand, it ended up giving rise to the idea, which was widespread in the past, that linearity is intrinsically 'elegant', because it is expressed in simple, concise formulae, and that a linear model is aesthetically more 'attractive' than a nonlinear one.

One consequence of this approach was that the aspects of natural phenomena that appeared to be more 'worthy' of the attention of scholars were actually the linear ones. In this way, the practice of considering linearity as elegant encouraged a sort of self-promotion and gave rise to a real scientific prejudice: mainly (or only) linear aspects were studied, as they were easier to describe. The success that was at times undeniably achieved in this ambit increasingly convinced scholars that linearization was the right way forward for other phenomena that adapted badly to linearization, but nonetheless that they attempted to treat with linear techniques, sometimes obtaining good results, but other times obtaining bad, if not totally unsuccessful ones.

However, an arbitrarily forced aesthetic sense led them to think (and at times still leads us to think) that finding an equation acknowledged as elegant was, in a certain sense, a guarantee that nature itself behaved in a way that adapted well to an abstract vision such as mathematics.[22]

---

[22] In this regard the following case is exemplary. On 2 January 1928, the theoretical physicist Paul Adrien Maurice Dirac presented an equation to the Royal Society of London that soon became one of the mainstays on which quantum physics is based, and for which he was awarded the Nobel Prize for physics in 1933 at the age of just 31. The solution of the equation envisaged the existence of a particle, at that time unknown, of an equal mass to that of an electron but with a positive charge: the positive electron (or positron). This, as was

On one hand, therefore, the relative simplicity of linearity and its effectiveness in some cases encourages us to use linear methods and to consider them elegant; on the other hand, the idea of associating the elegance of a mathematical description in linear forms to its effectiveness as a model ended up creating a prejudice of a philosophical–methodological nature. There was a tendency, therefore, to force nature towards linear mathematics, insisting that the former necessarily reflected the latter, rather than adapting the mathematical tool to nature, ignoring the fact that mathematics, in reality, is nothing more than the fruit of our mental processes. This meant that scholars sought linear evolutionary trajectories because they were often what was observed in a first approximation, because they are easier to treat and because they provide valid predictions for systems that are close to conditions of stable equilibrium.[23]

The limitations of linear models

Just as eighteenth century scientists believed in a deterministic world, regulated like clockwork, scientists in the nineteenth century and in the first half of the twentieth century believed in a linear world (Gleick, 1987; Stewart, 1989). A long tradition, therefore, has been created on the terrain of linear modelling, while precious little attention has been focused, more or less in all branches of science and until relatively recently, on nonlinear modelling, both due to the increased mathematical difficulty its treatment objectively entails, and due to the widespread aesthetic and methodological prejudices cited above. The reflection on the use of linear mathematics as a tool of investigation should be extended and put into a more general

---

acknowledged shortly afterwards, was the actual discovery of antimatter, from a theoretical point of view, in the solution of a linear differential equation with partial derivatives; this occurred four years before the positron was observed experimentally in 1932 by Carl David Anderson, as an effect of the interactions of cosmic rays in the atmosphere (Anderson received the Nobel prize for physics in 1936 for this experimental discovery, he too at the age of 31). Dirac, much later, confirmed that it was his 'acute sense of beauty' that enabled him to obtain the equation and to convince him that it was correct (Dirac, 1982). In this case, Dirac was right, because the experimental observation confirmed his theory, but other times, 'the sense of beauty' of a technique or of a mathematical form has led scholars to construct theories that have then proved to be failures, as in the case, for example, of Landau and Liftshitz's theory of turbulence (see p. 43–44 following).

[23] In this regard, James Gleick observes how, in the mid-twentieth century, there was no lack of scientists who understood 'how nonlinear nature's own soul is. Enrico Fermi, once exclaimed: "It does not say in the Bible that all laws of nature are expressible linearly!" ' (Gleick, 1987, p. 71).

framework that regards the relationship between mathematics, all of mathematics whether linear or not, and sciences, a framework in which the most constructive approach is based not on forcing nature towards mathematics but just the opposite: forcing mathematics towards nature.

Without entering into the merits of the foundations of mathematics, we would like to make some observations. Mathematics is an abstract discipline, the result of the speculations of the human mind, even if very often it has originated and been encouraged by problems connected to reality (Changeux and Connes, 1989; Granger, 1993; Boncinelli and Bottazzini, 2001; Lolli, 2002). It involves entities that are *objective*, but they are not *real* in the usual sense of the term (Giusti, 1999). The use of the tool of 'mathematics' has certainly demonstrated itself to be one of the most effective methods that we can apply to describe many of the phenomena observed. However, we must not forget the fact that, if it comes naturally to us to use mathematics because it 'works', this doesn't necessarily mean that nature (that is concrete) *has to totally* follow *our* mathematics (which is abstract). All we can say is that, if the evolution of the human mind has led to the development of what we call 'mathematics', this means that mathematics is, in some way, one of the most effective tools that man possesses to operate successfully in the environment (we will return to this point in chapter 31).

We have to be careful not to let ourselves get carried away in our attempts to dogmatically and assertively impose mathematics and its criteria on the description of phenomena, as much to those of natural science as to those of social science. In doing this, we actually run the risk that the criteria currently used in mathematics (formal coherence, logical rigour, coherence and completeness of postulates, etc.) may transform themselves, when applied to science, into real methodological prejudices.

In science too, just as in other fields, we may end up considering as 'elegant' all that is in harmony with the norm, with deep-rooted points of view or with expectations. Thus the tendency to apply linear models has led to a *desire* to recognize the presence of cycles, namely of various types of periodicity, in dynamic processes, even in situations where they do not exist.

A well-known example of this was the first complete theory formulated by Lev Davidovich Landau and Evgeny Mikhailovich Lifshitz (1959) regarding the phenomenon of turbulence in fluids, a problem that had been tackled without success for over two centuries. According to Landau (Nobel prize for physics in 1962) and Lifshitz (after the death of the master and colleague, Landau prize, in the Soviet Union), the onset of turbulence in a fluid was described as being due to the successive linear superposition of periodic waves that were generated in the movement of the fluid in certain specific conditions. The theory of Landau and Lifshitz, the conception of which was truly ingenious and which was technically very well constructed, was accepted even in the absence of in-depth experimental validation, because it appeared that it could work, in line with the expectations of the current era, and because it was 'elegant', since it was based on linear mathematical techniques that had been recognized for some time and used with great skill (essentially Fourier series and integrals). It was only later, thanks to a closer examination of the so-called Taylor–Couette flow, i.e. of the dynamics of a viscous fluid between two coaxial cylinders rotating at different speeds (a relatively simple experiment conducted by Geoffrey Ingram Taylor in 1923, long before the work of Landau and Lifshitz), that Landau and Lifshitz's model was rejected and a totally different theory was formulated, a theory conceived by David Ruelle and Floris Takens making use of new techniques and new mathematical concepts, such as that of chaos (Ruelle and Takens, 1971; see Bayly, 1990, and Ruelle, 1991).[24]

The same applies to the tendency to want to recognize the presence of periodicity in economics, and more generally in social sciences, as a structural unchangeable characteristic, where historic series of data also reveal repeated structural changes in addition to the presence of cyclical phenomena (Barkley Rosser, 1991; Mandelbrot, 1997ab).

Linear dynamical models are able both to provide stable solutions and to predict situations of instability, but the solutions of such models can be reduced to just four very general types, some examples of which will be illustrated in the remaining paragraphs of this section: (1) stable oscillations, (2) exponentially growing (explosive) oscillations, (3) asymptotical stability, (4) limitless (explosive) growth. These can account for short or medium term changes, but are not able to encompass the whole scope of possible long-term developments within the field of natural science and, even more so, given the increased conceptual difficulties of modelling, within that of social sciences.

---

[24] There is a well-known saying according to which turbulence is a 'cemetery of theories' (Ruelle, 1991). Numerous mathematicians and physicists, long before Landau and Lifshitz (from Euler and Daniel Bernoulli in the eighteenth century, to Poincaré and to Werner Heisenberg in the twentieth century) had tackled the subject, using classic mathematical techniques, but without achieving significant results. Interested readers can find in-depth discussions of the problems linked to Taylor–Couette flow and an illustration of the (still imperfect) current theories on the origin and on the description of turbulence in Pomeau and Manneville (1980); Manneville and Pomeau (1980); Swinney (1983, 1985); Mullin (1991); Chossat and Iooss (1994); Meyer-Spasche (1999).

It is important to observe that linear models enable us to formulate precise predictions about the future of a system, while nonlinear models, in general, only enable such with a degree of approximation which worsens the further into the future we try to predict; essentially, the predictions that nonlinear models provide can, at most, after the event has occurred, tell us why the evolution was of a certain type, why it followed a certain course characterized by particular changes in direction, rather than others, but predictions before the event are more unreliable the larger the intervals of time being considered. The dynamics of natural systems, just as those of social systems, are not linear; therefore, the more distant the future, the more unpredictable they are. We cannot make long-term predictions, for example, about turbulence that manifests itself not only in the movement of a fluid, but also, for instance, in the financial markets, in the changes of mass opinions and in demographic and macroeconomic developments.

Interest in long-term modelling started a few decades ago, at the end of the 1960s, with the development of the theory of dynamical systems. The first models that belonged to this new field were mostly simple numeric simulations that specifically focused on the effects that feedback cycles exercised on a system's stability. This, in particular, was made possible, and actually stimulated, by the increasing proliferation of electronic computers and by the enormous development of their power of calculation. These models, in reality, were no longer made up of systems of differential equations, equations that are by definition in continuous time, but by programs for computers, that essentially consisted of recursive and discrete calculations. In contrast with linear models, these dynamical models were able to generate evolutionary schemes other than the four schemes typical of linear models and to predict, for certain parameter values, the possibility of structural changes. An example of one of the simplest and most renowned nonlinear models is that of the logistic map (May, 1976; Feigenbaum, 1978), which is the version in discrete time of Verhulst's continuous law of growth. While the general integral of the differential version in continuous time can be calculated without difficulty and gives rise to a continuous and regular curve (see Chapter 20), the version in discrete time shows a very particular, articulated and varied evolution (see Chapters 21 and 22).

We will now return to our analysis of the model of the oscillations of a pendulum that we have examined up until now only in linear approximation. We will see in Chapters 7–10 how, abandoning the assumption of linearity, this model can be improved and can come closer to the object that we want to describe.

7 The nonlinearity of natural processes: the case of the pendulum

The nonlinear pendulum (Model 3 without friction, and Model 3′ with friction)

A closer analysis of the dynamics of the pendulum, still starting from Newton's law, leads to a different formulation of the equation of motion, less rough in its approximation and more articulated in its details. Ignoring the approximation introduced in Chapter 5 regarding the consideration of small oscillations only, we have, depending on whether we do or do not take friction into account, equation (7.1) and (7.2) respectively:

$$ml\ddot{\vartheta}(t) + mg\sin\vartheta(t) = 0 \qquad (7.1)$$

$$ml\ddot{\vartheta}(t) + kl\dot{\vartheta}(t) + mg\sin\vartheta(t) = 0 \qquad (7.2)$$

The similarity between equations (7.1) and (7.2) and between equations (7.2) and (5.4) is immediately apparent. The only difference, in both cases, is that the component of the force of gravity that recalls the pendulum towards the position of equilibrium, when the pendulum is distanced from it, represented by the last terms in the sums in the first members, is no longer proportional to the angle that the pendulum forms with the vertical, but to its sine. It is actually the presence of the term $\sin\vartheta(t)$, in this case, that makes the equation nonlinear with respect to the unknown function $\vartheta(t)$.

In the case that we are discussing, Figure 5.1 that describes the geometry of the system is transformed into Figure 7.1.

If we replace the function $\sin\vartheta$, that appears in (7.1) and (7.2) with its development in a Taylor series around the value $\vartheta = 0$, i.e.

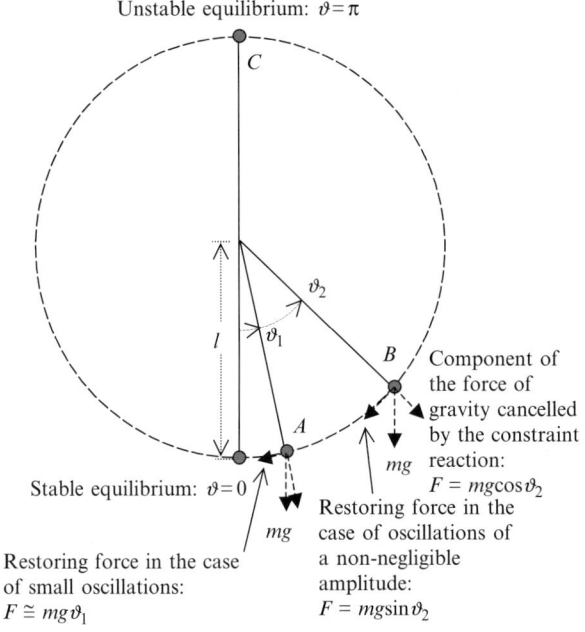

Figure 7.1 Diagram of a pendulum that makes oscillations of a non-negligible amplitude.

$$\sin(\vartheta) = \vartheta - \frac{\vartheta^3}{3!} + \frac{\vartheta^5}{5!} - \frac{\vartheta^7}{7!} + \ldots$$

we see that for small values of ϑ, i.e. for oscillations of an amplitude close to 0, we can ignore the terms of an order higher than first, which become increasingly smaller as the exponents that appear there increase. Equations (7.1) and (7.2) can thus be traced back to (5.2) and (5.4) respectively, i.e. they can be linearized. In this way, for small angles, the linear case appears to be a limit case of the general case.

Non-integrability, in general, of nonlinear equations

As we have said, while it is always possible to solve a linear differential equation, a nonlinear differential equation can only be solved in exact terms, i.e. without using approximations, in a limited number of cases only: the list of functions in which the solutions are usually expressed (polynomials, exponentials, etc.) is too limited to be able to adapt itself to the huge variety of nonlinear differential equations that we could be faced with in practice. Often, the attempt to write general solutions leads us to use integrations by series or integral functions, but methods of this nature are difficult to treat, are not always effective and can rarely be read intuitively. We need to focus, therefore, on other methods to qualitatively describe a phenomenon, after the differential equation has been formulated, methods that do not require the calculation of the general solution of the equation itself or methods that give an approximated solution (Jordan and Smith, 1977; Bender and Orszag, 1978). Each individual case should be studied in its own right; there are no general methods that can be applied to all the situations that we meet in models.

Given the historical importance of the problem of oscillations, and given the relative simplicity of the model that we are using as an example, we will see, for this case only, how it is possible to get around the difficulty of the lack of explicitly exact solutions[25] and make, in any event, several observations on the dynamics of the system.

We would like to immediately clarify that, even though we do not obtain exact solutions, it is still always possible to integrate (7.1) and (7.2) numerically, obtaining approximated solutions with arbitrary precision, at least in principle, for any set of initial conditions. What we are interested in is not so much giving life to a model, whose evolution is followed step by step (which we will do, however, for the logistic map in chapters 21 and 22), but rather in being able to formulate some general predictions on the future state of the system, at least where possible, even without a formula that provides the exact evolutionary trajectory of the system in question.

We will return to this point shortly. Before embarking on a discussion of oscillating systems, however, in chapter 8 that follows, we would like to make a digression of a more technical nature, where we will better clarify the two concepts that we have already used on bases that are little more than intuitive: the concepts of dynamical system and of phase space.

---

[25] Technically, the solution of (7.1) and (7.2) is ascribable to the calculation of so-called elliptic integrals, a class of integrals that is frequently applied in mathematical physics, that cannot be expressed by means of algebraic, logarithmic or circular functions, but that can only be calculated approximately, using an integration by series.

8 Dynamical systems and the phase space

What we mean by dynamical system

Let us reconsider a point that was introduced in chapter 2, but specifying the terms better: we want to define what is meant by dynamical system. When we describe the evolution of a system we use several state variables, i.e. a set of magnitudes, chosen appropriately, functions of time, the values of which define everything we know about the system *completely* and *unambiguously*; let us indicate them, just to be clear, with $x_i(t)$. Thus the evolution of a system is given by the evolution of the set of n state variables, each of which evolves according to a specific deterministic law. In this sense, we can speak of a dynamical system, identifying a system of this nature with the system of equations that defines the evolution of all of the state variables.

A dynamical system, therefore, is defined by a set of state variables and by a system of differential equations, of the following type:

$$\frac{d}{dt}x_i = F_i(x_1,\ldots,x_n,t) \quad i = 1,\ldots, n \qquad (8.1)$$

In this case the dynamical system in continuous time is called a flow. Alternatively it can be defined by a set of state variables and by a system of finite difference equations, which make the system evolve at discrete time intervals:

$$x_i(k+1) = f_i(x_1(k),\ldots, x_n(k)) \quad i = 1,\ldots, n \qquad (8.1')$$

In this second case, the value of the variable x_i at instant $k+1$ depends on that of all of the n variables at instant k; a dynamical system defined thus is called a map (occasionally also cascade) (Ansov and Arnold, 1988).

A mathematical model is usually written in the form of a dynamical system, i.e. in the form of (8.1) or (8.1'). In addition to these, there are also other forms in which dynamical systems and models can be formulated: for example, cellular automata, which are essentially dynamical systems in which, similar to finite difference equations, where time is considered discrete, the set of values that the state variables can assume is also assumed to be discrete (Wolfram, 1994), or in the form of partial derivative equations or integral equations.

Speaking in general terms, as we have already observed in Chapter 7, rather rarely can we integrate a dynamical system such as (8.1) analytically, obtaining a law that describes the evolution of the set of state variables. The set of all of the possible dynamical systems (8.1) is in fact much more vast than the set of elementary functions such as sine, cosine, exponential, etc., that are used, in various ways, to integrate equations and to provide solutions expressed by means of a single formula (this is called at times 'in a closed form') (Casti, 2000).

We almost always need to use numerical integrations that transform the continuous form (8.1) into the discrete form (8.1'), or we need to use mathematical tools that enable us to study the qualitative character of the dynamical system's evolution (8.1). This has given rise to a sector of modern mathematics expressly dedicated to the study of dynamical systems and to the development of specific techniques for this purpose.

The phase space

The phase space (also called the space of states) is one of the most important tools used to represent the evolution of dynamical systems and to provide the main qualitative characteristics of such, in particular where the complete integration of a system is impossible. It is an abstract space, made up of the set of all of the possible values of the n state variables $x_i(t)$ that describe the system: for two variables the space is reduced to a plane with a pair of coordinate axes, one for $x_1(t)$, the other for $x_2(t)$; for three variables we would have a three-dimensional space and a set of three Cartesian axes, and so on for higher dimensions. In the case of the oscillation of a pendulum, for example, two state variables are needed to define the state and the evolution of the system: the angle $\vartheta(t)$ and the angular speed $\dot{\vartheta}(t)$.

Everything we know about the state of a dynamical system is represented by the values of the n coordinates in the phase space: knowing the values of the state variables is like knowing the values (dependent on time) of the coordinates obtained 'projecting' an object that we call a dynamical system on each of the n axes. The evolution of the system is given by the evolution of these variables in their entirety, in the sense that the law of dynamics (the law of motion) that defines the evolution of the system also defines the evolution over time of each of the n state variables. All that we need to know are the coordinates of the starting point, i.e. the initial values of the state variables. At times, however, it is more useful and easier to do the contrary, i.e. describe the system by obtaining its law of evolution from the law of evolution of each of the state variables considered individually. We should also add that a dynamical system is deterministic if the law of dynamics is such as to generate a single state consequent to a given state; on the other hand, it is stochastic, or random, if, consequent to a given state, there are more possible states from among which the dynamical system, in some way, can choose according to a probability distribution.

The phase space is a useful tool to represent the evolution of a dynamical system. Let us clarify this point with some basic examples. Let us consider a very simple case of dynamics, the simplest of all: a material point in uniform rectilinear motion in a direction indicated by a sole spatial coordinate x (Figure 8.1). In this case, (8.1) becomes simply: $\dot{x} = $ constant.

The state of this system is characterized by a single spatial magnitude: $x(t)$, the coordinate of the point as a function of time (hereinafter, for the sake of simplicity, we will not write the explicit dependence of x on time). Uniform rectilinear motion means that the increase of x starting from an initial value x_0, i.e. the distance travelled, is directly proportional to the time t elapsed from the moment in which the mobile point was in x_0. We represent the dynamics of the point on a set of three Cartesian axes, in which the magnitudes x (the position), t (time) and $\dot{x}$ (speed) are shown on

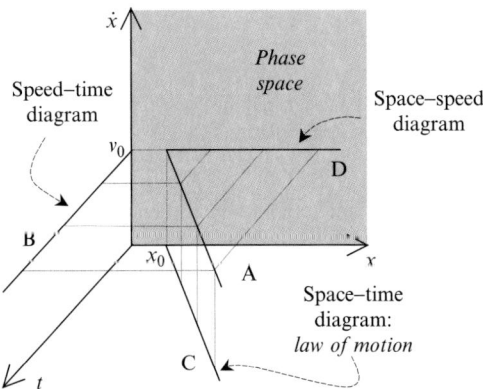

Figure 8.1 Dynamics of uniform rectilinear motion represented in a three-dimensional space x, $\dot{x}$, t and its projections on three planes.

the coordinate axes, as represented in Figure 8.1. In a case such as this, in which the motion is described by a sole coordinate, the plane $x\dot{x}$ is, precisely, the phase space (with just two dimensions: the phase plane).

It is then possible to consider the three diagrams that describe the uniform rectilinear motion: the space–time diagram (curve C in Figure 8.1), the speed–time diagram (curve B) and the space–speed diagram (curve D), as if they originated projecting a curve (curve A) on three planes xt, $\dot{x}t$ and $x\dot{x}$ in the three-dimensional space $x\dot{x}t$. It is immediately apparent that, in this case, this curve can only be a straight line, straight line A; a straight line, in actual fact, is the only curve that, projected on any plane, always and invariably gives a straight line.[26]

The same way of representing dynamics, applied to a case of motion that accelerates steadily with respect to a single variable x, gives Figure 8.2. Again in this case, in which (8.1) becomes $\dot{x} = kt$, the trend of the variables described by some of the known formulae of kinematics is immediately apparent.[27]

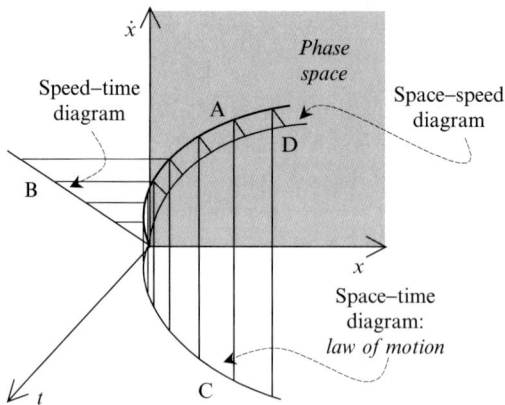

Figure 8.2 Dynamics of uniform accelerated motion represented in a three-dimensional space x, $\dot{x}$, t and its projections on three planes.

---

[26] More strictly speaking, we would like to remind readers that the equation that defines uniform rectilinear motion is $\ddot{x} = 0$; its integration is immediate and provides some of the laws of kinematics. In fact, a first integration with respect to time gives $\dot{x} = v_0$, the straight line B in Figure 8.1. A further integration gives $x = v_0 t + x_0$, curve C (law of motion). The speed, constant over time, is also constant in space, from which we obtain the horizontal straight line D in Figure 8.1 (space–speed diagram in the plane of phases). v_0 and x_0, respectively initial speed and initial coordinate, appear as constants of integration.

[27] The law of motion, in this case, is $\ddot{x} = a$, with a constant. By integrating once with respect to time and, for the sake of simplicity, assuming that the integration constants, i.e. the initial coordinate and the initial speed are nil, we have $\dot{x} = at$, curve B of Figure 8.2 (speed–

The usefulness of the phase space with respect to a space–time (the law of motion) or speed–time diagram lies in the fact that in the phase space the variables that appear are not represented as a function of time; time is only indirectly present, as the independent variable with respect to which space is derived.

This enables us to obtain a representation of the system's states in what we could call an atemporal way, that is to say in a perspective in which time does not play the primary role of the system's 'motor element', but rather the secondary role as a reference for the speed at which a spatial magnitude varies. In other words, the system is represented only with respect to the state variables (a bit like saying 'with respect to itself'), without the introduction of variables, such as time, that do not characterize the system itself, and in this sense, are not strictly necessary.

Therefore, the information shown in a phase space diagram is better, because it contains as much information as that provided by two space–time and speed–time diagrams, but it is expressed more succinctly in a single diagram. Obviously, this only applies to autonomous differential equations, in which time is not expressly present; in the case of non-autonomous systems, on the contrary, time is one of the variables which define the phase space.

The trajectories traced by a system in the phase space are commonly called *orbits*.

To conclude this digression into the phase space, we can add an initial definition of the concept of the stability of a system, a concept that we already briefly mentioned in our discussion of the pendulum, using the phase space as a tool to represent the dynamics; we will be discussing the concept of stability in further detail in Chapters 17–19. We will be looking at the concept of a system's stability again in Chapter 12, when we discuss the evolution of a system of two interacting populations.

If a system is in some way distanced from equilibrium, as a result of said movement, it may demonstrate different dynamics, depending on the type of law that governs it: (1) the system may evolve, remaining close to the abandoned equilibrium configuration, which, in this case is a stable equilibrium; (2) the system may tend to return to the abandoned equilibrium, which, in this case is a asymptotically stable equilibrium; (3) the system may distance itself further still from the equilibrium, which in this case is known as unstable equilibrium.

---

time diagram); by integrating a second time, we have $x = at^2/2$, curve C (law of motion). By eliminating time from the last two equations, we obtain the expression of speed with respect to space, i.e. the trajectory in the phase space: $\dot{x} = \sqrt{2ax}$, i.e. curve D (space–speed diagram).

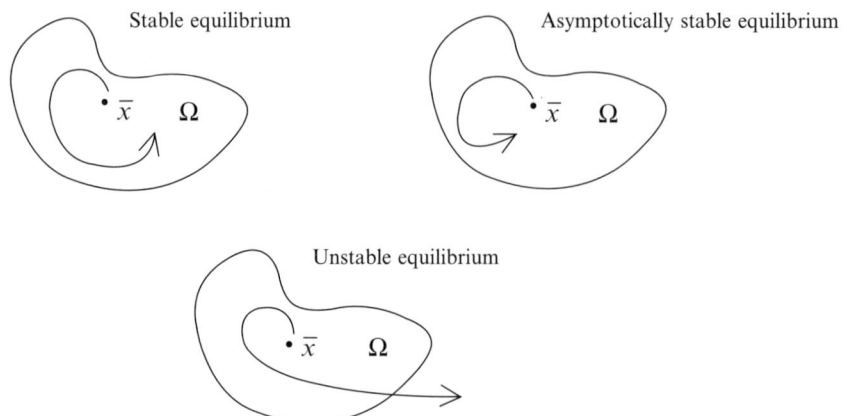

Figure 8.3 Depiction of stable, asymptotically stable and unstable equilibrium.

More precisely, we can indicate the set of the values of the coordinates of a point of equilibrium as $\bar{x}$: if the system's orbit remains indefinitely confined to a neighbourhood Ω of $\bar{x}$ (Figure 8.3), then $\bar{x}$ is a point of stable equilibrium (this is the case, for example, of the *non-damped* oscillations of a pendulum around the rest position). If on the other hand, as time tends towards infinity, the successive states of the system always tend increasingly towards $\bar{x}$, then $\bar{x}$ is an asymptotically stable point of equilibrium (this is the case, for example, of the dampened oscillations of a pendulum around a position of equilibrium). Lastly, if no neighbourhood of $\bar{x}$ can be found that satisfies what has been defined, then $\bar{x}$ is an unstable point of equilibrium of the system (this is the case, for example of a pendulum placed vertically, with the mass above the centre of rotation, as in case C in Figure 7.1, in which $\vartheta = \pi$).

Oscillatory dynamics represented in the phase space

Let us now return to the dynamics of the pendulum according to the linear model (Model 1: small oscillations without friction) that we described in Chapter 4 (hereinafter we will refer to the linear model by its proper name in mathematical physics: 'harmonic oscillator'). The considerations that we will make are quite technical: on one hand they can be considered as an example of the use of the concepts that we have introduced so far (the phase space, stable/unstable equilibrium, etc.); on the other, they actually enable us to introduce other concepts (attractor, unstable trajectories, etc.) that we will also find in cases of less basic dynamics.

As we know, the solution of the equation of motion (5.2) is the law of motion (5.3), in which $\omega^2 = g/l$. They are both shown below:

$$ml\ddot{\vartheta}(t) = -mg\vartheta(t)$$

$$\vartheta(t) = \vartheta_0 \cos \omega t$$

Equation (5.2) is a second order equation that can be either fully (i.e. twice) integrated with respect to time or just once. By integrating once only, with some intermediate steps, we thus obtain a first order differential equation, a so-called first integral of motion:[28]

$$\frac{1}{2} ml^2 \dot{\vartheta}^2 + \frac{1}{2} mgl\vartheta^2 = \text{constant} \qquad (8.2)$$

The first integral (8.2) expresses the well-known fact that during the oscillation of a system, there is one magnitude that remains constant over time, and is therefore called a 'constant of motion': we will indicate it with E, and we will call it the system's total energy. In the case in which the harmonic oscillator is in actual fact the linear pendulum we discussed earlier, we can demonstrate that the two terms that are summed to the first member of (8.2) are, respectively, the kinetic energy and the potential energy of the pendulum, and the total energy is simply the mechanical energy. A system in which mechanical energy is a constant of motion is called *conservative*.[29]

---

[28] More precisely, given a differential equation of order $n \geq 2$, all relationships between the independent variable, the unknown function and its first $n - 1$ derivatives are called first integrals. The study of the first integrals of the equations of motion enables us to identify constants of motion that can be as fundamental to the description of the phenomena, as the complete solution of equations of motion. It is precisely from the examination of first integrals that in the eighteenth century, the formulation of important general principles (or rather theorems) for the conservation of mechanical systems was achieved, such as the theorem of the conservation of mechanical energy, of linear momentum and of angular momentum, a theorem that played a fundamental role in the birth and the development of deterministic concepts in the science of the era (see Chapter 3).

[29] The fact that a constant of motion, mechanical energy, *exists* is the well-known theorem of the conservation of mechanical energy of a material point, one of the greatest conquests of mathematical physics and, in general, of scientific thought of the eighteenth century. This *theorem* represents a specific case of the general *principle* of energy conservation (if non-conservative forces are acting), fundamental to all branches of physics. We observe that we have discovered that in oscillations, a constant of motion *exists*, without fully integrating the equation of motion (5.2), therefore without calculating the real law of motion (i.e. the law that links space to time).

In the phase space defined by the variables ϑ and $\dot{\vartheta}$ (in this case, therefore, it is a phase plane), the curves represented by (8.2), as E varies, constitute a set of ellipses, the centre of all of which is at the origin of the coordinate axes and the lengths of whose semiaxes are, respectively, $\sqrt{2E}$ and $\sqrt{2E}/\omega$. Therefore, the dimensions of the orbit in the phase space depend on energy E. In Figure 8.4 we show several different values of E.

Each of the ellipses traced is the set of the points in the phase space that define states of equal total energy. The arrows indicate the direction in which the orbits are travelled with the passing of time.

As the system travels along one of the indicated orbits, the energy of the pendulum is transformed from kinetic to potential, then from potential to kinetic, then from kinetic to potential again and so on; the two forms of energy alternate periodically, maintaining their sum constant. As in the approximation considered, in which friction is not introduced, there is no dispersion of mechanical energy, because the latter is a constant of motion, the possible orbits are all isoenergetic and correspond to different quantities of total energy. Different orbits do not intersect each other, because a point of intersection, belonging to two different orbits, would result in two different values of the total energy (which is constant) for the same system.

Let us leave the harmonic oscillator and return to the linear pendulum subject to friction described in equation (5.4) (Model 2). With considerations similar to those described above, we discover, in this case, that the total mechanical energy (only the mechanical energy!) E is *not* maintained, but dissipates over time.[30] In this case the system is called dissipative; the

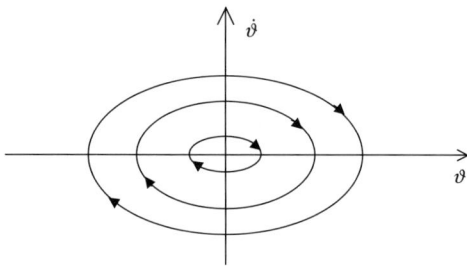

Figure 8.4 Representation of the motion of a harmonic oscillator in the phase space for different values of E.

---

[30] It must be absolutely clear to the reader that, when we speak of dissipated energy, we are only referring to mechanical energy (potential energy + kinetic energy), *not* to total energy (mechanical energy + thermal energy + energy of the electromagnetic field + mass energy +...). According to current scientific vision and knowledge, the total energy of a closed system is *always* conserved in *any* process.

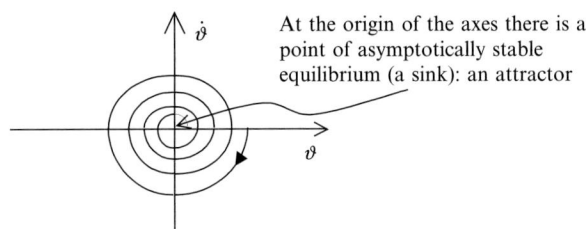

Figure 8.5 Representation of the motion of an oscillating system in the phase space that undergoes a damped oscillation in the presence of friction.

orbit in the phase space $\vartheta\dot\vartheta$ is now made up of a spiral with an asymptotically stable point of equilibrium at the origin of the axes (sometimes this point is called a 'sink'), i.e. at the angle $\vartheta = 0$ and with the pendulum still: $\dot\vartheta = 0$ (Figure 8.5).

We have therefore met a first, very basic example of a dissipative system, whose orbit in the phase space is an open curve. We will repeatedly come across dissipative systems characterized by non-periodic orbits in the second part.

Let us now consider the motion of the pendulum that makes oscillations of large amplitude in the simpler form described by (7.1), i.e. ignoring friction (undamped oscillations: Model 3). Equation (7.1) can be rewritten as follows:

$$m\ddot\vartheta + m\omega^2 \sin\vartheta = 0 \tag{8.3}$$

By integrating (8.3) once with respect to time, as we did with (5.2), we discover the existence of a magnitude that is maintained constant over time. Here again, as for (8.2), we are dealing with total mechanical energy E:

$$\frac{1}{2}ml^2\dot\vartheta^2 + mgl(1 - \cos\vartheta) = E \tag{8.4}$$

Just as in (8.2), (8.4) which represents the general case for any angles ϑ, the first addend to the first member is the pendulum's kinetic energy, while the second addend is the potential energy.

Oscillations of small amplitudes mean small energies and, as we have said, in this case (8.2) gives elliptic orbits. For oscillations of amplitudes that cannot be considered 'small', i.e. if linear approximation is not acceptable, the dynamics of the pendulum is no longer that provided by the model of the harmonic oscillator (5.2) and the orbits in the phase space are curves other than ellipses (8.2). We develop the term $\cos\vartheta$ in Taylor series around $\vartheta = 0$:

$$\cos\vartheta = 1 - \frac{\vartheta^2}{2!} + \frac{\vartheta^4}{4!} - \frac{\vartheta^6}{6!} + \ldots$$

If we ignore, for small ϑ, the terms above the second order, the equation (8.4) leads back to equation (8.2) and therefore we return to linear approximation. If, on the other hand, as the amplitude of oscillation ϑ increases, we progressively consider the subsequent terms of the series development of $\cos\vartheta$, the curves that (8.4) gives in the phase space defined by variables ϑ and $\dot\vartheta$ remain closed, but we get increasingly further away from the ellipse shape (8.2), with an increasingly marked difference as energy E grows. We can demonstrate that, in correspondence to a critical value of E, the orbits become closed curves formed by sinusoidal arcs in the upper half-plane and by sinusoidal arcs in the lower half-plane. For higher energy values, the orbits are not even closed curves: the pendulum no longer oscillates around a point of stable equilibrium, but rotates periodically.[31]

---

[31] Let us add some technical considerations that are not indispensable to the discussion presented in the text. Remembering that $\omega^2 = g/l$, we have from (8.4):

$$\dot\vartheta^2 = \frac{2E}{ml^2} - 2\omega^2 + 2\omega^2\cos\vartheta \qquad (8.5)$$

which, for small values of ϑ, is similar to the form of the harmonic oscillator:

$$\dot\vartheta^2 \cong \frac{2E}{ml^2} - \omega^2\vartheta^2 \qquad (8.5')$$

which is simply (8.2) rewritten expressing $\dot\vartheta^2$.

As the mechanical energy E (positive) varies, (8.5) represents a set of orbits in the phase space defined by the variables ϑ and $\dot\vartheta$ (Figure 8.6):

1. For $0 < E < 2ml^2\omega^2$, we have the *closed orbits* cited in the text: (i) in the limit case of small values of the angle of oscillation ϑ (the case of the harmonic oscillator), they become the ellipses given by (8.5'); (ii) in the general case of large values of ϑ, they are the closed curves (deformed ellipses) given by (8.5).
2. For $E = 2ml^2\omega^2$, the critical value of the energy cited in the text above, we have the case of *closed orbits* composed of arcs belonging to the two cosine curves that intersect each other on the axis of the abscissas of Figure 8.6. This is the limit case that separates the previous case from the following case: it corresponds to the motion of a pendulum whose energy is 'exactly' equal to that which would allow it to reach the vertical position above the centre of oscillation (point C in Figure 1.3) with nil speed and in an infinite time (the total energy E is 'exactly' equal to the potential energy of the pendulum in equilibrium above the centre of rotation: $U(\vartheta) = 2mgl$).
3. For $E > 2ml^2\omega^2$, (8.5) represents a set of cosine curves, symmetrical in pairs with respect to the axis of the abscissas, shifted higher (or lower) with respect to those of the previous case, the higher the value of E, and with amplitudes that decrease with the entity of the

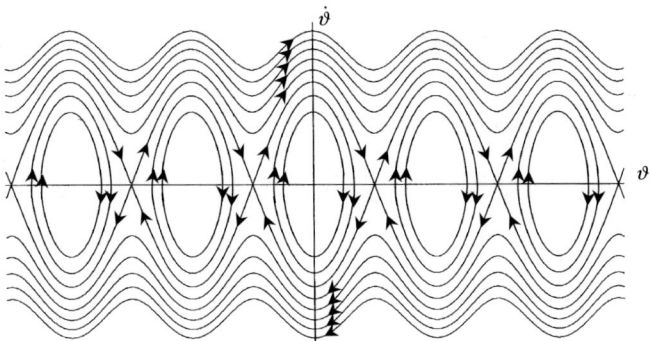

Figure 8.6 Orbits in the phase space of an undamped pendulum.

shifting. These are *open orbits* shown at the top (and bottom) of Figure 8.6: the pendulum has an energy E that is so great that it does not oscillate, but rotates periodically at a speed $\dot{\vartheta}$ that is not constant (the orbits are oscillating functions), but, on average, is greater the higher E is.

Extension of the concepts and models used in physics to economics

Jevons, Pareto and Fisher: from mathematical physics to mathematical economics

Models such as the oscillators, in particular the harmonic oscillator illustrated in chapter 8, constitute a set of mathematical models of physical systems that have been studied for some time and that represent some of the first and most important accomplishments of classical-era mathematical physics. In principle, models of this kind can be used to describe any oscillating system, with a few simple adaptations that don't affect their substance. In particular, the linear model of the harmonic oscillator is particularly suited to describing all cases of oscillating systems for which linear approximation is considered acceptable: these range from systems in physics, e.g. when we describe the movement of electrons in an oscillating electric circuit, or of the electrical or magnetic fields of an electromagnetic wave, to systems in other disciplines in which we believe that we can observe periodic oscillations over time, such as an economic system, a population, etc.

Taking a concept or a model that has enjoyed success in physics and applying it to social sciences was in actual fact one of the routes followed by physicalism, as we mentioned in Chapter 1. The general idea is that the nature of the real world (i.e. the universe and everything in it) conforms to the condition of being physical; when we observe oscillations of a certain empirically measured magnitude that remain constant over time, we can define other constants of motion (the first integrals), such as for example the energy in physical systems, giving them different meanings according to their context.

Between the last decades of the nineteenth century and the first decades of the twentieth century, considerations and concepts like those that we have mentioned so far, typical of a particular sector of mathematical physics called rational mechanics, have been transferred from physics to economics, traditionally the most quantitative of the social sciences, through real homologies. We have long been trying to build economic models that apply mathematical laws such as those we have illustrated so far, attempting to create a real mathematical economics on the model of rational mechanics.

The forefathers of marginalist neoclassic economics, the authors of this economic concept, attempted to literally copy the models used in physics, term by term, symbol by symbol, in a methodical and organic way, explicitly declaring their intentions of transferring the concepts and the methods used in physics to economics (Cohen, 1993). Economists such as William Stanley Jevons, Léon Walras, Vilfredo Pareto and Irving Fisher stated that their objective was to transform economics into a 'real' science, choosing physics as a reference.

There were various reasons for this: on one hand, physics was a science that they were all well acquainted with (Jevons graduated in chemistry at University College London, where, later, he became a professor of political economics, having been a student of the mathematician George Boole; Pareto graduated in civil engineering from Turin Polytechnic; Fisher had been a student, at Yale University, of Josiah Williard Gibbs, an eminent figure in mathematical physics and one of the fathers of theoretical thermodynamics); on the other hand, because physics, more than any of the other sciences, was very highly considered and enjoyed a general appreciation thanks to the significant successes it had achieved. However, above and beyond all of this was the important fact that physics was characterized, more than any other scientific discipline, by its extensive use of mathematics. Above all, this last aspect constituted the primary element, according to these economists, but not just them, to give a discipline such as economics, whose mathematical foundations were rather vague at the time, a more scientific character.[32]

Concepts were transferred from physics to economics in a number of different ways. Jevons, for example, sustained that in economics, the notion

---

[32] The economists cited in the text all belong to the marginalist school that reached its height of development during the last 30 years of the nineteenth century and of which the same Jevons, in Great Britain, and Walras, in France, were among the founders. For marginalists, the use of mathematics became common practice, although there were of course exceptions, such as, for examples, the marginalist economists from the Austrian school, firstly Carl Menger, who, unlike the marginalists of the French or British schools, considered the number that expresses the marginal value of a good not as a cardinal measure of the 'psychological pleasure' attributed to the good, but only as an indication of an ordinal nature, that could only be used to make comparisons, and not for quantitative calculations.

of value played the same role as that of energy in mechanics, and maintained that goods, in economics, can be attributed the same dimension that mass has in physics, and furthermore, that utility corresponds to Newton's gravitational force.

Pareto was convinced that there were close and surprising similarities between the equilibrium of a economic system and the equilibrium of a mechanical system. In his famous *Cours d'économie politique* (1896), he even went as far as drawing up a table to assist readers that had no knowledge of rational mechanics, in which he compared mechanical and economic phenomena, establishing strict parallels between the two disciplines.

Fisher also published a table in which he compared rational mechanics and economics, although he pushed the homology much further than Pareto. According to Fisher, for instance, the magnitudes used in economics, just as those used in physics, could be attributed scalar or vectorial properties. Fisher, for example, suggested the following analogies between physical magnitudes and economic magnitudes: particle–individual, space–basic products, force–marginal utility (both vectorial quantities), energy–utility (scalar quantities), work–disutility (scalar quantities; 'work' obviously is given the meaning it has in physics, i.e. the scalar product of a force applied to a body times the displacement made by the same). In any event, it should be noted that the analogies suggested by Fisher were criticized by other authors who were working in the same field. Their reasons were varied; for example, the fact that Fisher did not take due account of the principle of energy conservation that, applied to a closed economic system, would imply that the sum of total expenditure and of total utility has to be constant (Cohen, 1993).

Schumpeter and Samuelson: the economic cycle

The idea of applying mathematical techniques, that originated from classical physics, to economics, was developed and examined in further detail in the years that followed those of the first mathematical economists, even though under very different perspectives and without seeking or imposing strict parallels between physical and economic magnitudes, as Jevons, Pareto and Fisher had done in the past. In particular, these scholars concentrated their efforts on the idea of using a model to describe and reproduce the continually alternating occurrence of prosperity and depression that was observed in modern market economics, attempting to use models of various types of oscillations (linear, nonlinear, damped, forced, etc.) that had enjoyed so much success in physics.

The use of mathematical models of oscillations in economics therefore finds its most distant roots in the mathematical approach to the study of physical phenomena that originated in the seventeenth century, but that now abandons the assumption that there are strict homologies with physics. In this context, we no longer speak of magnitudes that simply *replace* those of physics, assuming the same role as the latter, but, for example, of a macroeconomic magnitude, usually gross domestic product or income, that oscillates over time according to a sinusoidal law, *in analogy* to the action of the harmonic oscillator; the concept is taken from physics, but the analogy with physics is now entirely formal.

From the end of the 1930s onwards, this new perspective generated a long series of mathematical models depicting the economic cycle, identified as a cyclical succession of phases (for example, as Joseph Schumpeter, 1939, suggested, the phases of prosperity, recession, depression and recovery), hypothesizing that there was a more or less sinusoidal mathematical law that could describe the regular patterns observed.

One of the first and fundamental models was the famous linear model formulated by Paul Samuelson (1939, 1947; see also Allen, 1959). Samuelson's model was made up of three equations: the consumption function, in which current consumption depends on the income of the previous period; the investment function, in which investment depends on the variation in income of the previous period; and the condition of equilibrium between production and demand in the goods market, to which a form of expenditure, called autonomous, is added, which is considered exogenous. The model, in its original form, produces a non-homogeneous second order linear equation, like the ones we presented in chapter 5 relative to the oscillations of a pendulum.[33] Samuelson's linear model envisages four possible dynamics: (1) divergence without oscillations, (2) divergence with oscillations, (3) convergence without oscillations, and (4) convergence with oscillations; two of these, (3) and (4), have exactly the same dynamics that are possible for a non-forced pendulum, those represented in Figures 8.4 and 8.5.

---

[33] In reality, the equation that Samuelson obtains is not a differential equation in continuous time, but a finite difference equation in discrete time, such as:

$$Y_t + aY_{t-1} + bY_{t-2} = A$$

where Y is the macroeconomic magnitude in question (for example income), that in the equation is considered at *three* different and successive points in time (indicated by $t-2$, $t-1$ and t), a and b are suitable parameters and A is autonomous expenditure. In our context, however, we can confuse the continuous and discrete nature of time, without this resulting in fundamental errors in the dynamics calculated.

Later, John Hicks (1950) made Samuelson's model nonlinear. The renowned, fundamental model of Goodwin–Kalecki–Phillips (Goodwin, 1951; Kalecki, 1954; Phillips, 1954, 1957) represented a further development of Samuelson's model in a nonlinear form, introducing new hypotheses. We will not be discussing the contents of these models; readers should refer to the cited references and to the wealth of literature on mathematical economics (for example, Allen, 1959).

Dow and Elliott: periodicity in financial markets

The models cited, from Samuelson's to the one developed by Goodwin–Kalecki–Phillips, which represent some of the most important milestones in the history of economic thought, were formulated making extensive use of mathematics and with clear analogies to the cyclical oscillations of systems in physical mathematics.

Alongside the cited models, other descriptions of certain specific economic systems in terms of oscillations were also developed, but without strict mathematical formulations. We are referring to models of the dynamics of the financial markets, i.e. the trend over time of the values of share prices and market indices. They are not mathematical models in the real sense of the word, and even less so deterministic models; however we mention them as examples of attempts to model human behaviour. A basic assumption of these models is that the trends in share prices and indices can be largely ascribed to dynamics that are based on human psychology; not the psychology of the single individual that operates on the market (the investor), but that of the mass of investors. We could even go so far as saying that efficient[34] and competitive financial markets, particularly share markets, are one of the contexts that best reflects mass psychology (Batten, 2000a).

Studies of this nature, aimed at highlighting recurrences in the trends of share prices, have mainly been developed in America, especially from when, at the end of the nineteenth century, the goods and financial markets have continued to increase in size. A pioneer of these theories was Charles Dow, who was the first to believe that he had identified recurrences in the form of cycles. In 1900, he actually wrote in the *Wall Street Journal*: 'The market is always considered as having three movements, all going at the same time.

---

[34] A market is said to be efficient when all information that is available instantaneously reaches all operators, is immediately processed and gives rise to a new value of the goods traded.

The first is the narrow movement from day to day. The second is the short swing, running from two weeks to a month or more; the third is the main movement, covering at least four years in duration' (cited in Batten, 2000a, p. 213).

Furthermore, Dow sustained that the stages of share price trends were characterized by three successive waves. The first is the rebound from the excess of pessimism from the decline that preceded it, the second reflects the real increase in earnings, while the third is simply caused by an overestimation of values. Moreover, again according to Dow, there is a time at which any upwards or downwards movement of the market is interrupted and is immediately followed by a counter-movement that recovers a part that can be between 2/5 and 3/5 of the previous movement, which is called *retracing*.

Around 30 years after Dow's observations, during the period immediately following the Wall Street stock market crash of 1929, the idea that the trend of the financial markets presented recognizable patterns that repeated themselves, almost as if they were real recurrent patterns, was taken up by Ralph Nelson Elliott, who developed and made the theory outlined by Dow more generalized and precise. According to Elliott, the collective behaviour of investors gives rise to phases of growth and decline of share prices and indices, which are manifested, in the graphs that report the values of share prices over time, in the form of periods of growth followed by periods of decline, which follow each other at more or less constant rhythms although with different amplitudes: these are known as *Elliott waves*.

According to Elliott's assumptions, in the graphs, characteristic phases of different periodicity can be identified, that provide a basis to describe the trends of the share prices in question, and more importantly, to formulate (an attempt at) forecasting (Elliott, 1946; see Frost and Prechter, 1990). In particular, Elliott's complete cycle is made up of five waves that drive share value upwards, followed by three waves that drive them back down and that represent the so-called correction of the previous phases. One of the most interesting aspects of this theory is that each of these waves can be broken down into a reduced time-scale, into an analogous cycle of eight waves, each of which can be further broken down in more or less the same way and so on. In this way, Elliott identified a sort of invariance of scale, which we would nowadays call a fractal structure, i.e. 40 years before Benoît Mandelbrot formally introduced the mathematical concept of a fractal object and applied it to the financial markets, using the concept of invariance of scale in a more detailed and complete way than Elliott had (Mandelbrot, 1997ab, 1999ab).

Elliott's pattern can be applied to any market that is characterized by efficiency, liquidity and fluidity in its price movements and by the competitiveness of the operators. It therefore finds its most natural application in the graphs of stock market indices, those of futures indices as well as the graphs of the most extensively traded shares.

At the basis of the Elliott wave principle is the assumption that the financial markets reflect a general principle of natural symmetry: symmetry that manifests itself, in this context, both in the succession of waves according to periodic patterns and in the invariance of the trends of share prices on different time-scales.[35] Elliott's deep-rooted conviction of the existence of this general principle of symmetry is the origin of the title he gave to his most famous book *Nature's Law. The Secret of the Universe* (1946), in which he fully illustrated the wave theory that he had developed in a series of previous articles.

We will now return, for the last time, to our discussion of the pendulum in a situation that is totally different to the ones examined up until now.

---

[35] The fact that there are patterns in the trends of share prices and indices, even though not truly periodical, but at least that repeat over time, is one of the fundamental postulates of the so-called technical analysis of the financial markets, of which Elliott waves, still today, are one of the most important elements.

10 The chaotic pendulum

The need for models of nonlinear oscillations

In this part we have discussed oscillations at some length. Whether we are examining a pendulum or a mass connected to a spring, an electrical current in a specific circuit or a wave broadcast by an antenna, economic cycles (real or assumed) or the listing of a stock, or further still, in a more general and abstract sense, the condition of dynamic equilibrium of a system that is interacting with other systems, the mathematical model of oscillations represents a basic schema that can be adapted, by making the appropriate corrections, to fit the case in question. A wide number of variants of the model of motion of an oscillating system can be constructed. These depend not so much on the nature of the specific system being studied, but rather on the different external conditions in which the system finds itself, or, to put it better, that we wish to be taken into account. The oscillations described by a linear model, for example, can be applied to the dynamics of a pendulum as well as to a mass attached to a spring, a current in an electric circuit, etc., as long as the disturbances to the state of stable equilibrium are small. This is precisely what the model does: it abstracts the common characteristics, in this case the dynamics of the motion close to the condition of stable equilibrium, ignoring the other characteristics that define the specific oscillating system. If the dynamics envisaged by the linear model do not successfully adapt to the description of the oscillations observed, this may be due to the fact that the oscillating system is not close to its stable equilibrium, or even that the oscillating system is so disturbed that it cannot find a stable equilibrium. In this case a new model of nonlinear oscillations, one of the many variants that can be conceived starting from the basic model of the harmonic oscillator, must be used to provide an acceptable description of the dynamics observed.

The case of a nonlinear forced pendulum with friction (Model 4)

Of the numerous possible variants of the model of an oscillator, we will illustrate a further one, without entering into its mathematical details, given its singular nature and importance to our discussions on chaos in the second part of the book. We will be referring to the case of a forced pendulum, i.e. a pendulum subjected to the action of another force in addition to that of gravity, that is free to rotate a 360° angle; we are specifically interested in the case in which the oscillation pivot, under the action of an external force, makes sinusoidal oscillations along a vertical axis. In this way the pendulum moves due to the force of gravity, friction and the force of a sinusoidal oscillation, acting on the pivot of the pendulum. The dynamics of the system can be expressed by the following equation (again, for the sake of simplicity we will not include the dependence of ϑ on time):

$$ml\ddot{\vartheta} + kl\dot{\vartheta} + mg(1 + A\phi^2 \cos \phi t)\sin\vartheta = 0 \qquad (10.1)$$

where A indicates the maximum amplitude of the vertical oscillation of the pivot, and ϕ is the ratio between the oscillation frequency of the pivot and the oscillation frequency of the pendulum; compare (10.1) with (5.4) for the other symbols used.

This is a nonlinear and non-autonomous system, in which the resulting acting force depends explicitly on time through the term $A\phi^2 \cos \phi t$. Equation (10.1) differs from (7.2), applicable to the case of nonlinear unforced oscillations, due to the presence of the product $A\phi^2 \cos \phi t$ in the term that gives the restoring force. A system described in this way was discussed by Rayleigh (1902) and has been studied in more depth in recent years (Leven et al., 1985).[36] The pendulum thus described is illustrated in Figure 10.1.

We will not enter into the technical details of the dynamics that this system may manifest, which are rather complex. We will limit ourselves to pointing out that this pendulum, in certain circumstances, represents an initial example of a type of dynamics that we will be discussing at length in the second part: chaotic dynamics. The peculiarity of this system, in fact, is that for certain values of parameters A and ϕ, the dynamics become unstable and substantially unpredictable. The fact that the force applied

---

[36] Alternatively, and without making significant changes to equation (10.1), we could also consider the case of oscillations of the pivot along a horizontal axis. Obviously, the most general case is that in which the pivot moves according to any law.

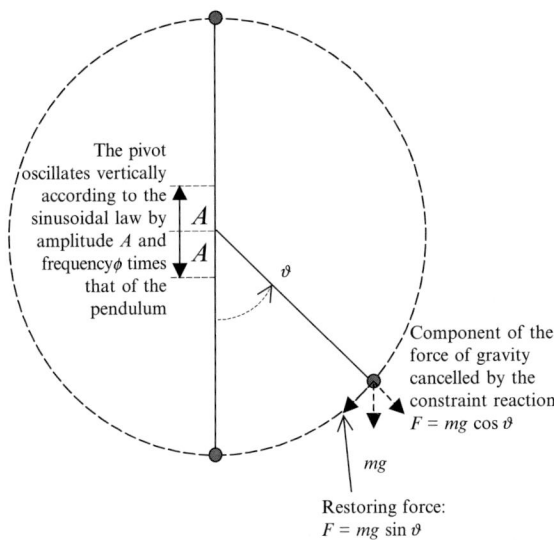

Figure 10.1 The case of a nonlinear pendulum forced on the pivot.

to the pivot is variable and that it acts with different intensities and in a different way at each phase of motion of the pendulum can cause different effects, such as a brusque change in the oscillation amplitude, or a sudden complete rotation, or similarly a sudden change in the direction in which the pendulum is moving. In general, the dynamics of this system are therefore unpredictable, even though the forces in play are both known: the tangential component of the force of gravity, that acts on the oscillating mass, and the sinusoidal force of known amplitude and frequency, that acts on the pivot.

In addition to the above, it is interesting to note that it can be shown (Beltrami, 1987) that for specific values of the amplitude and frequency of the force applied to the pivot, A and ϕ respectively, the pendulum can even find a position of stable equilibrium upside down, i.e. with the mass above the oscillation pivot. Obviously, as we have seen in chapter 9, a position of equilibrium with the oscillating mass above also exists for an unforced pendulum, which, however, unlike the case that *can* take place in the forced pendulum, is unstable equilibrium, that cannot be achieved in practice merely by applying a force that moves the mass that is still in a position of stable equilibrium.[37]

---

[37] It can be demonstrated that, by putting the oscillating mass of a unforced pendulum, still at the bottom, in motion, by applying a force (a blow), of a suitable intensity, the mass reaches a position of unstable equilibrium, at the top, with a decreasing speed that becomes

We will end our discussion of models of oscillating systems here; further and more in-depth examination of these and other dynamical systems can be found in the wealth of literature dedicated to this topic; for example Jordan and P. Smith (1977); Arnold (1983, 1992); Beltrami (1987); Anosov and Arnold (1988); Perko (1991); Moon (1992); Bellomo and Preziosi (1995); Alligood, Sauer and Yorke (1996).

nil when the mass stops, but in an infinite amount of time. *In practice*, this means that an unforced pendulum cannot reach a state of unstable equilibrium.

11 Linear models in social processes: the case of two interacting populations

Introduction

After having presented, in the paragraphs above, several versions of one of the simplest and most important models in mathematical physics, the model of the oscillator, which has served as a basis for developments and extensions into other sectors, we will now look at another example of modelling that has become a classic theme in social science: the case of a system of two interacting populations.

A number of different theories have been adopted in the approach to this subject which have given rise to a numerous family of models. In general these are models that are essentially of theoretical interest: they do not represent anything more than an initial attempt to model extremely complex phenomena such as social ones (in this context, naturally, the term social is intended in a broad sense, referring to any group of individuals, identifiable according to certain parameters, that interact with another group of individuals).

The simplest hypothesis on which a model can be formed is that in which the two populations interact with each other in a manner that can be expressed in a purely linear form; this hypothesis, however, is unrealistic, as we have already pointed out several times for other types of linear models. The linear model of two populations interacting with each other that we will illustrate in this part of the book, constitutes a very schematic example of how an extremely simplified mathematical model can be constructed to describe phenomenologies that belong to sciences that, at least in the past, have been little mathematized, and of how mathematical techniques developed in certain contexts (differential equations in mathematical physics) can be usefully transferred to other environments.

In the picture that we will be illustrating, we will not enter into the mathematical details; for a more in-depth exploration of the technical aspects, we recommend the reader consult specific literature on the topic of differential equations and dynamics systems, such as for example Zeldovich and Myškis (1976); Elsgolts (1977); Jordan and Smith (1977); Bender and Orszag (1978); Arnold (1983, 1992); Beltrami (1987); Anosov and Arnold (1988); Perko (1991); Moon (1992); Epstein (1997); Bellomo and Preziosi (1995); Hannon and Ruth (1997ab); Riganti (2000).

The linear model of two interacting populations

Let us consider two populations P_1 and P_2 evolving over time; the number of members of P_1 and P_2 are respectively $n_1(t)$ and $n_2(t)$. The two populations interact with each other in a linear manner, that is to say in such a way that the instantaneous speeds of change of $n_1(t)$ and $n_2(t)$, i.e. their first derivatives with respect to time, $\dot{n}_1(t)$ and $\dot{n}_2(t)$, are directly proportional to a linear combination of the same values of $n_1(t)$ and $n_2(t)$. For the sake of clarity, imagine the case of two populations that collaborate for their reciprocal development in the presence of unlimited natural resources. We can therefore deduce that the higher the number of individuals in each population the faster the growth of said population.

Let us now assume that the speed of growth of population P_1, indicated by $\dot{n}_1(t)$, is proportional to the number of individuals n_1 of the same population, through coefficient k_{11}, and also to the number of individuals n_2 of the other population, through coefficient k_{12}. In other words, we assume that the speed of growth of P_1 is given by a linear combination of the numbers of individuals of P_1 and P_2. The same is valid for P_2. In formulae, this translates into the following system of first order homogeneous linear differential equations with constant coefficients:

$$\begin{cases} \dot{n}_1(t) = k_{11}n_1(t) + k_{12}n_2(t) \\ \dot{n}_2(t) = k_{21}n_1(t) + k_{22}n_2(t) \end{cases} \quad (11.1)$$

The (real) values of the coefficients k_{ij} of the linear combination in the second member determine the contribution that *each* of the two populations, individually, makes to the speed of growth of *each* population. The values of coefficients k_{ij}, therefore, express the reciprocal relationship between the growth (or decline) of the two populations.

Depending on the sign of each of the coefficients k_{ij}, we have a constructive action if $k_{ij} > 0$, or destructive action if $k_{ij} < 0$. More specifically, k_{12}, in the first of the equations (11.1), and k_{21}, in the second of the equations (11.1), relate the speed of growth of a population to the number of individuals in the other. For example, a positive value of k_{12} indicates the positive contribution by P_2 to the speed of growth of P_1, in the sense that the higher the number of individuals in P_2, the faster population P_1 grows. Coefficients k_{12} and k_{21}, taken separately, therefore represent the effect of the cooperation (if both are positive) or of the competition (if both are negative), on P_1 by P_2 and on P_2 by P_1 respectively, or of the parasitism of a population on another (if k_{12} and k_{21} have opposite signs).

The coefficient k_{11} in the first of the equations (11.1) and the coefficient k_{22} in the second of the equations (11.1) indicate the effect of the number of individuals on the speed of growth of the population to which they belong. Therefore, if $k_{11} > 0$, population P_1 sustains itself and grows exponentially (à la Malthus), for example following an increasing amount of reproductive activity of the number of individuals n_1 of population P_1; if on the other hand $k_{11} < 0$, the speed of growth of P_1 decreases in proportion to the growth of the number of individuals n_1 in population P_1, for example due to the effect of internal competition within population P_1, caused by a lack of food resources (overpopulation crises).

Some qualitative aspects of linear model dynamics

Without resorting to the actual integration of (11.1) (so-called quadrature), we can often form a qualitative picture of the dynamics of a system by examining the state of equilibrium of the dynamics. The understanding of the behaviour of the system in relation to states of this nature, once identified, can actually enable us to predict the general behaviour of the system in any other condition: in actual fact the system evolves by following trajectories that distance themselves from the (stationary) states of unstable equilibrium and come closer to the (stationary) states of stable equilibrium. It is as if we were studying the dynamics of a ball that is rolling on ground that has rises and depressions; more than the slopes between them, where the dynamics are obvious and easy to describe, we would be interested, in order to have a clear picture of the general dynamics, in identifying the peaks of the rises, as points of unstable equilibrium (repellers) and the bottoms of the depressions, as points of stable equilibrium (attractors).

Using the metaphor of the pendulum again, a subject that has been discussed at some length in the previous paragraphs, a point of stable equilibrium is that occupied by the pendulum in a vertical position under the pivot, while at a point of unstable equilibrium the pendulum occupies a vertical position above the pivot.

The states of equilibrium, whether stable or unstable, are therefore the fundamental references of the system's dynamics. This justifies the interest taken in their identification and in the study of the system's dynamics in the neighbourhood of the equilibrium points.

Establishing the states of equilibrium of the system of equations (11.1) is not difficult. Based on the definition of derivative, they are those in which the first derivatives $\dot{n}_1$ and $\dot{n}_2$ are both equal to zero (for the sake of simplicity, from now on we will not indicate the dependence of n_1 and n_2 on time):

$$\begin{cases} 0 = k_{11}n_1 + k_{12}n_2 \\ 0 = k_{21}n_1 + k_{22}n_2 \end{cases} \qquad (11.1')$$

Excluding the trivial case in which all the k_{ij} are null, the *only* solution of the system is given by $n_1 = n_2 = 0$, which thus identifies the *only* point of equilibrium of the linear model (11.1); this is the situation in which the populations are 'empty', i.e. they contain no individuals. The question that we now ask ourselves is: how do the dynamics of the two populations behave with respect to the point of coordinates $n_1 = n_2 = 0$? Do the evolutions of P_1 and P_2, tend towards extinction (point of stable equilibrium, attractor point) or towards unlimited growth (point of unstable equilibrium, repeller point)? We observe that, as $n_1 = n_2 = 0$, the *only* state of equilibrium of the linear model, we cannot have the extinction of only one of the populations: in fact if we assume that $n_1 = 0$ and $\dot{n}_1 = 0$, we see that equation (11.1) gives $n_2 = 0$ and $\dot{n}_2 = 0$.

Before discussing the solution of system (11.1), we need to make another observation: we are attempting to find a link between the numbers n_1 and n_2 of the individuals of P_1 and P_2, starting from (11.1), by eliminating time from it; the function that we obtain will be the orbit of the system of the two populations in the phase space. Eliminating time from the two derivatives in equations (11.1), we obtain, after several steps:

$$\frac{dn_2}{dn_1} = \frac{k_{21}n_1 + k_{22}n_2}{k_{11}n_1 + k_{12}n_2} \qquad (11.2)$$

Equation (11.2) illustrates, point by point, the position of the straight line tangent to the curve, and therefore the direction of the system's orbit.[38] Equation (11.2), in other words, shows what the *instantaneous* variation of

---

[38] To be thorough, together with (11.2) we should also consider the inverse, i.e. the derivative of n_1 with respect to n_2:

the number of individuals of P_2 is with respect to that of the number of individuals of P_1, whereas in (11.1) the instantaneous variations of the number of individuals of P_1 and P_2 with respect to time, were considered separately.

Equation (11.2) can be calculated only if $n_1 \neq 0$ and $n_2 \neq 0$. The point of coordinates $n_1 = 0$ and $n_2 = 0$, in which the derivative (11.2) does not exist, is called the *singular point*. But $n_1 = n_2 = 0$ are also the only values that make both the derivatives of (11.1) equal to zero, which leads us to identify the states of equilibrium of model (11.1) with its singular points (this happens for any system, not just linear ones), and, in practice, to turn our attention to the study of solutions of the system of equations (11.1) in the neighbourhood of the only singular point.[39] We can claim, using a paragon that is slightly forced, but serves to clarify the situation, that the singular points of an equation are a bit like the points of support that hold a suspended chain: they determine the general form of the chain (the orbit in the phase space); what we are particularly interested in is studying the location of the supporting points, because they determine the form that the chain takes when suspended between one point and another.

In the discussion that follows in this chapter, we will turn our attention towards the study of the solutions of the system of equations (11.1) in a neighbourhood of the only singular point. We would like to state now that variables n_1 and n_2 should be understood in a general sense, not limiting ourselves to attributing only the meaning of the number of individuals of a populations in its strictest sense to them, but assuming that they can also have negative values.

$$\frac{dn_1}{dn_2} = \frac{k_{11}n_1 + k_{12}n_2}{k_{21}n_1 + k_{22}n_2} \qquad (11.2')$$

The existence of (11.2) alone, in fact, is not sufficient, for example, to describe closed orbits, in which there are at least two points in which the curve has a vertical tangent. The existence of (11.2'), together with (11.2), guarantees that all points of the curve can be described using a formula that links n_1 to n_2, and that the curve is continuous and has a unique tangent at any point

---

[39] To be precise, we note that at the singular points of a differential equation, the conditions set by the well-know theorem of the existence and uniqueness of the solutions of a differential equation are not satisfied, conditions that, furthermore, are sufficient but not necessary. Therefore, it *could* occur, as we will see in the discussion that follows in this chapter, that more than one (from two to infinity) trajectory of the system's dynamics passes through the singular points. This means, for example, that the system evolves towards the same final state of stable equilibrium, starting from any point on the phase plane, following different orbits, as happens with a damped pendulum that tends to stop, regardless of the initial oscillation amplitude.

The solutions of the linear model

A discussion, even if not complete nor rigorous, of the solutions of system (11.1) is useful because it enables us to highlight general characteristics and elements that can help us better understand certain aspects of the dynamics generated by more elaborate models. Some aspects of our discussion can be generalized, without making too many changes, to linear models with more than two interacting populations, and also, at least in part, and more or less along the same lines, to nonlinear models (which is actually much more interesting).

The theory of differential equations shows that the solutions of the system of linear equations (11.1) are made up of linear combinations of exponentials such as:

$$\begin{cases} n_1 = \alpha_1 e^{\beta_1 t} + \alpha_2 e^{\beta_2 t} \\ n_2 = \alpha_3 e^{\beta_1 t} + \alpha_4 e^{\beta_2 t} \end{cases} \quad (11.3)$$

In order to understand the evolutive dynamics of the two populations, the values of β_1 and β_2 are particularly important, because they determine the *nature* of the singular point: specifically its stability or instability characteristics. In fact, depending on whether the exponentials are real and increasing or real and decreasing or complex, we have, as we will see, a situation in which P_1 and P_2 distance themselves from a condition of equilibrium (instability) or tend towards a state of asymptotic equilibrium (stability) or make periodic oscillations.

The values of the coefficients α of (11.3) depend on the two integration constants and are less important than the values of β_1 and β_2. For now we will limit ourselves to observing that it can be demonstrated how, in reality, to define the dynamics of a system in the neighbourhood of the singular point we do not need to establish four independent coefficients α, because it is sufficient to determine the ratios α_1/α_3 and α_2/α_4, once β_1 and β_2 have been established, and therefore the nature of the singular point has been established.

The theory of differential equations shows that coefficients β_1 and β_2 can be found by solving a simple second degree algebraic equation, called the characteristic equation of (11.1), that can be written using the coefficients of the same (11.1):

$$\beta^2 - (k_{11} + k_{22})\beta + (k_{11}k_{22} - k_{12}k_{21}) = 0 \quad (11.4)$$

Now we need to examine the different cases that the signs of the coefficients k_{ij} of (11.1) give rise to. Depending on the signs of the k_{ij}, β_1 and β_2 can be real or otherwise. If they are real we can identify the cases in which β_1 and β_2 are negative (Case 1A), positive (Case 1B) or one positive and one negative (Case 1C). If β_1 and β_2 are not real, two possible cases can be identified: either β_1 and β_2 are imaginary conjugates (Re β_1 = Re β_2 = 0, Im β_1 = −Im β_2) (Case 2A), or β_1 has a real part common to β_2, i.e. β_1 and β_2 are complex conjugates (Re β_1 = Re $\beta_2 \neq 0$, Im β_1 = −Im β_2) (Case 2B).

Firstly, however, we would like to state three properties of the second degree equations to which we will be referring in the discussion that follows; we will refer to these properties indicating them as Point 1, Point 2 and Point 3 respectively:[40]

Point 1. The coefficient of (11.4), indicated by $-(k_{11} + k_{22})$, is equal to the sum of the roots β_1 and β_2 the sign of whose real component has been changed (obviously this does not mean that the roots of the equation are necessarily k_{11} and k_{22}).

Point 2. The term $(k_{11}k_{22} - k_{12}k_{21})$ is the product of the roots β_1 and β_2.

Point 3. The characteristic equation (11.4) has real roots β_1 and β_2 only if the discriminant of the coefficients, that is indicated by Δ, is positive or nil, i.e. if it is:

$$\begin{aligned}\Delta &= (k_{11} + k_{22})^2 - 4(k_{11}\,k_{22} - k_{12}\,k_{21}) \\ &= (k_{11} - k_{22})^2 + 4k_{12}k_{21} \geq 0\end{aligned} \quad (11.5)$$

Real roots of the characteristic equation: both populations grow or both extinguish themselves

Case 1A Real negative roots: $\beta_1 < 0$, $\beta_2 < 0$.

---

[40] In basic algebra it can be demonstrated that in the second order equation (in which $a \neq 0$):

$$ax^2 + bx + c = 0$$

there are the following relations between coefficients a, b, c (real or complex) and roots x_1 and x_2 (real or complex):

$$(-x_1) + (-x_2) = b/a \text{ and } (-x_1) \times (-x_2) = c/a$$

The characteristic equation (11.4) has real and negative roots (for Point 1) if: (1) k_{11} and k_{22} are both negative or (2) only one is, but, in absolute value, it is higher than the other; at the same time, in both cases the following must be true: $k_{11}k_{22} - k_{12}k_{21} > 0$ (for Point 2).

Let us interpret these conditions.

In case (1), k_{11} and k_{22} both negative means that there is internal competition within each of the two populations for the exploitation of the (limited) resources: as we can see from (11.1), in fact, an increase in population P_1 impacts the speed of growth of both P_1 and P_2 which decreases, because the derivatives $\dot{n}_1$ and $\dot{n}_2$, in this case, are both negative.

In case (2), on the other hand, we have real negative roots if there is a level of competition within only one population that is so high that it compensates for any (small) self-sustainment capacity of the other population. Let us suppose, for the sake of clarity, that there is internal competition within P_1, and therefore that $k_{11} < 0$, while P_2, on the other hand is self-sustaining: $k_{22} > 0$. As long as $k_{11}k_{22} - k_{12}k_{21} > 0$ (Point 2), the term $-k_{12}k_{21}$ must now be positive, in order to compensate for the product $k_{11}k_{22}$ which is negative: k_{12} and k_{21} therefore must have opposite signs. This means that one of the two populations benefits from the growth of the other, but at the same time, damages its growth.

In this case (2) we not only have k_{11} and k_{22} that indicate internal competition within a population and self-sustainment of the other, but also the interaction between P_1 and P_2 and that between P_2 and P_1, considered globally in the product $k_{12}k_{21}$, represents a relationship that we can call parasitic, for example of P_1 with respect to P_2. In other words, we find ourselves in a situation in which the dominant characteristic of the system is the internal competition, which, even if present in only one population, is large enough to inevitably lead *both* populations towards extinction. On the other hand, as we have already observed, in a linear system, either both populations grow or both populations extinguish themselves; there is no possibility that just one population survives over the other which completely extinguishes itself.

An example will clarify which type of system can give real negative roots. If we give the following values to the coefficients of (11.1):

$k_{11} = -10$ (high internal competition in P_1)
$k_{12} = +3$ (moderate gain by P_1 due to cooperation with P_2)
$k_{21} = +2$ (moderate gain by P_2 due to cooperation with P_1)
$k_{22} = -5$ (high internal competition in P_2)

system (11.1) thus becomes:

$$\begin{cases} \dot{n}_1 = -10n_1 + 3n_2 \\ \dot{n}_2 = 2n_1 - 5n_2 \end{cases}$$

for which the characteristic equation (11.4) is:

$$\beta^2 + 15\beta + 44 = 0$$

A simple calculation shows that, in this example, the characteristic equation has the following roots:

$$\beta_1 = -4 \text{ and } \beta_2 = -11$$

The high values of the internal competition indicated by the values of k_{11} and k_{22} in this example are such that they globally exceed the benefits of the cooperation given by the positive coefficients k_{12} and k_{21}, and therefore the populations tend to extinguish themselves.

The fact that the evolution of the system leads inevitably to the extinction of both populations, regardless of their initial values, is reflected in the fact that exponents β_1 and β_2, in this Case 1A, are both negative and give rise to solutions (11.3) that are linear combinations of exponentials that both decrease over time. This is a condition of asymptotically stable equilibrium, in which the origin of the axes acts as an attractor point, and is known as a *stable nodal point* (Figure 11.1).

In general, it can be demonstrated that the values of β_1 and β_2 determine the speed of convergence of the orbits towards the singular point, while the ratios α_1/α_3 and α_2/α_4 [equal to the two arbitrary constants of integration of the system (11.1)] determine the direction the orbits take to reach the stable nodal point, i.e. the inclination of the straight line in Figure 11.1 tangent to all the orbits at the singular point.

Case 1B Real negative roots: $\beta_1 > 0$, $\beta_2 > 0$.

The characteristic equation (11.4) has real and positive roots if the following conditions are satisfied: $k_{11} > 0$, $k_{22} > 0$ (for Point 1) and, at the same time, $k_{11}k_{22} - k_{12}k_{21} > 0$ (for Point 2).

Still with reference to (11.1), the conditions imposed by Point 1 tell us that each population, individually, grows indefinitely, as long as there are unlimited resources. The condition imposed by Point 2 tells us that the self-sustainment of both populations, considered globally in the product $k_{11}k_{22}$, is

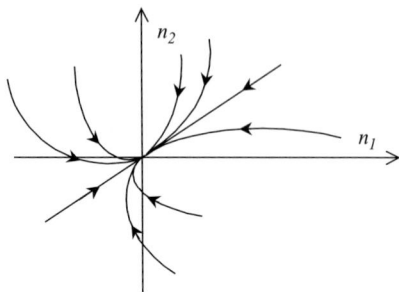

Figure 11.1 Dynamics of n_1 and n_2 in Case 1A: extinction of the populations, asymptotically stable equilibrium; the singular point is called *a stable nodal point*.

sufficient to compensate for any effect generated by the interaction of each population with the other, the combined value of which is represented by the product $k_{12}k_{21}$, regardless of the specific type of interaction between the populations. The product $k_{11}k_{22}$ is positive like in Case 1A, but, and this is the only difference between the two cases, k_{11} and k_{22} are now both positive.

A numerical example will clarify the situation that occurs in this case. If, for example, we give the following values to the coefficients:

$k_{11} = +10$ (high level of self-sustainment of P_1)
$k_{12} = +3$ (moderate gain by P_1 due to cooperation with P_2)
$k_{21} = +2$ (moderate gain by P_2 due to cooperation with P_1)
$k_{22} = +5$ (high level of self-sustainment of P_2).

system (11.1) thus becomes:

$$\begin{cases} \dot{n}_1 = 10n_1 + 3n_2 \\ \dot{n}_2 = 2n_1 + 5n_2 \end{cases}$$

for which the characteristic equation (11.4) is:

$$\beta^2 - 15\beta + 44 = 0$$

which has the following roots:

$$\beta_1 = +4 \text{ and } \beta_2 = +11$$

With all four coefficients k_{ij} positive, it is evident that the evolution can only be growth, as no term acts as a restraint to growth. Observe, however, that in this Case 1B, growth is given above all by the self-sustainment of the two populations individually and not by their cooperation (the product of the terms of the self-sustainment is greater than the product of the terms of their interaction: $10 \times 5 > 3 \times 2$). The case of growth driven mainly by cooperation, that occurs only if $k_{11}k_{22} - k_{12}k_{21} < 0$, is illustrated in Case 1C which follows.

In a situation of this nature, in which all of the elements contribute to the growth of the populations, (11.3) gives P_1 and P_2, which, regardless of their initial values, both tend to increase over time. This is a condition of unstable equilibrium, in which the origin of the axes acts as a repeller point, and is known as an *unstable nodal point* (Figure 11.2).

Note how the dynamics in Case 1B envisage orbits of the same form as those obtained in Case 1A, with the only difference that now the system is not directed *towards* the singular point, but distances itself *from* the singular point. Here again, the ratios α_1/α_3 and α_2/α_4 determine the direction the orbits take from the unstable nodal point.

Case 1C   Real roots of opposite signs: $\beta_1 < 0 < \beta_2$.

The characteristic equation (11.4) has real roots of opposite signs only if the product of the roots is negative, i.e. $k_{11}k_{22} - k_{12}k_{21} < 0$ (for Point 2). This occurs only if the product $k_{12}k_{21}$, which represents the combined effect of the interaction between P_1 and P_2, is positive and is greater than the product $k_{11}k_{22}$, which represents the combined effect of the self-sustainment or of the internal competition of both P_1 and P_2.

A situation of this type may arise if there is a very high level of cooperation between the two populations, in which each of them benefits considerably from its interaction with the other ($k_{12} > 0$, $k_{21} > 0$), so much so

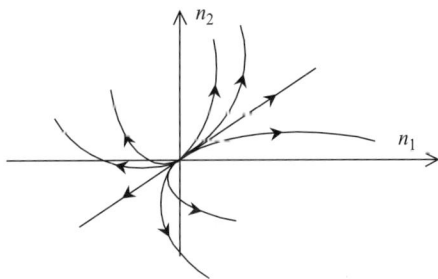

Figure 11.2   Dynamics of n_1 and n_2 in Case 1B: unlimited growth of the populations, unstable equilibrium; the singular point is called *an unstable nodal point*.

that the cooperation becomes the dominant element at the origin of the system's dynamics. The benefit brought by cooperation, in this case, is such that it exceeds the contribution to a population's speed of growth generated by the number of individuals in the same, regardless of its type and regardless of the fact that this contribution is positive (if $k_{11} > 0$ and $k_{22} > 0$, self-sustainment of the two populations), negative (if $k_{11} < 0$ and $k_{22} < 0$, internal competition within the two populations), or nil.[41]

An example of a linear system that gives rise to a situation of this nature is obtained with the following values of the coefficients:

$k_{11} = +5$ (moderate level of self-sustainment of P_1)
$k_{12} = +9$ (high gain by P_1 due to cooperation with P_2)
$k_{21} = +6$ (moderate gain by P_2 due to cooperation with P_1)
$k_{22} = +2$ (low level of self-sustainment of P_2).

The system thus becomes:

$$\begin{cases} \dot{n}_1 = 5n_1 + 9n_2 \\ \dot{n}_2 = 6n_1 + 2n_2 \end{cases}$$

for which the characteristic equation (11.4) is:

$$\beta^2 - 7\beta - 44 = 0$$

which has the following roots:

$$\beta_1 = -4 \text{ and } \beta_2 = +11$$

In the above example, the roots are real, because the effect of the interaction between the two populations P_1 and P_2 is positive ($9 \times 6 > 0$), but the most important aspect is that the given effect of the interaction between P_1 and P_2 exceeds the effect of the self-sustainment of the two populations ($9 \times 6 > 5 \times 2$), i.e. the overall effect of the cooperation is greater than the overall effect of the self-sustainment. This latter aspect is what really

---

[41] From an algebraic point of view, if we want to be thorough, the situation illustrated in the text is not the only one that can give two real values with opposite signs for β_1 and β_2. For example, the same result could be achieved by changing the signs of all four k coefficients with respect to those described in the text. We have limited ourselves to describing the situation indicated, because its interpretation is the most immediate in the terms that we have established, i.e. the application of the model to population dynamics, and we will ignore the other situations whose interpretation is less immediate, and whose significance is almost exclusively mathematical.

distinguishes Case 1C with respect to Case 1B: both cases involve the growth of the two populations, but in Case 1B this is the result of self-sustainment, while in Case 1C, the cooperation between the two populations is what 'drives' the growth of both.

In Case 1C, the solutions (11.3) of the linear model (11.1) are a linear combination of one increasing exponential and one decreasing exponential. The singular point at the origin of the axes is now an unstable point, but is a different type from that of Case 1B; it is called a *saddle point* (Figure 11.3).

It can be demonstrated that there is *one* direction, namely one *straight line*, along which the saddle point acts as if it were an attractor, and *another straight line* along which the instability, i.e. the 'force' with which the saddle point repels the system, is greatest. The two straight lines [two singular integrals of system (11.1)], are illustrated in Figure 11.3, and their equations are respectively:

$$n_2 = \frac{\alpha_1}{\alpha_3} n_1 \quad \text{and} \quad n_2 = \frac{\alpha_2}{\alpha_4} n_1$$

The first straight line indicates the *only* direction in which the linear system evolves over time following an orbit that brings it *towards* the saddle point, which is seen by the system (and only in this case) as a point of stable equilibrium and not as a repeller point. In all other directions, the saddle point is a point of unstable equilibrium. In particular, in the direction indicated by the second straight line, the 'force' with which the saddle point repels the orbit, and consequently the speed with which the system travels along the same orbit, is greater with respect to all other directions (see Figure 11.3). This second straight line is of particular interest to us because it is an asymptote for the system's orbits. In a certain sense, we can say that this straight line 'attracts' all of the system's orbits, which, after a

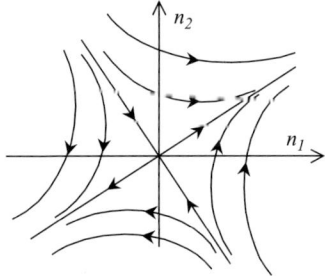

Figure 11.3 Dynamics of n_1 and n_2 in Case 1C: unstable equilibrium; the singular point is called *a saddle point*.

sufficient period of time, end up as being approximated by this same straight line. The first straight line, on the other hand, the one that leads towards the saddle point, also an asymptote for the orbits, from a geometrical standpoint, acts in the opposite way to the second straight line in terms of dynamics: it repels the orbits that arrive in its direction, as shown in Figure 11.3. Both an unstable nodal point and a saddle point (with the exception of one direction), therefore, 'repel', so to speak, the orbits, but one saddle point differs from an unstable nodal point due to the fact that the effect of the repulsion that it exercises on them varies according to the direction, i.e. according to the orbit that follows the dynamics of the system.

In summary, the values of the ratios α_1/α_3 and α_2/α_4 determine the two asymptotes, while the values of β_1 and β_2 (one positive, the other negative) determine the specific curve that these two asymptotes have.[42]

Lastly, it can be shown that the straight line that drives the orbits away from the saddle point is more inclined towards axis n_1 or axis n_2, the higher the value of the self-sustainment, respectively, of P_1 or P_2 (i.e. the larger k_{11} is with respect to k_{22}, or respectively, the larger k_{22} is with respect to k_{11}). This restates the fact that the population that has the highest level of self-sustainment determines the general dynamics of the system, drawing, so to speak, the growth of the other population into its dynamics, which, over time, ends up growing at a speed that is directly proportional to that of the 'stronger' population in terms of self-sustainment. In a certain sense, therefore, it is as if the growth of the population of the two that has the higher self-sustainment coefficient 'draws' the growth of the other.

We will now examine Cases 2A and 2B whose characteristic equation (11.4) has roots β_1 and β_2, which are not real.

Complex conjugate roots of the characteristic equation: the values of the two populations fluctuate

Case 2A Imaginary conjugate roots β_1 and β_2: $\mathrm{Re}\beta_1 = \mathrm{Re}\beta_2 = 0$.

The characteristic equation (11.4) has imaginary conjugate roots if two conditions are satisfied simultaneously. The first condition is that the dis-

---

[42] The understanding in mathematical terms of this unique case of stability is simple: if we assume that $\alpha_2 = \alpha_4 = 0$, in (11.3) only two negative exponentials remain (remember that β_1 is the negative root and β_2 is the positive one). In the same way, by assuming $\alpha_1 = \alpha_3 = 0$, we have the condition of maximum instability of the singular point, because only the positive exponentials remain. The two straight lines are asymptotes common to all of the orbits characterized by the same values for ratios α_1/α_3 and α_2/α_4.

criminant Δ of the characteristic equation (11.4) is negative; in this case, instead of (11.5) of Point 3, we have:

$$\Delta = (k_{11} - k_{22})^2 + 4k_{12}k_{21} < 0 \qquad (11.6)$$

Equation (11.6) dictates that $4k_{12}k_{21}$ is negative and that $(k_{11} - k_{22})^2$, which is positive because it is a square, is not large enough to give a positive result if summed to $4k_{12}k_{21}$. This implies that k_{12} and k_{21} must have opposite signs, and thus that the interaction between P_1 and P_2 is advantageous to only one of the populations and not to the other.

The second condition that must be satisfied is that the sum of the roots β_1 and β_2 is nil, which, for Point 1, can occur in two cases: (1) $k_{11} = k_{22} = 0$, i.e. the two populations are not characterized by self-sustainment nor by internal competition; or (2) $k_{11} = -k_{22}$, i.e. the intensity, so to speak, of the effect of the self-sustainment of a population is equal and opposite to that of the effect of the internal competition of the other population.

To clarify, let us suppose, for example, that P_2 is the population that benefits from its interaction with $P_1 (k_{21} > 0)$, while, on the contrary, P_1 is the population that is damaged by its interaction with $P_2 (k_{12} < 0)$, to the extent that product $k_{12}k_{21}$ is negative, and that P_2 is the population characterized by internal competition ($k_{22} < 0$), while P_1 is self-sustaining ($k_{11} > 0$).[43] The situation that the values of the coefficients generate is that typical of a relationship that, in biology, is defined as predation of population P_2 over P_1, or also as parasitism of P_2 towards P_1.

It is clear that in a prey–predator relationship, the extinction of the prey population P_1 due to an excess of predation, would also lead to the extinction of the predator population P_2, as the growth of the number of predators depends, to a large extent, on coefficient k_{21}, namely on the number of available prey. The prey, for their part, would tend to multiply in an unlimited manner, in the absence of internal competition, if the predators did not intervene and reduce their numbers. A dynamic is thus established, according to which the number of prey decreases due to the action of the predators ($k_{12} < 0$), but when such reduction becomes excessive, even the predators suffer due to the limited amount of food resources, which creates internal competition within the population and hinders its growth ($k_{22} < 0$). A restricted number of predators, however, would leave the prey free to multiply, as the latter do not have any internal competition

---

[43] Our observations in note 41 on p. 82 are also valid in this case; as well as the situation that we have described here, whose interpretation is the most immediate, there are other configurations of the coefficients k that can give imaginary conjugate roots β_1 and β_2 but which are of less interest in this context.

($k_{11} > 0$); the growth of the prey stimulates, in turn, the growth of the predators ($k_{21} > 0$), which once again restrain the unlimited growth of the prey, and so forth.

Using Euler's formula:

$$e^{\text{Re }\beta + i\text{Im }\beta} = e^{\text{Re }\beta}(\cos(\text{Im }\beta) + i\sin(\text{Im }\beta))$$

we see, after several steps, that in Case 2A in question, which has pure imaginary roots (Re $\beta = 0$), (11.3) becomes:

$$\begin{cases} n_1 = \alpha_1 \cos(\text{Im }\beta t) + \alpha_2 \sin(\text{Im }\beta t) \\ n_2 = \alpha_3 \cos(\text{Im }\beta t) + \alpha_4 \sin(\text{Im }\beta t) \end{cases} \quad (11.7)$$

where $\beta = |\beta_1| = |\beta_2|$.

The equations (11.7) describe the dynamics of a system in which the two populations fluctuate periodically between a maximum and a minimum value, each of which according to a sinusoidal law (Figure 11.4). It can be demonstrated that both the phase difference between the two sinusoids and their amplitudes depend on the (real) values of the coefficients α of the equations (11.7).

Once time has been eliminated, the equations (11.7) generate, on the phase plane, ellipses whose centre is the singular point. The values of the four coefficients α determine both the amplitude of the axes of the ellipses, and their inclination with respect to the coordinate axes. Figure 11.5 shows the dynamics corresponding to that of Figure 11.4. In this specific case, the coefficients α that are not 0 directly give the length of the semiaxes of the

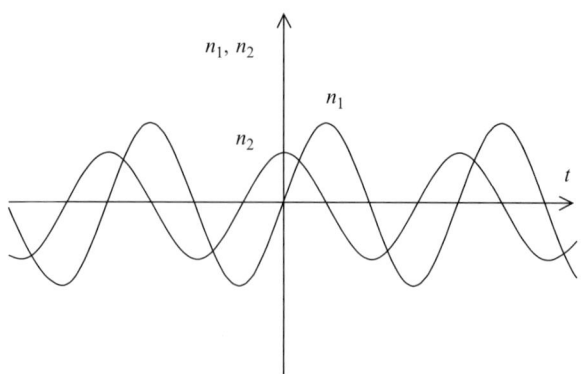

Figure 11.4 Trend of n_1 and n_2 as a function of time t in Case 2A.

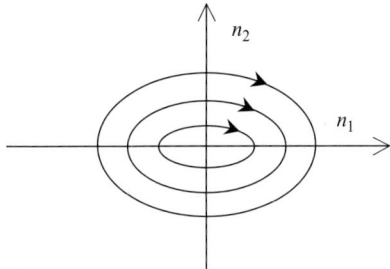

Figure 11.5 Dynamics of n_1 and n_2 in Case 2A; the singular point is called a *centre*.

ellipses. The projections on the two axes of the successive states in Figure 11.5 give the trend over time of the values n_1 and n_2 shown in Figure 11.4.

The evolution of the system over time, represented by the motion of a point that travels along an elliptic orbit like that shown in Figure 11.5, is similar to the one we illustrated previously in our discussion on oscillating systems in Chapter 10.

We are therefore dealing with stable orbits around a singular point that, as we know, is situated at the origin of the axes in the phase plane and that now is not characterized as an attractor nor a repeller; an indifferent point of equilibrium which is called a *centre*.[44]

The system of the two populations, therefore, does not evolve in any manner towards an asymptotically stable final state, but demonstrates a periodically repetitive situation in which the number of prey and the number of predators continually fluctuates, with the recurrence of the same values. The predation relationship means, in the linear hypothesis, that the prey–predator system makes neither damped nor amplified oscillations, demonstrating close similarities to that of a non-damped linear pendulum,[45] described in the model of the harmonic oscillator (Chapter 8).

---

[44] Remember that we are still describing an abstract model, in which the magnitudes n_1 and n_2 have been interpreted as the values that two (abstract) populations could have, and which also include negative values. In a real case, the state variables n_1 and n_2 should be reinterpreted in light of the specific situation to which they refer. Obviously, a population, in the strictest sense, cannot have negative values: attributing the numerical significance of members of a population to n_1 and n_2 in a strict sense, would imply, from a mathematical point of view, at least a shift of the axes in Figures 11.1–11.7, so that the singular point is not the state in which both populations are zero, but another pair of positive values. For example, identifying a centre in the pair of coordinates $N_1 = 50$, $N_2 = 100$ (i.e. if P_1 and P_2 oscillate around these two respective values) would lead us back to the dynamics described in Case 2A above, assuming $n_1 = N_1 - 50$ and $n_2 = N_2 - 100$.

[45] To establish the correspondence between the prey–predator system and the pendulum, it is sufficient to replace n_1 with the angle and n_2 with the speed of the pendulum.

Consider, for instance, the linear model of a set of two interacting populations, which have the following coefficients:

$k_{11} = +1$ (moderate level of self-sustainment of P_1)
$k_{12} = -2$ (high damage of P_1 caused by its interaction with P_2)
$k_{21} = +3$ (high gain by P_2 due to its interaction with P_1)
$k_{22} = -1$ (moderate level of internal competition within P_2 with the opposite value to that of the self-sustainment of P_1).

The system thus becomes:

$$\begin{cases} \dot{n}_1 = n_1 - 2n_2 \\ \dot{n}_2 = 3n_1 - n_2 \end{cases}$$

for which the characteristic equation (11.4) is:

$$\beta^2 + 5 = 0$$

which has the following roots:

$$\beta_1 = \sqrt{5}i \text{ and } \beta_2 = -\sqrt{5}i$$

We can see from the above that the system's coefficients satisfy the conditions of Case 2A: $k_{11} = -k_{22}$, and furthermore, $k_{12}k_{21} < 0$.

Let us interpret the dynamics of the example. Population P_1 grows, due to the effect of self-sustainment, at a speed proportional to the number of individuals $n_1 (k_{11} = 1)$, which indicates ease of reproduction in the absence of competition for food resources. The speed of growth of P_1, however, decreases as the number of individuals n_2 of P_2 increases (as $k_{12} = -2$), which can be seen as the effect of the predation of P_2 to the detriment of P_1. On the other hand, the number of individuals in P_2 grows at a speed proportional to the number of individuals in $P_1 (k_{21} = 3)$, and this reintroduces the predation relationship of P_2 over P_1, but decreases as the number of individuals n_2 increases ($k_{22} = -1$), which indicates competition for (limited) resources within P_2.

We can observe that if coefficients k_{11} and k_{22} didn't have opposite values as in the above example, the situation could change. If, for example, the self-sustainment of P_1 (the prey) was greater than the internal competition within P_2 (the predators), i.e. if $|k_{11}| > |k_{22}|$, and if the difference $|k_{11}| - |k_{22}| > 0$ was such as to compensate for the effect of the interaction between the two populations, then condition (11.6) would no longer be satisfied and we would return instead to (11.5):

$$\Delta = (k_{11} - k_{22})^2 + 4k_{12}k_{21} \geq 0$$

which would result in (11.4) having real roots β_1 and β_2, and therefore we would return to one of the Case 1 situations. Furthermore, if the difference between the self-sustainment of the prey and the internal competition of the predators was not zero, but was not sufficient to satisfy (11.5), then the resulting situation would be that described in Case 2B below.

Case 2B Complex conjugate roots: $\beta_{1,2} = \text{Re}\,\beta \pm \text{Im}\,\beta$.

In this case, the roots β_1 and β_2 of the characteristic equation (11.4) have the same real component $\text{Re}\,\beta \neq 0$, and an imaginary component $\text{Im}\,\beta \neq 0$, which appears with opposite signs in the two roots.

We obtain a situation of this type if (11.6) is satisfied, which guarantees the non-reality of β_1 and β_2 (Point 3) and if, at the same time, we have $k_{11} + k_{22} \neq 0$ (Point 1: the sum of the roots β_1 and β_2 must not be zero). We will interpret this situation as we did for Case 2A. The first condition imposes that $(k_{11} - k_{22})^2$ is small, so as not to exceed in absolute value $4k_{12}k_{21}$, which must be negative so as to yield $\Delta < 0$, as in Case 2A. The second condition means, on the other hand, that the self-sustainment of a population, say P_1, has a different intensity to that of any internal competition within the other population P_2.

In Case 2B, the solutions of the system (11.1), with respect to (11.7), have an extra real exponential component, that is increasing or decreasing over time depending on the values of the real components of roots β_1 and β_2. They thus become:

$$\begin{cases} n_1 = e^{\text{Re}\,\beta t}(\alpha_1 \cos(\text{Im}\,\beta t) + \alpha_2 \sin(\text{Im}\,\beta t)) \\ n_2 = e^{\text{Re}\,\beta t}(\alpha_3 \cos(\text{Im}\,\beta t) + \alpha_4 \sin(\text{Im}\,\beta t)) \end{cases} \qquad (11.8)$$

Equation (11.8) represents an evolutive dynamics in which n_1 and n_2 fluctuate as in Case 2A, but in which their maximum amplitudes are no longer constant, and actually decrease exponentially if $\text{Re}\,\beta < 0$, or increase exponentially if $\text{Re}\,\beta > 0$. In the first case, the orbits trace spirals that go *towards* a point of stable equilibrium, called a *stable focal point* (Figure 11.6); in the second, they trace spirals that *distance themselves*

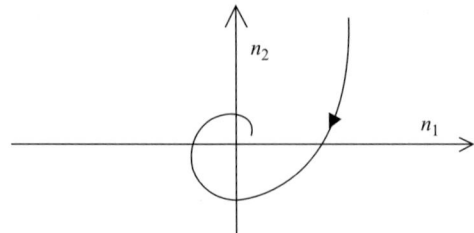

Figure 11.6 Dynamics of n_1 and n_2 in Case 2B, with Re $\beta < 0$: stable equilibrium; the singular point is called a *stable focal point*.

from a point of unstable equilibrium, called an *unstable focal point* (Figure 11.7).[46]

Let us consider two examples of this dynamic.

The first example is represented by the system in Case 2A, which we modify by making $k_{11} = 1/3$, and leaving the other coefficients unchanged; we establish, therefore, that the self-sustainment of the prey is less than the internal competition between predators. We thus obtain the following system:

$$\begin{cases} \dot{n}_1 = \frac{1}{3}n_1 - 2n_2 \\ \dot{n}_2 = 3n_1 - n_2 \end{cases}$$

whose characteristic equation is:

$$\beta^2 + \frac{2}{3}\beta + \frac{17}{3} = 0$$

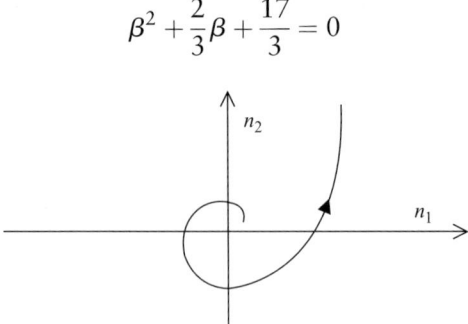

Figure 11.7 Dynamics of n_1 and n_2 in Case 2B, with Re $\beta > 0$: unstable equilibrium; the singular point is called an *unstable focal point*.

---

[46] Both the focal point and the nodal point can be stable or unstable. From a geometric point of view, the difference between a stable/unstable focal point (Figures 11.6 and 11.7) and a stable/unstable nodal point (Figures 11.1 and 11.2) lies in the fact that at the nodal point there is a straight line tangent to the orbits passing through it, that can clearly be defined and calculated; in the focal point, on the contrary, there is no tangent to the orbits at all.

which has the following roots:

$$\beta = -\frac{1}{3} \pm \frac{5\sqrt{2}}{3}i$$

The second example is still made up of the system of Case 2A, modified by supposing $k_{22} = 1/3$, leaving the same coefficients as the example in Case 2A; we thus establish that the self-sustainment of the prey is greater than the internal competition between predators. The system therefore becomes:

$$\begin{cases} \dot{n}_1 = n_1 - 2n_2 \\ \dot{n}_2 = 3n_1 - \dfrac{1}{3}n_2 \end{cases}$$

whose characteristic equation is:

$$\beta^2 - \frac{2}{3}\beta + \frac{17}{3} = 0$$

which has the following roots:

$$\beta = +\frac{1}{3} \pm \frac{5\sqrt{2}}{3}i$$

In the first example, solutions (11.8) indicate a tendency towards the extinction of the two populations, which fluctuate with amplitudes that decrease exponentially over time (Figure 11.8). This is due to the lack of

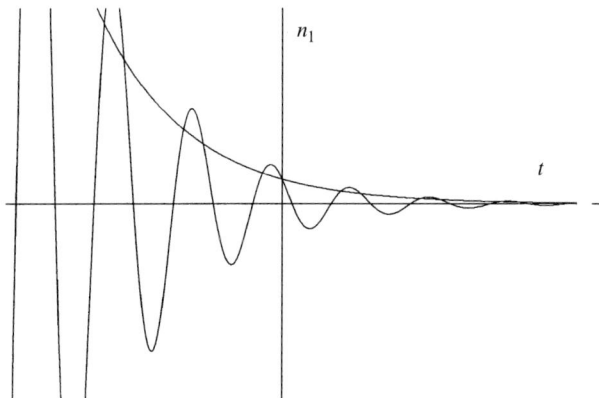

Figure 11.8 Dynamics of one of the two populations with respect to time t, in the first example, with $\alpha_3 = \alpha_4 = 1$. The figure also shows the decreasing exponential exp(Re βt), with $\beta = -1/3$.

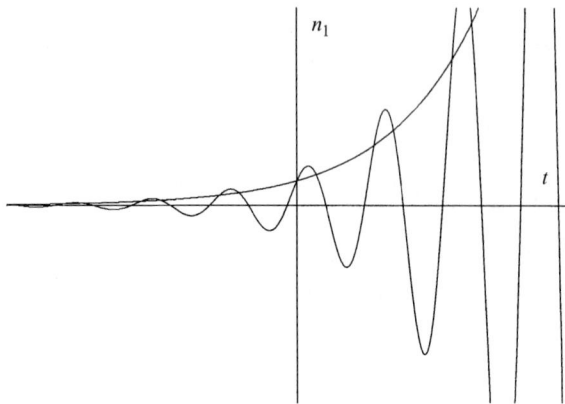

Figure 11.9 Dynamics of one of the two populations with respect to time t, in the second example, with $\alpha_1 = \alpha_2 = 1$. The figure also shows the increasing exponential $\exp(\text{Re } \beta t)$, with $\beta = 1/3$.

self-sustainment of the prey which, added to the predation relationship, does not permit either the prey or the predators to grow, or even to fluctuate around an average value. In the second example, on the other hand, the prey is characterized by self-sustainment that is sufficient to cause a progressive amplification of the fluctuations in the number of individuals of the two populations (Figure 11.9).

We will end our description of the linear model here. We will now be looking at (chapter 12) a well-known nonlinear prey–predator model and at some developments that have resulted from this.

12 Nonlinear models in social processes: the model of Volterra–Lotka and some of its variants in ecology

Introduction

Around the mid-1920s, the American biologist Alfred Lotka (1925) and the Italian mathematician Vito Volterra (1926) independently proposed a nonlinear model of the dynamics of a prey species interacting with a predator species, which although still rather abstract, like the linear model (11.1) described in 11, was more realistic than the latter. The model, which is very well known and which has been extensively applied in a variety of areas, is known by the name of the Volterra–Lotka model.[47]

---

[47] Alfred Lotka had been working on self-catalytic processes since the first decade of the twentieth century, proposing in 1920 a well-known model of three chemical reactions (Epstein, 1989), whose equations were used as a basis for the Volterra–Lotka model in question. Vito Volterra, one of the most remarkable figures in mathematics and science in Italy in the early decades of the twentieth century, was inspired by the observation that, during the course of the First World War and in the years immediately following, the frequency with which cartilaginous (or selachian) fish, usually predators, such as small sharks or rays, were caught in the ports of Fiume, Trieste, and Venice, went from fairly stable values of around 10 and 12 per cent, to higher values, reaching, depending on the port and the year, values of around 20 per cent and, on occasion close to 30. The biologist Umberto D'Ancona, Volterra's son-in-law, who had been working for several years on fishing statistics in the Adriatic, came up with the hypothesis of a relation between the reduction in fishing caused by the raging war, and the consequent restoration of a natural equilibrium between prey and predators, which had been altered previously by intense fishing activity. D'Ancona asked his father-in-law if his hypothesis could be confirmed in mathematical terms. Volterra formulated the model in question [equations (12.1) on page 94], with which he identified the existence of a dynamic equilibrium corresponding to the situation observed during and after the war. He then demonstrated that the insertion in the model of a reduction in the number of cartilaginous fish and of the number of other fish at the same rates (the effect of fishing activity carried out indiscriminately for cartilaginous fish and other fish, which favoured neither the prey nor the predators, as occurred before the war), effectively shifted the natural

Setting aside the linear model discussed in the previous pages, although significant above all on a theoretical level, the Volterra–Lotka model was one of the first models formulated with the specific aim of describing the actual interaction between an animal prey species and an animal predator species in mathematical terms. Not only does it have a historical value, even today it is the fundamental starting point for countless series of other more refined prey–predator models which feature a wealth of elements, used to describe the dynamics of reciprocally interacting populations in continuous time. The general fundamental principles of the model are still considered valid today and many theoretical biologists use them as a reference for modelling in a field known as mathematical biology (Jost, Arino and Arditi, 1999). In addition to mathematical biology, the Volterra–Lotka model has been successfully applied, as we will see in chapter 13, to the areas of urban and regional science (Pumain, Sanders and Saint-Julien, 1989; Pumain, 1997).

The basic model

The original form of the Volterra–Lotka model, which we will call the basic model, describes the evolution of two populations of different species, P_1 and P_2, that interact in a closed ecosystem. Population P_1 belongs to an animal prey species that has n_1 individuals, who have an unlimited food supply and therefore are not in any nature of internal competition. Population P_2 belongs to an animal predator species that has n_2 individuals, who feed only on the prey represented by population P_1. Prey P_1 are present in limited quantities and, therefore, the carnivores are in competition between themselves for the food. One could imagine, for example, the prey to be herbivores that have an unlimited quantity of grass, while the predators could be a species of carnivore that feed on the above-mentioned herbivores.

The equations of the basic model (Beltrami, 1987; Arnold, 1992; Çambel, 1993; Israel, 1996; Epstein, 1997) are:

$$\begin{cases} \dot{n}_1 = k_{11}n_1 - k_{12}n_1n_2 \\ \dot{n}_2 = k_{21}n_1n_2 - k_{22}n_2 \end{cases} \quad (12.1)$$

(dynamic) equilibrium between prey fish and predator fish, reproducing the data observed. Volterra's model, thus formulated, confirmed D'Ancona's hypothesis (Israel, 1986, 1996).

in which the coefficients k_{ij}, that are all positive, have the following meanings:

k_{11} growth rate of the prey
k_{12} mortality rate of the prey due to predation
k_{21} reproduction rate of the predators (how many predators are born for each prey consumed)
k_{22} mortality rate of the predators

A fundamental characteristic of the model is that the two populations interact in a nonlinear way, due to the presence in the two equations of the products of the two variables n_1 and n_2. The nonlinear term $n_1 n_2$ represents the frequency with which a prey and a predator encounter one another which, according to the law of compound probability, is assumed to be proportional to the product of the total number of prey multiplied by the total number of predators, through coefficients k_{12} and k_{21}.

In the absence of encounters between prey and predator, i.e. when $k_{12} = k_{21} = 0$, each species would develop according to a Malthusian schema: the prey, which possess unlimited quantities of food, would grow exponentially (k_{11} is positive) due to a lack of encounters with predators, while the number of predators would decrease exponentially ($-k_{22}$ is negative) due to a lack of food, because they never meet the prey. The presence of a positive coefficient k_{11} in the equation which gives the speed of growth $\dot{n}_1$ of the number of prey n_1, the first of the equations (12.1), indicates that said speed of growth is proportional to the number of existing prey.

In the first of the equations (12.1), there is a term with a negative coefficient $-k_{12}$; this means that the speed of growth, in reality, decreases proportionally to the frequency of encounters, represented by the product $n_1 n_2$. On one hand, in fact, the more numerous the prey, the higher the probability each prey has of surviving and reproducing; on the other, the more numerous the prey, the higher the probability of encounters with predators. In the first equation, therefore, there are two types of feedback, one positive, proportional to k_{11}, and one negative, proportional to $-k_{12}$.

The second of the equations (12.1) gives the speed of growth of the number n_2 of predators. It also contains a positive term, k_{21}, and a negative term, $-k_{22}$, but the positions with respect to the first equation are exchanged. This means that the speed of growth of the predators is directly proportional to the ease with which the predators procure their food, which depends on the frequency of their encounters with the prey, here also represented by $n_1 n_2$. However, this time, unlike the case of the prey, as $k_{21} > 0$, the speed of growth of the predators n_2 increases as the number of prey n_1 increases, and thus as the product $n_1 n_2$ increases, but decreases due to the effect of competition between predators, proportionally to their number ($-k_{22}$ negative), due to the limited food supply represented by the prey. Therefore, also in this second equation there are two types of feedback: a positive one, proportional to k_{21}, and a negative one, proportional to $-k_{22}$.

Basically, we have a mechanism according to which an increase in the number of prey provokes, through an increase in the frequency of encounters with predators, an increase in the number of the latter. The increase in the number of predators, however, leads to a decrease in the number of prey, which negatively influences the growth of the number of predators. At a certain point, the number of predators stops increasing and starts to decrease due to the lack of food, which leads to a reduction in encounters between predators and prey, with a consequent increase of the latter, which again causes a growth in the number of predators, and so on. Note how these aspects of dynamics recall, at least partly, those of the linear model, in the case of imaginary conjugate roots, discussed as Case 2A in chapter 11.

The numerical integration of the model is not complicated and leads to interesting results, at least in theoretical terms. The model gives rise to cycles in the phase plane $n_1 n_2$, i.e. to an oscillatory dynamics of the number of the individuals of the two populations, like those shown in Figures 12.1 and 12.2 (the two figures illustrate dynamics similar to those shown in Figures 11.4 and 11.5 regarding Case 2A of the linear model, described in chapter 11).

The model is interesting at a conceptual level, but there are a number of problems regarding its real effectiveness as a way to describe interactions

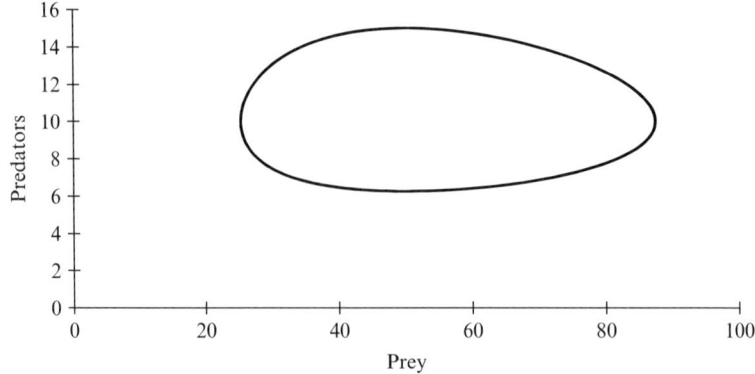

Figure 12.1 Numerical integration of the dynamics of the Volterra–Lotka model. The values of the parameters of the model assumed for the integration are: $k_{11} = 0.1\ k_{12} = 0.01\ k_{21} = 0.001\ k_{22} = 0.05$.

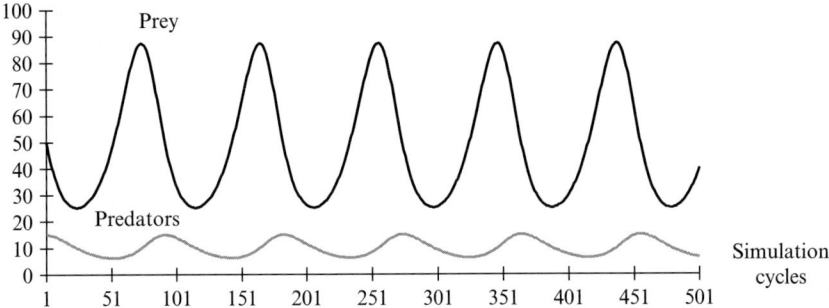

Figure 12.2 Numerical integration of the trend of prey and predator populations in the Volterra–Lotka model. The values of the parameters are the same as those in Figure 12.1.

between species in a real environment. A fundamental characteristic of the dynamics envisaged by the basic model is that there are no points of stable equilibrium; the populations of the prey and the predators evolve according to endless cycles, without ever tending towards a situation of stability. The equilibrium is only dynamic; it can even be demonstrated that this occurs for any set of values of the parameters.

With regard to the cyclicity of the populations, although some cases of cyclicity have actually been observed in nature, such as for example the populations of hares and lynx, the trends of which over time were recorded in Canada between 1847 and 1903 by the Hudson's Bay Company (Figure 12.3),

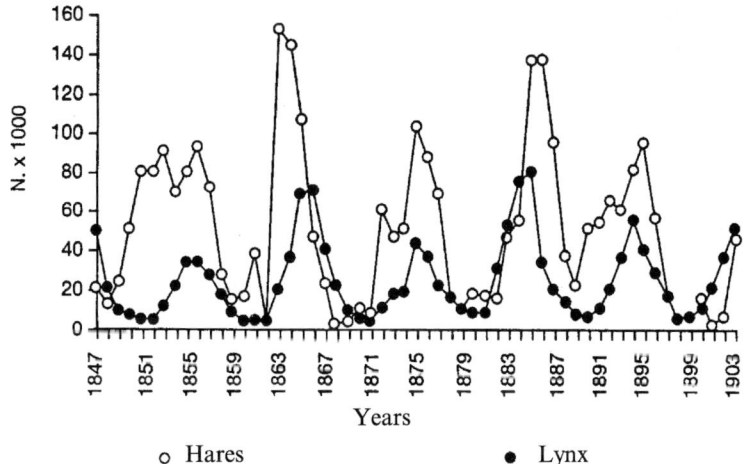

Figure 12.3 Trend of the populations of hares and lynx recorded by the Hudson's Bay Company.
Source: Israel (1996, p. 53 of the Italian edition of 1997).

this does not, in any event, appear to be the most common case. The basic assumptions of the model are evidently not realistic: for example, it does not consider any type of competition within the population of prey that is assumed to grow without any limits, if predators were not present; furthermore, it assumes that the predators are never satiated, as the quantity of food they feed from is directly proportional to the number of existing prey. It is no surprise, therefore, that it does not show any type of asymptotic stability. Moreover, the basic model does not envisage the possibility of the extinction of the populations, which, on the other hand, is a frequent result in laboratory tests of simple prey–predator systems, which is something that appeared right from the first experiments carried out by the Russian biologist Georgyi Frantsevich Gause (1934, 1935). As the Volterra–Lotka model only envisages periodic dynamics, extinction was explained as the result of random fluctuations of the populations which manifest themselves when the numbers of individuals is low, i.e. when the trajectory traced by the prey–predator system (see Figure 12.1) passes close to the origin of the axes.

It is evident that the basic form of the Volterra–Lotka model, schematic as it is, is of great theoretical and historical value, but it does not appear to be sufficient to model a predation relationship in a general and realistic way. On one hand, the model needs other theoretical elements to be added to it, that need to be defined based on the specific context to which they refer; on the other, the assumption that the number of individuals in a populations is directly proportional to the speed with which it grows, in the case of the prey (coefficient k_{11}), or with which it decreases, in the case of the predators (coefficient k_{22}), is a very simple and schematic assumption, useful as a working hypothesis, but too rigid to be realistic. Appropriately defined functions need to be introduced in some way, that make the model more detailed, flexible and varied for the dynamics it envisages. This is exactly what we will be doing.

Before illustrating later versions of the model, however, we need to introduce a new type of attractor that we will be encountering, to describe a type of dynamics that is not present in any of the systems illustrated thus far: the limit cycle.

A non-punctiform attractor: the limit cycle

Let us re-examine the dynamics of a forced pendulum, subject to friction. This time the pendulum is activated by a mechanism that keeps it in motion, the escapement, connected to a weight, that periodically supplies the energy that the pendulum loses due to friction, as occurs with a pendulum clock. Therefore, if the pendulum is still, the escapement, under the thrust of the weight, sets it in motion, making it oscillate at its characteristic frequency and with a specific oscillation amplitude; if, on the other hand, the pendulum is oscillating with too large an amplitude, perhaps because it has been set in motion by a very large initial impulse, the friction damps the motion, until an equilibrium is reached between the energy lost due to friction and the energy supplied to the pendulum by the weight. The result is the establishment of a stable oscillatory motion, a closed orbit in the phase space travelled periodically, as occurs with a linear pendulum without friction.

Now, however, there is a fundamental difference with respect to the previous models. The linear pendulum that we saw (Models 1 and 2) can find an infinite number of orbits, one for each value of the total energy that characterizes the oscillation (see chapter 8): different initial states of energy distinguish different periodic orbits. On the other hand this is not true for the pendulum we have just described: it always ends up in one periodic orbit only, in dynamic equilibrium between lost energy and absorbed energy. If the system is not in that orbit, over time it tends to reach it, and once it has reached it, it will travel along it periodically.

The closed orbit 'attracts' the dynamics of the oscillating system towards itself, so to speak, starting from whatever state it finds itself in. The closed orbit is called the *limit cycle*; it is a real attractor, not unlike a point attractor in a damped pendulum or in a linear system. At a point attractor, the system is still in a state of stable equilibrium; in a limit cycle, on the contrary, the system is in a state of dynamic equilibrium; it is not still, although, in any event, the condition is one of stability.

A limit cycle can have a large variety of forms, depending on the system and the values of the parameters, but it is always a closed, stable curve. In general in dissipative systems characterized by a single variable, for which, therefore, the phase space has only two dimensions (the variable in question and its first derivative with respect to time), in addition to the point attractor, the case of an attractor represented by a limit cycle could thus arise: a closed, continuous curve, that can be described using ordinary techniques of mathematical analysis, even though not always of elementary expression (Figure 12.4).

When the system has reached its limit cycle, it no longer preserves the memory of the previous states, nor of the initial states from which the evolution started. The situation, from this point of view, is similar to that of a point attractor, like in the damped pendulum; in both cases, future evolution is predictable, because it is always the same, regardless of the initial state, but the memory of the past is lost.

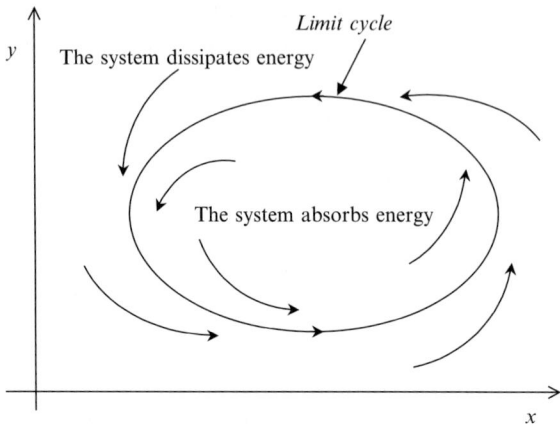

Figure 12.4 The limit cycle in the phase space.

The establishment of a limit cycle assumes that the system, on one hand, is losing energy but, on the other, it is receiving it from an external source, in a dynamic equilibrium between the dispersion and acquisition of energy. The system behaves like a dissipative system if the energy it is supplied with is less that the energy it loses (in this case the orbit reaches the limit cycle from the outside), or like a system that absorbs energy if the energy supplied is more than that lost due to friction (in this case the orbit reaches the limit cycle from the inside).

Limit cycles are typical, in general, of certain forced oscillating systems, such as the case of the pendulum described; the establishment of a limit cycle in the dynamics of a system is a frequently encountered phenomenon, that can be observed every time a system oscillates in a condition of dynamic equilibrium between lost energy (for example, due to different types of friction) and the energy absorbed from a source.[48] A typical case of

---

[48] There are several very well-known mathematical models whose solutions, in certain conditions, are limit cycles. One of the first was formulated following the study of a mechanical phenomenon: a particular type of nonlinear oscillating system, subject to an external force dependent on time. The equation is known as the Duffing equation (1918) and can have different forms; one of the most common is as follows:

$$\ddot{x} + k\dot{x} + \alpha x + \beta x^3 = \Gamma \cos \omega t$$

in which k, α, β are parameters that define the oscillator and $\Gamma \cos \omega t$ is an external force (if the oscillator is made up of a spring, then β is a parameter linked to the hardness of the spring). The solution of this model (also known as the double-well oscillator), for certain sets of parameter values, is a limit cycle.

a limit cycle can be found, for example, in the dynamics of a heart beat (Van der Pol and Van der Mark, 1928), even though it has been observed recently that the physiology (not the pathology) of the heart rate usually has more chaotic features (Goldberg, Rigney and West, 1994).

Carrying capacity

As we have said, numerous variants of the basic Volterra–Lotka model have been conceived, in which various corrective terms have been introduced. We will cite a few of the simpler ones, although we make no claims to have covered the subject thoroughly.

To start with, let us consider the model in which, like in the basic model, we have a population P_1 of prey, which, however, now have a limited food supply, and a population P_2, of predators, which feed on the population of prey P_1. Here again n_1 and n_2 indicate the number of individuals of the two populations. The reciprocal interaction between prey and predators, like in the basic Volterra–Lotka model, is demonstrated by the two terms containing the product $n_1 n_2$. The equations are as follows (Beltrami, 1987):

$$\begin{cases} \dot{n}_1 = k_{11} n_1 \left(1 - \dfrac{n_1}{A}\right) - k_{12} n_1 n_2 \\ \dot{n}_2 = k_{21} n_1 n_2 - k_{22} n_2 \end{cases} \quad (12.2)$$

A second, equally famous model, is presented by the equation of Van der Pol (1927), who describes the periodic motion of the electrical charges in an oscillating electrical circuit, according to assumptions that are very similar to those described in the text regarding the pendulum clock: there is a resistance that makes the electrical charges lose energy (which corresponds to the friction) and an amplifier that supplies them with energy (which corresponds to the weight–escapement mechanism of the pendulum clock). Balthazar Van der Pol established that the dynamics of the electrical charges are described by a nonlinear second order equation:

$$\ddot{q} + \mu(q^2 - 1)\dot{q} + q = 0$$

where the unknown function $q(t)$ is the number of electrical charges, $\dot{q}(t)$ is the intensity of the electrical current and μ is a parameter that depends on the characteristics of the circuit.

The system's nonlinearity lies in the term $q^2 - 1$: when $q < 1$, the system tends to reach the limit cycle from the outside (it dissipates energy); when $q > 1$, the system tends to reach the limit cycle form the inside (it absorbs energy), more or less like what happens with the pendulum clock when the oscillation amplitude is, respectively, smaller or larger than that corresponding to the stable oscillation. In the limit $\mu \to 0$, the Van der Pol equation becomes the same as that of the harmonic oscillator [equation (5.2)] that, therefore, constitutes the limit for negligible resistance or amplifications (Israel, 1996).

As in the basic model, the coefficients k_{ij} are all positive.

Equation (12.2) differs from (12.1) due to the presence of the term $(1 - n_1/A)$ in the equation that describes the dynamics of the prey. The interaction between prey and predators takes place in accordance with the same mechanism described for the basic model, but now, with regard to the dynamics of the prey, the presence of the term n_1/A in brackets that is multiplied by k_{11}, has the effect of limiting the values of n_1 slowing down the speed of growth, which becomes smaller the closer n_1 gets to the value A.

When n_1, initially smaller that A, grows over time, the value of the brackets $(1 - n_1/A)$ decreases, distancing itself from the unit, and $\dot{n}_1$ decreases with it. The brackets would become equal to zero, if n_1 reached the value of A, totally cancelling, in this way, the contribution of n_1 to growth, that originates from n_1 itself. Term A therefore takes on the meaning of the maximum value of the number of individuals of P_1 permitted by the available food and in a certain sense, plays the role of a limiting device of the speed of growth. In this context, it is interpreted as being due to the limited availability of food supplies and is called the *carrying capacity* of the environment. The presence of the factor $(1 - n_1/A)$, in addition to the constant k_{11}, therefore, changes the growth pattern that n_1 would follow in the absence of predators, which now is no longer Malthusian, as in the linear model and in the basic Volterra–Lotka model, but is a new type, called logistic.[49]

The carrying capacity has been used in another variant of the basic model that describes the dynamics of two separate populations, no longer prey and predators, that compete for the same limited resources. As a consequence of the limited availability of resources, the individuals are now in competition with one another within the same population, and with the individuals of the other population, according to the following model:

$$\begin{cases} \dot{n}_1 = k_{11} n_1 \left(1 - \dfrac{n_1}{A}\right) - k_{12} n_1 n_2 \\ \dot{n}_2 = k_{21} n_2 \left(1 - \dfrac{n_2}{B}\right) - k_{22} n_1 n_2 \end{cases} \quad (12.3)$$

in which the coefficients are all positive; A and B are the carrying capacities of the two populations (Beltrami, 1987).

---

[49] The concept of carrying capacity and the law of logistic growth will reappear and will be discussed at length when in Chapters 20–22 we will be illustrating the Verhulst model and the logistic map. For now, let us just say, anticipating the basic elements, that the law of logistic growth (function of time, continuous and derivable on the entire real axis) describes the trend of a magnitude that grows over time at a speed that is initially low, then grows and reaches a maximum value and subsequently decreases tending to zero, as time tends towards infinity.

Functional response and numerical response of the predator

The basic Volterra–Lotka model has been enhanced, besides the introduction of the carrying capacity, also by the addition of further terms that consider, for instance, the possibility of the prey fleeing from their usual habitat, or by the addition of a second population of prey as an alternative to the first for the same population of predators, or again still with the addition of various kinds of exogenous agents (climatic, biological, etc.) or other.

With the aim of demonstrating the fertility of the idea at the origin of the Volterra–Lotka model, even without going into the technical details, we would like to illustrate some important variants of the model that have been developed, enhancing and making the basic model more flexible with elements suggested by observations.

A first variant which is widely used to describe the prey–predator interaction was proposed by Robert May (1973). May made various changes to the basic model. Firstly, he assumed a logistic growth for the prey, as in models (12.2) and (12.3). Secondly, he assumed that the predators could satiate themselves, which led him to replace the product $k_{12}n_1n_2$, which in the equation for the prey in the basic model gives the decrease of the prey due to their interaction with the predators, with the term:

$$\frac{\beta n_1}{\alpha + n_1} n_2$$

where the positive constant β, expresses the per capita feeding rate of the predators[50] and α is an appropriate constant which is also positive.

With regard to the predators, May assumed that the limited availability of prey simply leads to a logistic growth for them as well, and that, therefore, the number of predators does not depend on the probability of

---

[50] This is justified by the fact that for large values of n_1 we can approximate:

$$\frac{\beta n_1}{\alpha + n_1} \sim \beta$$

When there are large numbers of prey, therefore, the term which gives their decrease depends *only* on n_2, the number of predators, through the constant of proportionality β, and no longer on $k_{12}n_1n_2$, i.e. on the probability that the two populations will encounter each other. β can thus be interpreted as the number of prey that each of the n_2 predators consumes in an infinitesimal time interval, i.e. as the feeding rate of each predator.

their encounter with the prey, which in the basic model was represented by the term $k_{21}n_1n_2$, which now disappears. A peculiarity of May's model is that the carrying capacity of the predators is not assumed to be constant, but proportional to the number n_1 of prey by means of a factor B.

The equations of May's model are as follows:

$$\begin{cases} \dot{n}_1 = k_{11}n_1\left(1 - \dfrac{n_1}{A}\right) - \dfrac{\beta n_1}{\alpha + n_1}n_2 \\ \dot{n}_2 = k_{21}n_2\left(1 - \dfrac{n_2}{Bn_1}\right) \end{cases} \qquad (12.4)$$

It can be demonstrated (Beltrami, 1987) that this model can have a limit cycle as its solution, unlike the basic Volterra–Lotka model which, as we said, only leads to cyclical, and not limit cycle solutions.

Another variant of the basic model, one of the most well-known and most used in mathematical biology, is that of Crawford S. Holling (1959). Holling, while studying several cases of predation by small mammals on pine sawflies, observed that the predation rate was not constant, but grew as the number of prey increased; therefore not only the number of predators grew as that of the prey grew, but each predator increased its own consumption rate following the growth in the number of prey. To describe this particular predation relationship, Holling introduced two new terms, called *the functional response* and *the numerical response of the predator*.

The functional response of the predator is the capture rate of prey for each individual predator, a rate that is now no longer assumed to be constant, as in the models that we have examined up until now, but depends on the number of prey. Holling assumes that a predator divides its time into two parts: the time employed in searching for the prey, called the *searching time*, indicated by t_s, and the time employed in digesting the prey, called the *handling time*, indicated by T_h, which is equal to the handling time t_h required for a single prey, multiplied by the number of prey H that have been captured, i.e. $T_h = t_h H$. Holling obtained the following expression for the functional response of the predator, which in the model plays the role of the mortality rate of the prey per individual predator:[51]

---

[51] The reasoning is as follows. Assuming that capture is a random process, the number of prey H captured by a single predator is given by the product of the searching time t_s by the density of prey d per unit of area, multiplied by the surface area a of the territory in question: $H = adt_s$. From this last formula, we obtain: $t_s = H/ad$, while, as we said in the text, $T_h = t_h H$. As the time t of the predator is $t = t_s + T_h$, we can write $t = H/ad + t_h H$, from which we can deduce H, the total number of prey captured by a predator in time t:

$$H = adt/(1 + adt_h)$$

$$\text{Functional response of the predator} = \frac{k_{12}n_1}{1 + k_{12}t_h n_1}$$

The functional response of the predator, multiplied by the number of predators n_2, provides the total number of prey captured by all predators in a unit of time.

The numerical response of the predator is the predators' growth rate. The basic idea is that, for each prey captured, there is an increase in the number of predators according to a constant of proportionality c, called the *conversion efficiency*, which, basically, counts how many offspring each individual predator has for each prey that it eats (for example, for every ten prey consumed, a predator is born).

The dynamics of the predators is the effect, thus, of the simple sum of two components. The first component is the growth in the number of predators (the numerical response of the predator), which occurs according to a constant of proportionality c (conversion efficiency):

$$\text{Numerical response of the predator} = \frac{c k_{12} n_1}{1 + k_{12} t_h n_1}$$

The second component is the decrease in the number of predators that takes place according to a mortality rate which is assumed to be constant; the number of predators that die, therefore, is proportional to the number of predators and is expressed by the product $k_{22}n_2$, where k_{22} is positive.

Holling's model thus takes on the following form:

$$\begin{cases} \dot{n}_1 = k_{11} n_1 \left(1 - \dfrac{n_1}{A}\right) - \dfrac{k_{12} n_1}{1 + k_{12} t_h n_1} n_2 \\ \dot{n}_2 = \dfrac{c k_{12} n_1}{1 + k_{12} t_h n_1} n_2 - k_{22} n_2 \end{cases} \quad (12.5)$$

Without entering into the technical details of the model, we will limit ourselves to pointing out the wealth of the states of Holling's models. In fact, if it is numerically integrated choosing different values for the param-

---

Let us now assume that ad, the product of the density of the prey per unit surface multiplied by the surface covered by the predator, is proportional to the total number of existing prey n_1, i.e. $ad = n_1 k_{12}$, and let us divide H by t: in this way we obtain the capture rate of a predator shown in the text, namely the functional response of the predator.

The model that Holling suggested for the functional response H remains very popular among ecologists: it is often called the 'disc equation' because Holling used paper discs to simulate the area examined by predators. Mathematically, this model is equivalent to the model of enzyme kinetics developed in 1913 by Lenor Michaelis and Maude Menten.

eters, it can give rise to states of asymptotic equilibrium or to damped oscillations of the number of individuals or, further still, to limit cycles.

A third variant of the basic model, which is also very popular, is the Holling–Tanner model (Tanner, 1975) which simply combines several elements of May's model (12.4) with others of Holling's model (12.5). To be more precise, the equation of the prey is the same as the one that appears in Holling's model, while the equation of the predators is that of May's model (12.4):

$$\begin{cases} \dot{n}_1 = k_{11}n_1\left(1 - \dfrac{n_1}{A}\right) - \dfrac{k_{12}n_1}{1 + k_{12}t_h n_1} n_2 \\ \dot{n}_2 = k_{21}n_2\left(1 - \dfrac{n_2}{Bn_1}\right) \end{cases} \qquad (12.6)$$

In Figure 12.5 we show an example of the dynamics envisaged by the Holling–Tanner model integrated numerically: the choice of the values of the parameters indicated in the figure gives rise to a state of asymptotic equilibrium (the arrow indicates the direction followed over time).

In general, the variants developed from the basic Volterra–Lotka model (12.1) currently used in mathematical ecology, all adopt the following type of canonical schema:

$$\begin{cases} \dot{n}_1 = n_1 f(n_1) - n_2 g(n_1, n_2) \\ \dot{n}_2 = n_2 c g(n_1, n_2) - k n_2 \end{cases} \qquad (12.7)$$

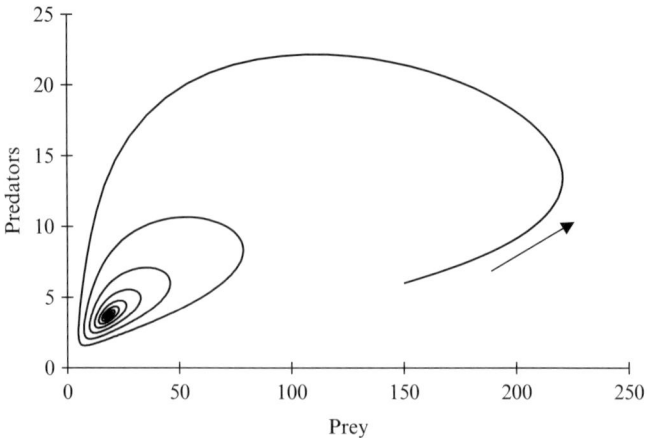

Figure 12.5 Dynamics of a Holling–Tanner prey-predator model (12.6). The values of the parameters are: $k_{11} = 0.2$, $A = 500$; $k_{12} = 0.1$, $t_h = 0.5$, $k_{21} = 0.1$, $B = 0.2$; initial values: prey $= 150$, predators $= 6$.

where $f(n_1)$ is a function of the growth of the prey, $g(n_1, n_2)$ is a functional response of the predators, c is the conversion efficiency and k is the mortality rate of the predators.

The stability properties of a model such as (12.7) have been examined by Christian Jost, Ovide Arino and Roger Arditi (1999). Assuming certain particular forms for f and g, they particularly studied the properties of the critical point corresponding to the extinction of both populations, reaching the conclusion that the dynamics of the two populations, according to their model, can present a saddle point, or even an attractor, at the origin of the axes, depending on the values of the parameters. In other words, the model studied by Jost, Arino and Arditi reproduces Gause's experimental observations, which we mentioned on p. 98, and therefore envisages that, unlike that which occurs in the basic Volterra–Lotka model, which is not able to result in extinction, in certain circumstances, the extinction of both populations arises after several fluctuations in the number of individuals in both.

13 Nonlinear models in social processes: the Volterra–Lotka model applied to urban and regional science

Introduction

As we mentioned at the beginning of Chapter 12, several variants of the Volterra–Lotka model have been used to model urban and regional systems. The dynamics obtained generally lead back to those obtained with the ecological models illustrated in Chapter 12. Nevertheless, it is evident that ecological models can only be applied through homologies, i.e. purely formal equivalences conducted between the biological sphere and that of urban and regional science;[52] meanings relating to other scientific sectors can under no circumstance be strictly attributed to the magnitudes typical of urban and regional science. In urban and regional modelling, the relations between prey and predators in ecology can be likened, for example, to the dynamic interaction between demand (the predator) and supply (the prey) of a good (also called *stock*), and we assume that it is possible to establish a

[52] Regional science studies social, economic, political and behavioural phenomena with a spatial dimension. In regional science, a region is, usually, a geographical area smaller than the State in which it is found: it might be a city, a country or province, a group of countries or provinces or even, sometimes, a State, according to the scale of the phenomenon studied. Regions may even extend over governmental/political boundaries, as, for example, when the issue under study relates to a labour market area or a watershed. In 1954 Walter Isard (emeritus professor at Cornell University) founded the RSAI, Regional Science Association International (see www.regionalscience.org), an interdisciplinary, international organization focusing on understanding the growth and development of urban, regional, and international systems; by that, regional science became formally recognized as an interdisciplinary field of scientific research which uses jointly the methods and the knowledge developed in different fields such as economics, geography, sociology, demography, statistics, planning, design and others.

relationship in the form of one of the variants of the Volterra–Lotka model between said magnitudes.

Model of joint population–income dynamics

Transfers of formulations of this type to urban and regional analysis have been carried out, for example, by Dimitrios Dendrinos (1980, 1984a-c, 1992) and by Dendrinos and Henry Mullally (1981, 1985). The basic idea at the heart of the model that these scholars propose is that the macro dynamics of spatial units (whether such are urban areas, regions, or countries) can be described by means of the dynamics of two key development indicators: population and average per capita income. Instead of the prey and the predator, in models of this nature, which generally refer to urban areas, the variables used are the average income available in the urban area in question, which constitutes supply (the prey), and the population of the urban area 'hunting' the income, which constitutes demand (the predator). The model, conceived in this way, expresses the joint population–income dynamics that develops in a closed area, i.e. without external flows, following continual adjustments between the demand (population) and the supply (income) in the various locations within the area in question (Dendrinos, 1992).

The model simulates the evolution of a population in a given geographic area, a population that is indicated by variable y, and the level of average income indicated by x, according to the basic hypothesis that an average income x available in an area, which is higher than in other areas, will attract population, giving rise to immigration to said area; but a population that grows in an area, because it is attracted by the average income available, provokes a reduction in said income, thus diminishing the attraction of said area. A dynamic-evolutive mechanism can immediately be recognized in this population–income relationship, similar to that of the prey–predator system illustrated in chapter 12:

$$\begin{cases} \dot{x} = ax - bxy \\ \dot{y} = -cy + dxy \end{cases} \quad (13.1)$$

in which constants a, b, c and d are all positive. In the equation that gives the income dynamics, the first of the equations (13.1), the term ax expresses the positive effect that the income has on the growth of the income itself (to use a popular expression 'wealth generates wealth'). The term $-bxy$, on the other hand, expresses the negative effect that the increase in population has on the growth of average income: this effect corresponds to that which occurs in the basic model (12.1) when the growth in the number of prey slows down or becomes negative due to the high number of predators.

In the equation that gives the population dynamics, the second of the equations (13.1), the term $-cy$ expresses the negative effects that the above-mentioned congestion, reducing the attractiveness of the area, exercises on the speed of growth of the population. In the basic model, it corresponds to the fall in the number of predators caused by internal competition due to the limited amount of available prey. The term $-cy$ expresses a sort of spontaneous decline in the population, contrasted by the term $+dxy$, which describes the growth of the population due to the immigration caused by the average available income. The term dxy basically expresses the attractiveness of the labour market (Paelinck, 1992, 1996) and corresponds, in the basic model, to the growth of the predators cause by predation.

Model (13.1), which is substantially identical to the basic model (12.1), shows the same dynamics as the latter: only periodic cycles, similar to those produced by the harmonic oscillator (chapter 8). The population–income dynamics can be made more rich and varied by adding a quadratic term to model (13.1), in addition to the already existing nonlinear terms. This enables us to take into account phenomena due to productivity and congestion in a less rigid way than simple direct proportionality. A model generalized in this way was successfully applied to the city of Rotterdam, for which it reproduced annual data regarding population and income between 1946 and 1978 (Bachgus, Van Nes and Van der Zaan, 1985).

The ecological approach was adopted in the field of urban and regional science not only to describe population dynamics, but also to analyse competition between actors or businesses in a limited area. Volterra–Lotka equations have thus been used, in various competition models for urban areas between two types of land use or between two different types of urban actors. On occasion, attention has been focused on the space–time diffusion of technological innovation (Sonis, 1983ab, 1984, 1995; Camagni, 1985), on the evolution of the labour market (Nijkamp and Reggiani, 1990), on the competition between operators for one or two locations in an urban environment (Dendrinos and Sonis, 1986), on the evolution of a population with respect to the cost of land, and to the cyclical behaviour of the relative growth of an urban centre and its hinterland (Orishimo, 1987).

A number of variants of the Volterra–Lotka model have been used in applications of this nature. One often used in urban and regional science entails the introduction of carrying capacity; these are models that correspond to the ecological ones like (12.2) in which not only do both variables appear multiplied together, as in the basic model, but, and this is the fundamental aspect, one of the two variables now appears in a quadratic term as well (Nijkamp and Reggiani, 1992ab, 1998):

$$\begin{cases} \dot{x} = x(a - bx - cy) \\ \dot{y} = y(-d + ex) \end{cases} \quad (13.2)$$

Coefficients a, b and d are all positive and refer to the endogenous dynamics of the variables, while coefficients c and e, which are also positive, reflect the interaction between the two variables. One can easily transform (13.2) into the equivalent system (13.2'), identical in form to (12.2):

$$\begin{cases} \dot{x} = ax\left(1 - \dfrac{bx}{a}\right) - cxy \\ \dot{y} = exy - dy \end{cases} \quad (13.2')$$

Form (13.2') enables a comparison to be made with (12.2) and, more specifically, enables the carrying capacity of the prey population, indicated by A in (12.2) to be expressed explicitly through the a/b relationship. By comparing (13.2) with (13.2'), we note that a large carrying capacity, i.e. a high availability of resources in the environment, translates into a small value for b with respect to that of a, which signifies, in (13.2), a lesser importance of the second order term $-bx^2$ with respect to the linear term ax. The scarcity of resources (low carrying capacity), on the other hand, implies that the external environment exercises an action of higher intensity on the system; this more intense environmental feedback translates into an increased 'nonlinearity' of the model.

By making the two derivatives equal to zero, we obtain two states of equilibrium for model (13.2). One is trivial: $x = y = 0$, the other is a point on the xy plane whose coordinates are $\bar{x}$ and $\bar{y}$ as follows:

$$\begin{cases} \bar{x} = \dfrac{d}{e} \\ \bar{y} = \dfrac{a}{c} - \dfrac{b}{c}\dfrac{d}{e} \end{cases} \quad (13.3)$$

Figures 13.1 and 13.2 show an example of a type of dynamics envisaged by (13.2): a case of the orbit of the two variables in the phase plane xy and the corresponding dynamics of x and y as a function of time (represented by simulation cycles) respectively.

The curves shown in the figures were obtained by integrating system (13.2) numerically, using the same numerical method and the same values

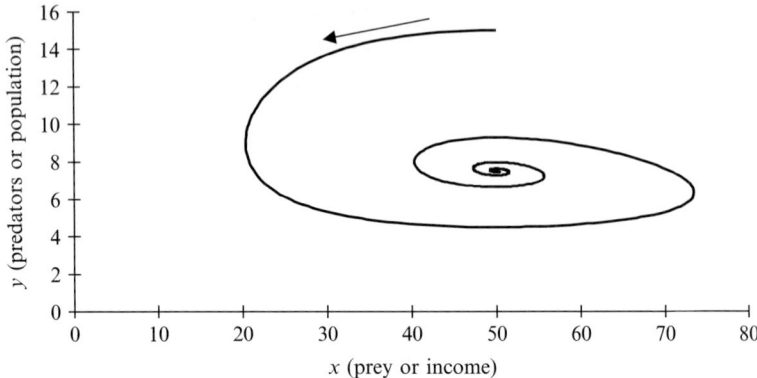

Figure 13.1 Dynamics of two variables x and y according to variant (13.2) of the Volterra–Lotka model. The values of the parameters are: $a = 0.01$, $b = 0.0005$, $c = 0.01$, $d = 0.05$, $e = 0.001$; initial values: $x = 50$ $y = 15$; final equilibrium values: $x = 50$ $y = 7.5$.

of the parameters that were used for Figure 12.1; stable equilibrium is obtained by making $b = 0.0005$. The presence of the quadratic term $-bx^2$ in the equation that describes the evolution of the prey (or of the income), as we can see by comparing Figures 13.1 and 13.2 with Figures 12.1 and 12.2, has the effect of the appearance of a state of asymptotic equilibrium for the dynamic evolution of the system. The dynamics described by this model are a convergent spiral type, with oscillations that dampen around an asymptotic state of equilibrium.

The value of parameter b, which is inversely proportional to the carrying capacity, is critical. It can be shown, in fact, that the equilibrium given by equations (13.3) is asymptotically stable for $b > 0$ (as that in Figure 13.1), as occurred in the linear model in Case 2B (chapter 10); equations (13.3)

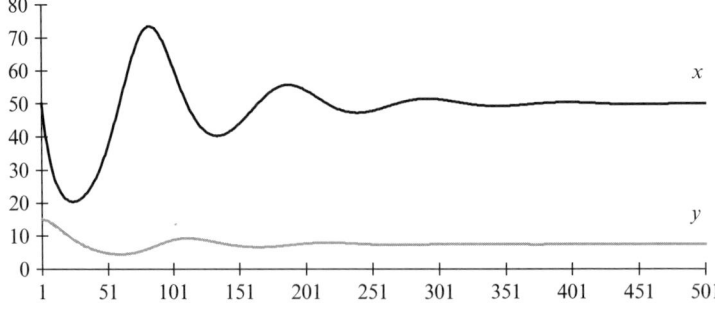

Figure 13.2 Trend of variables x and y as a function of time (the simulation cycles), according to variant (13.2) of the Volterra–Lotka model. The values of the parameters are the same as in Figure 13.1.

define, therefore, a stable focal point, like that shown in Figure 11.6. As b decreases, the speed with which the dynamics tends towards asymptotic equilibrium diminishes. It can be shown that, if the carrying capacity tends towards infinity (i.e. if b tends towards 0), the equilibrium becomes unstable: the orbit no longer converges at one point, but makes a stable cycle, as occurred in the linear model in Case 2A (chapter 10) (a singular centre type point, like that illustrated in Figure 11.5). For $b = 0$, there is just dynamic equilibrium: in fact, in this case, model (13.2) re-assumes the form of the basic Volterra–Lotka model (12.1), which envisages only closed orbits travelled cyclically and not points of equilibrium. If we also accept negative values for b, then we can see that the equilibrium defined by equations (13.3) is unstable, and that the dynamics of the model follow orbits that diverge from the point of equilibrium (13.3), which behaves, therefore, like an unstable focal point, like that shown in Figure 11.7.

The population–income model applied to US cities and to Madrid

The population–income model of ecological origin, founded on the prey–available income and predator–population associations, was used to describe the dynamics of 90 metropolitan areas in the United States between 1890 and 1980, even though the scarcity of available historical data (the so-called time series) does not enable us to ensure a good experimental basis to the model, as Denise Pumain observed (1997). Figure 13.3 shows a simulation of the evolutive dynamics described by Dendrinos (1992) for several urban areas of the United States. The continuous lines indicate the orbits simulated in the population–income phase plane, the dashed lines indicate forecasts and the state of stable equilibrium. It can be noted how the average available income in the Los Angeles area exceeds that of the New York area, in accordance with the forecasts of the US Bureau of Census, while, in contrast, the populations of the two areas tend to resemble one another.

The effectiveness of the ecological model applied to the field of urban and regional science becomes obvious when comparing the orbit simulated for the urban area of New York (Figure 13.3) with the empirical data available for 1930–80 (Figure 13.4), which describe the actual evolution of New York with respect to the population coordinates on the x axis, and average income on the y axis (population and income are expressed as percentages). The part of the calculated curve corresponding to 1930–80 accurately reproduces the empirical data for the same years.

13: The Volterra–Lotka model in regional science

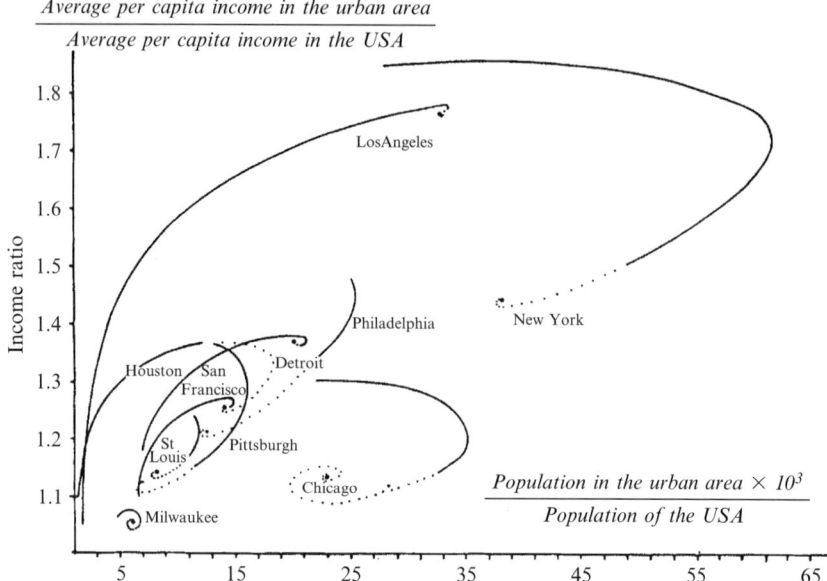

Figure 13.3 Simulation, with a Volterra–Lotka model, of the population–income dynamics of several metropolitan areas of the United States, between 1890 and 1980; the continuous lines are the simulated evolution, the dashed lines indicate forecasts and final equilibrium. *Source*: Dendrinos (1992, p. 120).

The great socio-economic transformations that occurred in the United States, above all from the end of the Second World War, have given rise to phenomena of suburbanization, internal migration, the increase in the size of the family nucleus, and of the wealth and the average age of the population. Moreover, they are the origin of the economic growth of the southern states, and of the growing importance of the service sector with respect to that of industry, and in particular, of the significant development of the electronics sector. These are none other than some of the many causes that have significantly influenced the spatial-sectorial structure of the United States.

The factors that characterize the US economy, in addition to the cited consequences, have also given rise to other economic and demographic changes. In particular, the impact of the socio-economic transformation of the two major metropolitan areas, New York and Los Angeles, has worked in opposite directions. Its effect has been a shift in the pole of attraction for the population from the New York area, and in general from north-eastern states, towards the younger and smaller area of Los Angeles, which was more economical and which had better opportunities of economic growth, in particular between the 1960s and the 1980s, as illustrated in Figures 13.3 and 13.4 (Dendrinos, 1992).

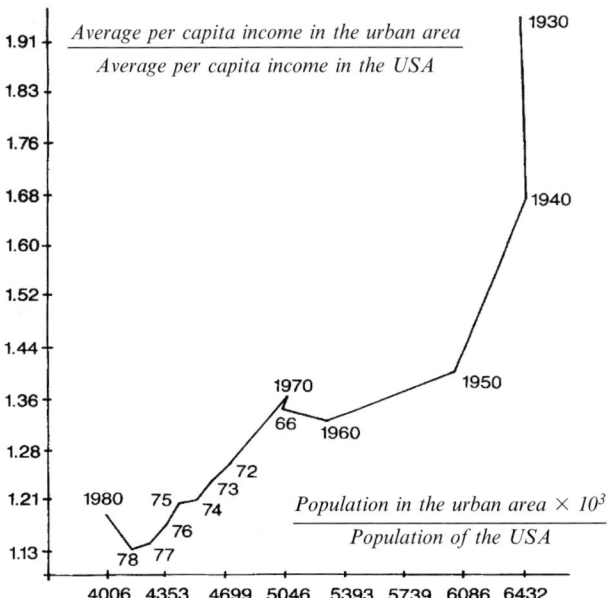

Figure 13.4 Population–income dynamics in the metropolitan area of New York between 1930 and 1980 (historic series).
Source: Dendrinos, (1992, p. 116).

A model like that of (13.2) was also used to analyse the development of the city of Madrid (Dendrinos, 1984a), obtaining the population–income dynamics shown in Figure 13.5. Figures 13.6 and 13.7 show, respectively, the average per capita income of Madrid, expressed as a percentage with respect to the average per capita income in Spain, and the population of Madrid, expressed as a percentage of the total population of Spain.

According to Dendrinos, empirical data compared with the simulation data of the model not only confirm the validity of metropolitan dynamics, but also allow us to conclude that the parameters responsible for the dynamic evolution of Madrid have very similar values to those applicable to a large number of metropolitan areas in the United States.

Dendrinos (1984c) used the following 'urban' variant of the Volterra–Lotka model for both the urban areas of the United States and for Madrid:

$$\begin{cases} \dot{x} = \alpha x(y - 1 - \beta x) \\ \dot{y} = \gamma y(x^* - x) \end{cases} \quad (13.4)$$

in which x is the value of the population, expressed as a percentage of total population, and y is the average per capita income available in the urban

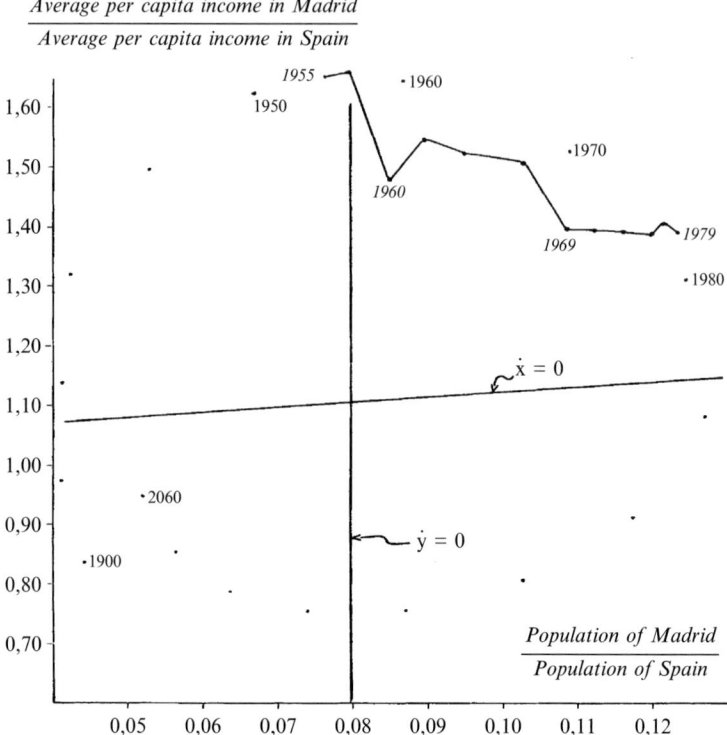

Figure 13.5 Population–income dynamics for the urban area of Madrid: comparison between simulated data, indicated by isolated points, and empirical data, indicated by points joined by the broken line; the intersection of the two lines is the state of equilibrium envisaged by the model.
Source: Dendrinos (1984a, p. 242).

area in question, expressed as a percentage of the average per capita income of the country. The parameter x^* identifies the carrying capacity of the metropolitan area expressed in relative terms with regard to that of the country.

The form of (13.4) is similar to that of (13.2): the first equation contains a first degree term in x, a second degree term in x and a second degree one with the product of the variables xy; the second equation contains a first degree term in y and a second degree term with the product xy. Model (13.4), therefore, also entails a stable state of equilibrium at a point, like model (13.2).

Parameters α and γ are the relative speeds that regulate the changes in population and income respectively. The population dynamics in the first

The population–income model 117

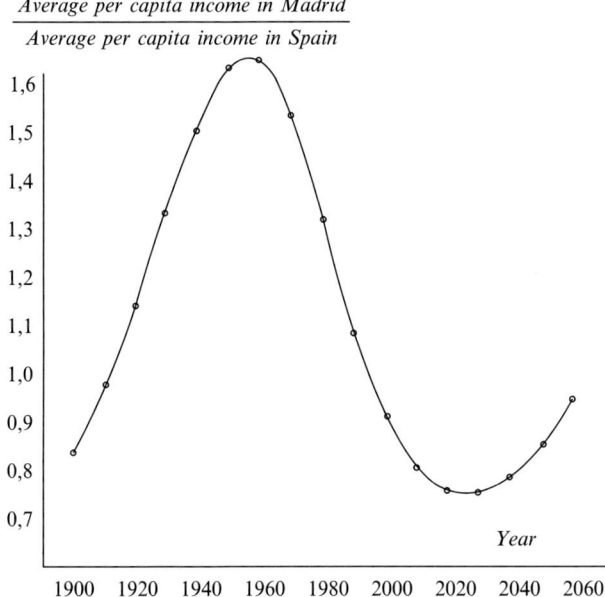

Figure 13.6 Simulation of the average per capita income of the urban area of Madrid (the dots on the curved line) and empirical data from historic series (the smaller dots).
Source: Dendrinos (1984a, p. 243).

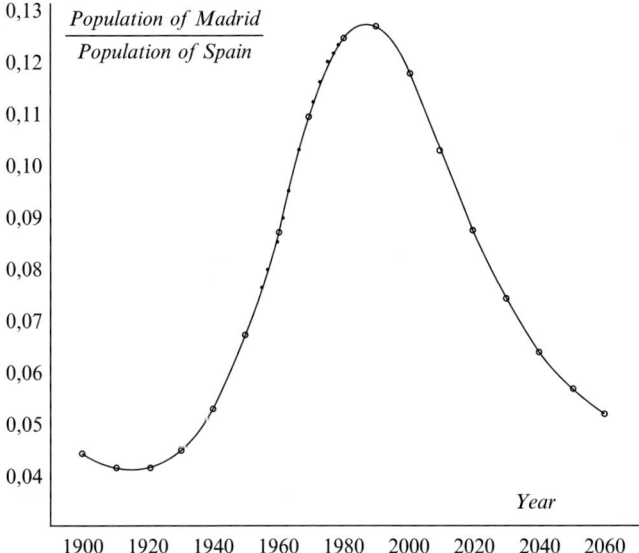

Figure 13.7 Simulation of the evolution of the population of the urban area of Madrid (the dots on the curved line) and empirical data from historic series (the smaller dots).
Source: Dendrinos (1984a, p. 244).

equation of (13.4) is the effect of two elements. The first is the term $\alpha x(y-1)$, which favours growth, for the entire time that the relative average per capita income y for the urban area (variable) exceeds the average of the whole country (which equals one unit). The second is the nonlinear term $-\alpha\beta x^2$, which represents the negative effect of agglomeration on population growth. Parameter β plays the role of friction factor: it is a term which, by opposing the effect of α, due to the negative sign with which it appears in the equation, acts by slowing down the movement of the population in question; the bigger the population, the more effect it has. The product $-\alpha\beta x^2$ constitutes a sort of force which, by opposing the action of $\alpha x(y-1)$, damps the system's dynamics. It can be shown that for $\beta > 0$, the systems evolves towards a state of stable equilibrium, like the attractor point at the centre of the convergent spiral in Figure 13.1; for $\beta < 0$, on the other hand, the system evolves amplifying its own oscillations, without tending towards a condition of stability.

We underline the fact that in the income equation, the second of (13.4), a quadratic y term is missing that would express the (positive or negative) effects of agglomeration on income. These effects are assumed to be negligible; the carrying capacity of income is therefore assumed to be infinite.

The symmetrical competition model and the formation of niches

In all of the variants of the basic Volterra–Lotka model that we have presented in the ecological sphere, with the sole exception of (12.3), we have been describing the interaction between two species, one of which, the predator, dominates the other, the prey. This is also true for the models that we have seen in regional and environmental spheres, in which the residents 'hunt' for the available income. The system, constructed in this way, presents an asymmetry with respect to the two species.

Another variant of the basic model can therefore be constructed, in which the roles of the two populations are symmetrical, in which, that is, the dynamics of the two populations are described by equations with the same form, like those of model (12.3) in an ecological sphere. In this case, it is a system that describes two populations of the same nature (or, in broader terms, that act according to the same mechanisms), that compete against one another for the limited resources available in the common environment. We therefore formulate a new variant of the basic model, introducing the carrying capacity for both populations. To this end, we add a second degree negative term to both system equations: $-bx^2$ and $-fy^2$, for the first and second populations respectively (Nijkamp and Reggiani, 1992ab, 1998):

$$\begin{cases} \dot{x} = x(a - bx - cy) \\ \dot{y} = y(d - ex - fy) \end{cases} \quad (13.5)$$

Parameters a and d are natural growth rates; $-b$ and $-e$ represent the use of the resources by the two populations, and are an index of the level of competition for resources within the population; $-c$ and $-f$ represent the sensitivity of the two populations to reciprocal competition and are an index of the level of competition between them.

A model such as (13.5) illustrates a dynamics, in certain aspects, similar to that of (13.2) and (13.4) (Pumain, Sanders and Saint-Julien, 1989).[53] It can be demonstrated (Smith, 1974) that, if the inequalities $d/e > a/b$ and $a/c > d/f$ are satisfied simultaneously, then (13.5) also results in the possibility of two states of stable equilibrium, that can be calculated, as in (13.2), by making the derivatives in (13.5) equal to zero. One solution is trivial: $x = y = 0$; the other is the point of coordinates $\bar{x}$ and $\bar{y}$ given by:

$$\begin{cases} \bar{x} = \dfrac{af - cd}{fb - ce} \\ \bar{y} = \dfrac{db - ea}{fb - ce} \end{cases} \quad (13.6)$$

In Figures 13.8 and 13.9 we provide an example of numerical integration of (13.5), similar to that shown in Figures 12.1–12.2 and 13.1–13.2. Note the similarity of the orbits in Figures 13.1–13.2 and 13.8–13.9, but the different coordinates of the points of equilibrium.

In reality, system (13.5) can also give rise to other very different orbits that differ again from those in the example illustrated in Figures 13.8 and 13.9, depending on the values of the parameters that characterize the system in question. For example, equilibrium can be achieved in a number of ways: the system can make repeated oscillations around the stability value before reaching it, following a convergent spiral shaped orbit in the phase space, or it can stabilize itself without oscillations, i.e. following an orbit that, we could say, 'aims' directly towards the state of equilibrium.

It can be seen that, if the two populations x and y are characterized by different carrying capacities, then they reach a constant value over time and

---

[53] Sonis and Dendrinos (1990), however, observe that the interpretation of the model (13.5) in terms of the income–population dynamics presents some difficulties with regard to internal consistency.

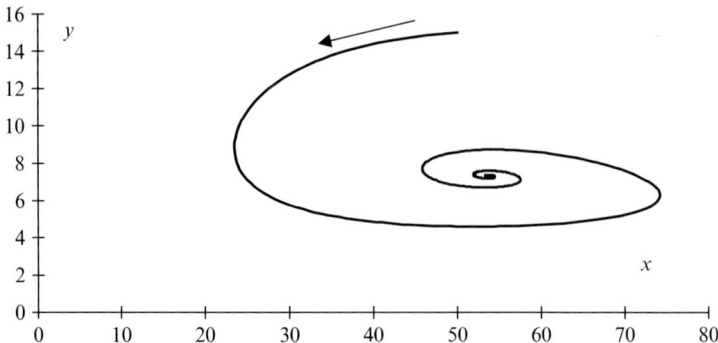

Figure 13.8 Dynamics of two variables x and y according to variant (13.5) of the Volterra–Lotka model. The values of the parameters are: $a = 0.1$, $b = 0.0005$, $c = 0.01$, $d = 0.05$, $e = 0.001$, $f = 0.0005$; initial values: $x = 50$ $y = 15$; final values of equilibrium: $x = 53.66$ $y = 7.32$.

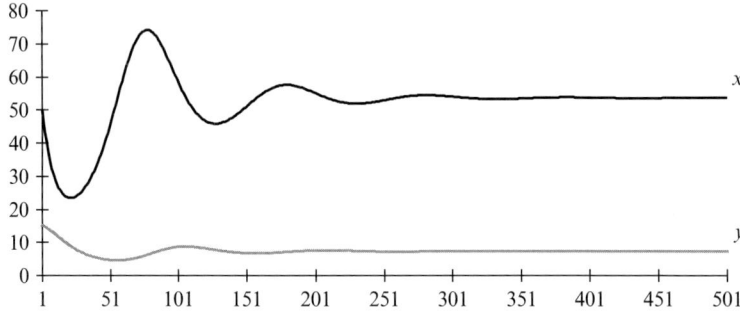

Figure 13.9 Trend of variables x and y as a function of time (simulation cycles), according to variant (13.5) of the Volterra–Lotka model. The values of the parameters, and the initial values are the same as those in Figure 13.8.

coexist; but if the carrying capacities are identical, the most efficient of the two will end up totally eliminating the other in accordance with Gause's principle of competitive exclusion.[54] The latter replicates one of the most important elements that characterize the economic theory of increasing returns, according to which the larger the market share of a company, the

---

[54] In 1934 Gause published *The Struggle for Existence*, a fundamental work in ecological literature, where the principle of competitive exclusion is discussed. This principle, of particular importance, is at the origin of the concept of ecological niche, according to which two different species that are not differentiated in their consumption of food resources cannot coexist indefinitely: either they differentiate themselves in some aspect or one of the two will eventually disappear.

higher the probability that said quota will increase and eventually absorb the market, eliminating competitors, and thus setting up a real monopoly (Bertuglia and Staricco, 2000).

Model (13.5) can be enhanced further, by replacing product xy with $x^n y^m (n, m \neq 1)$. The exponents n and m enable both the speed and the intensity of the change with which each of the two variables responds to a change in the other to be modified; this enables the interaction between x and y to become more rich and varied (it could be said that n and m introduce a form of partial elasticity to the interaction with respect to the dimensions of the species). In general, models like (13.5) are called niche models (Nijkamp and Reggiani, 1992b, 1998) and describe competitive situations between different populations, as well as cooperation and parasitism, depending on the signs of the coefficients c and e present in the terms that express the interactions between the populations (naturally, we would like to remind readers that 'populations' has a general meaning here, which should be specified on each occasion).

The concept of niche is particularly useful to describe the equilibrium towards which a process of evolution tends when it occurs in an environment in which the limited resources are exploited by various operators competing against one another, such as an undifferentiated set of prey for different species of predator, or an undifferentiated market, or a new market, for different goods manufacturers or service providers.

As we have said, in model (13.5), we can have a situation of equilibrium, in which the variables have the values indicated in (13.6). In this case we have the contemporary survival of all actors, in the same way as, in ecology, we have the contemporary survival of several species in a single environment. This happens if the species divide the limited available resources by means of a differentiation of interests with which the various parties (niches) approach the resources as a whole, therefore limiting interference between reciprocal sectors of interest. In contrast, according to model (13.5), it is possible that in an undifferentiated environment, i.e. where the two species have the same carrying capacity, no division into niches takes place, and this results in the inevitable disappearance of one of the operators.

The Volterra–Lotka model and its variants describe a situation in which two animal species that live in the same environment compete. These models, applied in an economic–spatial sphere, as we have said, have proved to be extremely useful to model any type of competitive scenario (see the study of three interacting species in May and Leonard, 1975, and within regional science, the study of the dynamics of a three-population system, conducted by means of simulations, in Dendrinos, 1991a).

For example, in an urban model, the services of two separate areas can be seen as being in competition with one another to attract users that live in a third area. The same can be said of a set made up of a larger number of services, or operators, generalizing the Volterra–Lotka model into N dimensions, with any value of N (Reggiani, 1997):

$$\dot{x}_i = \alpha_i x_i \left(k_i - \sum_{j=1}^{N} \beta_{ij} x_j \right) \qquad (13.7)$$

where x_i is the value of the population of species i ($i = 1, \ldots, N$), α_i is its growth rate, constants k_i are a function of the carrying capacity of species i, and coefficients β_{ij} represent the competitive coefficients between species i and j, measurable by the superimposition of the niches. For example, one can imagine describing the evolutive dynamics of three different types of transport systems along these lines (the predators: railways, roads, airways) that in an integrated transport network, compete with each other to 'capture' the limited available resources (the prey: the passengers and/or the goods).

Finally we can observe that (13.7), for $N = 1$, becomes simply:

$$\dot{x} = ax(1 - bx) \qquad (13.8)$$

known as the logistic function, which we will be looking at in Chapter 20.

We conclude our analysis of non-chaotic models here. In Part 2 we will be examining a topic that we have already mentioned several times in this first part: the stability of a system and deterministic chaos.

PART 2
From Nonlinearity to Chaos

If you say that all this, too, can be calculated and tabulated – chaos and darkness and curses, so that the mere possibility of calculating it all beforehand would stop it all, and reason would reassert itself, then man would purposely go mad in order to be rid of reason and gain his point!

Twice two makes four seems to me simply a piece of insolence. Twice two makes four is a pert coxcomb who stands with arms akimbo barring your path and spitting. I admit that twice two makes four is an excellent thing, but if we are to give everything its due, twice two makes five is sometimes a very charming thing too

Fyodor Mikhailovich Dostoevsky,
Notes From the Underground (1864)

14 Introduction

In Part 1 we looked at several examples of linear and nonlinear modelling in natural and social sciences. We also observed how nonlinear cases generally require much more sophisticated mathematical techniques, which still only enable us to obtain a partial description of the phenomena. In the linear case, on the other hand, we can obtain a complete description; if we have full knowledge of either the boundary conditions (i.e. the coordinates at different times) or the initial values of the dynamics (i.e. the coordinates and their derivatives), the system's future evolution is preordained and can be calculated.

We also observed, however, that linear models are not realistic and represent merely a mathematical approximation that is only acceptable when the system is very close to a situation of stable equilibrium. All models are, without a doubt, mathematical abstractions, but the hypotheses assumed for linear ones are the most drastically simplificative.

In Part 3 we will return to the question of the meaning of the abstractions made in mathematics, and more generally, the sense of mathematics and its methods. In this second part, however, we are going to examine a matter of fundamental importance in more depth: a nonlinear mathematical model can, at times, demonstrate very particular evolutive behaviour, giving rise to unstable evolutive trajectories that appear to be unpredictable, as if they weren't generated by any imposed deterministic law and that we call chaotic. Behaviour of this nature is not peculiar to abstract mathematical models alone: even real systems observed experimentally, in both natural and social sciences, on occasion may demonstrate behaviour that appears to be unstable and chaotic.

In this part, therefore, we will be focusing mainly on nonlinear models, describing some of the elements that characterize their dynamics with respect to those of linear models. We will begin by discussing, in Chapters 15 and 16 several rather technical aspects of deterministic chaos and some of the ordinate structures that sometimes manifest themselves in nonlinear dynamics: chaotic and strange attractors. We have already come across some types of attractors in Part 1, but they were ordinary attractors, whose structure is profoundly different to that of strange attractors, as we will see. In Chapter 17 we will clarify some of the differences between chaos in models and chaos in real systems. The onset or otherwise of chaos in models depends on the stability of an orbit in the phase plane, a concept that will be examined immediately afterwards, in Chapter 18. In Chapter 19 we will touch upon a very delicate matter regarding the recognition of chaos in data obtained from measurements of real systems. Then in Chapter 20 we will return to our discussion of growth models, looking at some properties of the logistic function. In Chapters 21 and 22 we will examine the logistic map, a variant of the logistic function in discrete time, an example of one of the simplest and most well-known dynamical systems that, in certain circumstances, manifests chaotic behaviour.

Our discussions will contain some technical mathematical details here and there, although we have no intention of providing formal demonstrations or entering into exhaustive discussions, favouring clear and simple explanations over technical ones, in order to enable the reader to better understand the key concepts behind the mathematical technicalities.

15 Dynamical systems and chaos

Some theoretical aspects

Before proceeding, we would like to remind readers of a concept that was stated in Chapter 8: by dynamical system we simply mean a system of equations that define the evolution of a set of state variables in discrete or continuous time. It is therefore important not to confuse the abstract mathematical concept of dynamical system, with that of system, which for the sake of clarity we will call real, and which instead refers to the empirical observation of phenomena and to the measurement of series of data. The mathematical models we are examining are written in the form of dynamical systems: therefore the properties of the models that we will be discussing should be considered properties of dynamical systems.

In the various types of oscillating systems described in Part 1, the existence of several geometric objects in the phase space, called attractors, was acknowledged. We came across them in a number of models: there was an attractor point in the damped oscillating systems and in some interacting population models, while, in certain variants of the Volterra–Lotka model, we encountered limit cycles. In general terms, an attractor is a particular region in the phase space (a subset of the phase space) that a dynamical system tends to reach during the course of its evolution. When the system, following its orbit, approaches an attractor, it does not escape from it unless external factors intervene and alter the system's dynamics.

A basin of attraction is associated to an attractor: a region in the phase space (it can even be the entire phase space, if the dynamics of the system envisage a single attractor) characterized by the fact that, starting from any one point in that region, i.e. from any state, the dynamical system evolves, always directing itself towards the attractor. If the attractor is a point, the system moves toward it in a condition of asymptotic stability; if the attractor is a closed line, called a limit cycle (see Chapter 2), the system tends asymptotically to travel along it with a constant periodicity.

If the system is relatively simple, such as a pendulum made up of one particle only, the oscillating mass, whose position can be obtained given the value of one single coordinate, then (see Chapter 11) the model that describes its evolution over time identifies the state of the system by a point in a phase space that is reduced to one plane, and its evolution can be represented by different forms of lines, depending on the mathematical model adopted and the values of the parameters. If, on the other hand, the system is more complicated and is made up of a large number of reciprocally interacting elements (such as a gas, a society of individuals, an economic system, etc.), then the model is more complicated. The system's state is now given by a set of points in a region in the phase space. Let us attribute a volume[55] to this region; so, if the system represented by the points contained in that region evolves over time conserving its total energy, then it can be shown that the region in question moves within the phase space, perhaps changing form as the system evolves, but the volume that it occupies remains constant. This is called a conservative, or non-dissipative dynamical system (and is the case of the pendulum without friction that oscillates periodically, without attractors). In dissipative systems (systems that dissipate energy into the environment, like a damped pendulum that over time tends towards a state of stable equilibrium), on the contrary, the points, that as a whole define the system's state, move towards a particular region in the phase space, the attractor, with the result that the volume in which the points are spread contracts during the course of the dynamical system's evolution.[56]

In other words, the dissipative system behaves as if the points that characterize it in the phase space 'slip' towards an attractor, like an object in a gravitational field that rolls along the sides of a valley to its bottom. The bottom of the valley is an attractor, while the valley represents, in this example, the basin of attraction (Figure 15.1).

In the case of a conservative dynamical system, every point in the initial region follows its own particular orbit, which depends only on the initial state and is not affected by the presence of any attractor. In the case of a dissipative system, on the other hand, all of the orbits tend towards a

---

[55] We will use the term 'volume' to indicate the extension of the region in question, without referring explicitly however to the three-dimensional case only; in the two-dimensional case, for example, volume is intended to mean area.

[56] This constitutes the content of a renowned theorem of the French mathematician Joseph Liouville: if in a dynamical system the energy is conserved (conservative system), then the volume of the subset in the phase space, made up of the points whose coordinates and momenta (mass multiplied by velocity) characterize the dynamical system, is a constant of the motion (Chavel, 1994).

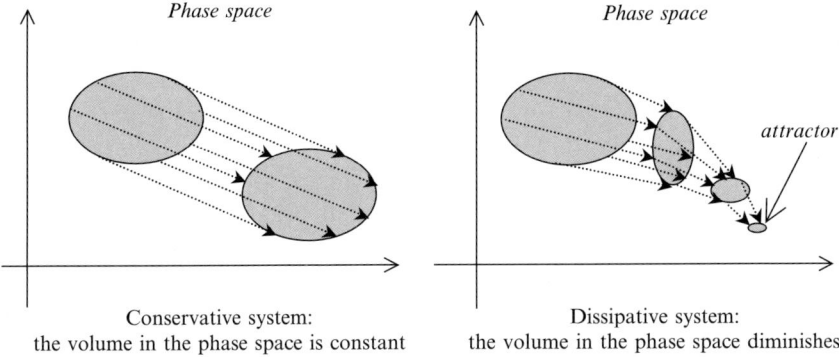

Figure 15.1 Evolution of a conservative dynamical system and a dissipative dynamical system in the phase space.

common attractor, which characterizes the general dynamics of said system. In the latter case, the solutions of the equations that describe the dynamical system tend towards a final stable configuration over time: apart from very particular situations, two orbits that start 'fairly' close together tend towards the same final configuration. The evolution of a system that behaves in this way is predictable; even if the starting state is only known through approximation, this does not affect the final state, which is always the same attractor.

The attractor can be an isolated point, such as that in the different models of the pendulum discussed in Part 1, with the exception of the chaotic model; it can be a limit cycle, i.e. a regular flat continuous curve with a tangent at each point, a curve in a three-dimensional space, but also a geometric object with more than three dimensions, depending on the number of state variables that make up the phase space; as we know, in fact, each state variable corresponds to one single dimension (i.e. one axis) of the phase space. There is also the case of an attractor made up of a set of non-continuous states, i.e. by a set of isolated points, as we will see in Chapter 21 in the logistic map. This can occur if the system is not described by differential equations in continuous time, but by maps,[57] i.e. by finite difference equations. An ordinary attractor (point or closed line) is

---

[57] Remember that a map is simply a function $f(x)$ in the phase space that, starting from state x_n, supplies the next state x_{n+1}. A map, therefore, is none other than a difference equation written isolating the $(n+1)$-th value of the variable in the first member, i.e. in the form: $x_{n+1} = f(x_n)$. Each differential equation can give rise to a map, increasing the independent variable by a unitary step (therefore not infinitesimal) that in dynamical models is time.

synonymous, therefore, with the stability of a dynamical system, with the reproducibility of the evolution, and with the return to the same conditions, starting from different initial conditions; that is to say, with the insensitivity to variations of the state assumed as initial or disturbances to the same, or to ambiguities in its definition or measurement.

Up until a few decades ago, it was believed that the only attractors possible for systems described by differential equations were made up of sets of isolated points or a closed line (a limit cycle) or even by a line that wraps itself around a toroidal surface,[58] attractors called simple or ordinary. In the wake of the pioneering work carried out by Poincaré, first, and later by Fermi, studies that incidentally were not immediately successful, the chance discovery of the existence of a new type of attractor (Lorenz, 1963; Smale, 1967; Ruelle and Takens, 1971), however, opened up entirely new perspectives.

There are real systems that manifest dynamical behaviour characterized by a strong dependence on the initial conditions. This means that small initial disturbances are increasingly amplified as the system evolves, giving rise to increasingly greater changes in the evolution. In the same way, there are models in the form of dynamical systems that give rise to diverging orbits. By integrating dynamical systems of this nature, we see that the subsequent states of the latter are not dispersed into increasingly larger regions in the phase space, but remain confined to limited regions, without, however, *ever* repeating the same sets of values of state variables, as happens on the contrary with limit cycles. Basically what happens is that these dynamical systems, evolving over time, repeatedly pass *close* to states they have already occupied, without ever displaying constant periodicity, tracing orbits that are continuous lines that never intersect each other and are of infinite length, even though, we will repeat, enclosed (we could say 'entangled') in limited regions in the phase space. The possible orbits of a dynamical system of this nature, even though all of an *almost* equal form, in reality depend on the initial state: they are all open lines, confined to the same region in the phase space, similar in form, but distinct from one another, which are travelled along at different speeds over time, with the effect that the states on one orbit and the corresponding states that we would have at the same points in time on another disturbed orbit, increasingly distance themselves. This is the well-known butterfly effect: even initial situations that are very close to one another may present a difference that becomes sizeable over time.

---

[58] A toroidal surface is the surface of a three-dimensional geometric object called a 'torus', whose form, in practice, is that of a ring or a doughnut.

Starting from an initial point, a dynamical system of this nature reaches an area in the phase space, within which it remains; its subsequent states construct a curved, open, infinite orbit, that is however permanently confined to that limited region in the phase space; an attractor of this type is called *chaotic*. A dynamical system whose dynamics are characterized by the presence of a chaotic attractor is also called chaotic.

Situations in which chaotic attractors are encountered can be generated by nonlinear models, which, in certain circumstances, can manifest evolutive dynamics that appear completely unpredictable, precisely because they lack ordinary attractors, that is to say they lack states of stable equilibrium (point attractors) or sets of states travelled periodically (limit cycles).[59]

Two examples: calculating linear and chaotic dynamics

Certain dynamical systems are characterized by the fact that small differences between the initial states, over time, result in increasing differences between subsequent states; this property is called *sensitivity to initial conditions*. We have already come across a model characterized by said sensitivity in Chapter 10, when we mentioned the chaotic pendulum (Model 4). Sensitivity to initial conditions means that a (small) disturbance to the initial conditions of an orbit, assumed to be undisturbed, generates another orbit, which we will call disturbed, which starts at a point in the phase space close to the initial one of the undisturbed orbit, but that develops over time, increasingly amplifying the difference with respect to the undisturbed orbit. This could lead to the points in the initial region in the phase space being spread over increasingly larger regions in the phase space, without giving rise to any attractor (Figure 15.2), or to the points being distanced from one another, but remaining confined to a limited region in the phase space thus indicating the presence of an attractor.

Let us consider an example of a very simple dynamical system: a mathematical law like (8.1'), called a linear map.

We assume:

$$x_{n+1} - 2x_n \qquad (15.1)$$

---

[59] Typical examples of dynamical systems of this type are forced oscillating systems, i.e. systems in which the 'natural' oscillatory dynamics are continually altered by an intervention from the external environment. The Duffing equation, that we mentioned in Chapter 12 (p. 100, note 48), is an example of a dynamical system that can display limit cycles, as we said in the cited note, but also situations of unpredictability (chaotic), depending on the values of the parameters.

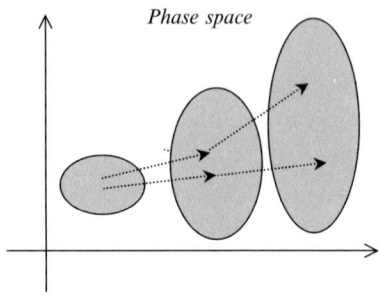

Figure 15.2 Evolution in the phase space of a dynamical system sensitive to the initial conditions, which does not have attractors.

Equation (15.1) is characterized by a single coordinate, x, whose value evolves over time simply by doubling itself. If we calculate two sequences starting from close initial values, the distance between two corresponding values of the two sequences doubles at each step; if, e.g. we (slightly) move the coordinates of the initial point of the evolutive trajectory, the latter does not return towards the initially undisturbed one, but gets increasingly further away.

Let us iterate the linear map (15.1), a first time, starting from $x_1 = 1$; and then a second time, starting from a point close to 1. Let us assume for example that $X_1 = 1.1$. The first terms of the two sequences are shown in Table 15.1; the values of the differences between corresponding terms are shown together.

The two sequences can be visualized as shown in Figure 15.3; the first sequence is represented by the points indicated above the real number axis, the second by the points shown below.

It is immediately apparent that the two sequences are unlimited and that, as the iterations progress, they generate numbers, represented by the points

Table 15.1 First iterations of the linear map (15.1) for two different initial values

n	x_n	X_n	$\Delta_n = X_n - x_n$
1	1	1.1	+0.1
2	2	2.2	+0.2
3	4	4.4	+0.4
4	8	8.8	+0.8
5	16	17.6	+1.6
6	32	35.2	+3.2

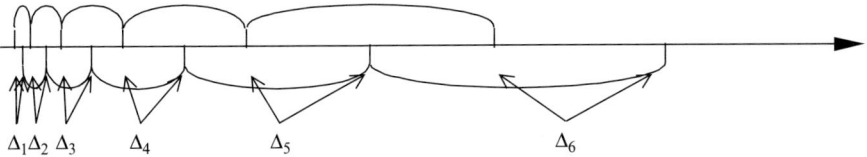

Figure 15.3 Iterations of the linear map starting from two totally separate values.

on the line in Figure 15.3, that do not remain confined to any limited part of the line [the phase space for map (15.1) is simply the line of real numbers]. The difference between the coordinates of the two points that are obtained after the same number of iterations, indicated by Δ_n, also grows without limit as n increases according to the same law (15.1).

Equation (15.1) is the recursive definition of a simple geometric sequence of common ratio 2, which does not in any way reflect the approximate idea of unpredictability that we intuitively associate with chaotic situations. The sequence generated by (15.1) is without doubt sensitive to the initial conditions, as the two trajectories increasingly diverge as the iterations progress, but the terms can be easily calculated not only using the recursive formula (15.1), but also using known non-recursive formulae provided in the elementary arithmetic for geometric sequences.

When we talk about chaos and chaotic dynamical systems other than the example we have just seen, we mean something more that a mere sensitivity to initial conditions. We can only talk about chaos if the orbits in the phase space remain confined, i.e. if the system does not evolve towards infinity for any variable, something which does happen, on the other hand, with the linear map (15.1); sensitivity to initial conditions is a necessary condition for chaos, but at the same time is not in itself sufficient.

In linear dynamics, there can be sensitivity to initial conditions and unlimited orbits, as in the cases of the open orbits discussed in Chapter 11, or limited orbits, but without sensitivity to initial conditions, as in the examples of oscillation discussed in Chapters 5 and 8 and in the cases of cyclical orbits discussed in Chapter 11. Linear dynamics, however, are not capable of generating orbits that are sensitive to initial conditions and that remain confined to a limited region in the phase space.

Sensitivity to initial conditions, in itself, is not an exclusive assumption of nonlinear dynamics; however, it plays a fundamental role in nonlinear cases. Briefly anticipating some elements of the logistic map, that we will be discussing in Chapters 21 and 22, we will use an example to better clarify what we mean by sensitivity to initial conditions and orbits confined to an area in the phase space, and in what context this gives rise to unpredictability.

Let us take an example of a logistic map and calculate two sequences starting from initial values that are close to one another, as we did for the linear map (15.1). Let us iterate, for example, the following map:

$$x_{n+1} = 3.7 x_n (1 - x_n) \tag{15.2}$$

once, starting from $x_1 = 0.5$ and a second time, starting from $X_1 = 0.6$. The numbers obtained for the first six iterations are shown in Table 15.2, together with the differences between the corresponding terms (we have limited ourselves to nine decimal places, those shown by a normal pocket calculator).

Several differences between the sequences originated by the initial 'close' values, generated by (15.1) and those generated by (15.2) are apparent. First of all, (15.1) generates unlimited sequences, while (15.2) generates sequences whose terms always fall between 0 and 1 (this is easy to check, by continuing with the calculation of subsequent terms). The ease of predicting the terms of the sequences generated by (15.1) right from the outset, moreover, is contrasted by the fact that the terms of the sequences generated by (15.2), even though generated by a deterministic law, do not demonstrate easily identifiable recurrences, such as those that we can 'guess' at a glance. In this context, we speak of sequence unpredictability of (15.2), referring to the substantial impossibility of identifying its law, if the latter is not already known. Lastly, the difference Δ_n between the corresponding terms of the 'close' successions also now has a nonlinear, unpredictable trend, whereas, for (15.1), Δ_n increased simply by doubling itself at each step.

Table 15.2 First iterations of the logistic map (15.2) for two different initial values

n	x_n	X_n	$\Delta_n = X_n - x_n$
1	0.5	0.6	+0.1
2	0.925	0.888	−0.037
3	0.2566875	0.3679872	+0.1112997
4	0.705956401	0.860518696	+0.154562295
5	0.626861975	0.444097199	−0.182764776
6	0.865452345	0.913437046	+0.047984701

A final observation of particular interest regards the sensitivity to initial conditions that originates from the unpredictability of the terms of sequence (15.2). Equation (15.2) defines the dynamics of a chaotic system on two grounds: (1) because it is characterized by unpredictability, in the context mentioned on the previous page, and (2) because subsequent states are confined to between 0 and 1, while this is not true for (15.1).

We will resume this topic in Chapter 16, returning to these two successions and further developing our discussions thereupon.

The deterministic vision and real chaotic systems

In real systems we talk of chaos when the prediction of the system's future evolution becomes impossible for anything beyond short periods of time, regardless of the precision with which the initial conditions are measured, even if the mechanisms that control its evolution are known and can be transformed into a model. Sensitivity to initial conditions, also in the case of real systems, means that it becomes increasingly difficult to make predictions about the system's future evolution the further into the future we wish to look, due to the amplification of measurement errors that originate from approximations in measuring the state variables.

Improving the precision of measurements can, in some way, extend the time interval within which predictions can be considered meaningful, the so-called predictability horizon. However, as the predictability horizon typically only grows according to the logarithm of measurement precision (see Chapter 18), even though we may make considerable efforts to improve observations, in practice we obtain only modest results regarding the predictability of the system's future evolution.

At least until questions related to quantum uncertainty are introduced,[60] the limit to measurement precision might be thought to be imposed by purely practical difficulties. Improving measurement precision, in principle, would enable new decimal places to be added to the number obtained from each measurement. However, in no way would this enable the predictability

---

[60] The uncertainty principle is one of the fundamental principles of quantum physics. Formulated by Werner Heisenberg (1927), it maintains the existence of specific pairs of magnitudes, called conjugates (for example, the x coordinate of a particle and its speed in the direction of the x coordinate, or energy and time), that cannot be measured simultaneously with arbitrary precision; the measurement precision of one of the two magnitudes is inversely proportional to the measurement precision of the other magnitude, conjugated to it. This principle can be ignored for magnitude values typical of everyday experience and is relevant only with regard to atomic or subatomic phenomena.

horizon to be extended to infinity, as this would require 'absolute' precision at the time of the measurement of the initial state and the maintenance over time of the same 'absolute' precision with regard to the definition of the system's subsequent states.

First of all, the idea of 'absolute precision' in determining a system's state is an abstract concept (or even intrinsically wrong), completely lacking in meaning in practice, absolutely unachievable in any measurement, at least (but not only) due to the constraints posed by the principle of quantum indetermination. At most, increasingly improving the precision of measurements enables us to relegate the approximation towards a decimal place that is increasingly further from the point, but not to eliminate it and obtain absolute precision. If we use the approximate result obtained from the measurement of the present state as an input for a model that simulates the evolution of a real system to calculate its future states, then, if the model is an unstable dynamical system, the error introduced by the approximation does not remain confined, over time, to the figure on which the approximation was initially based, but increasingly amplifies itself.

We can never have *thorough* knowledge of reality, because we cannot measure any of the state variables with *absolute* precision, even if we assume that we have identified all of the state variables needed to fully define a system (which, if the system is complicated, is by no means necessarily easy, or may even be impossible). The (small) inevitable approximations in measurements may hide small immeasurable (i.e. we cannot know them) differences between the initial states of the orbits. In general, we are not at all sure whether these small immeasurable differences will remain small during the course of the dynamical system's evolution or whether they will tend to disappear, or, on the other hand, whether they will progressively grow. It depends on the mechanisms that control the system's evolution. Similarly, we are not at all sure whether small disturbances that intervene during the system's evolution have effects that, in the long term, become significant, substantially changing its evolution. This is exactly the fundamental point that Laplace was unaware of and where classical determinism erred: the small, negligible approximations (or imprecisions or disturbances, however you wish to call them) that intensify over time and cease to be negligible. This is therefore the unpredictability that characterizes an unstable dynamical system (i.e. a model), which is sensitive to initial conditions. This concept is in complete antithesis to the Laplacian deterministic vision; as accurate as our knowledge of the present state may be, there may be intrinsic instability in the evolution of the dynamical system (model), which limits the predictability horizon, i.e. the interval of time within which the predictions provided by the model, i.e. from our representation of the situation, can be considered realistic.

The question of the stability of the solar system

Laplacian determinism was also the result of a series of studies and calculations regarding the problem of describing the movements of the planets. The idea that *small* errors (or approximations or disturbances) could *only* generate *small* consequences appeared in classical science, not only because it was intuitive, but also following the success of Newtonian physics applied to several aspects of celestial mechanics. Studies regarded in particular the problem of two bodies in gravitational interaction: Sun–Earth, Earth–Moon, etc., namely the initial historical problem from which classical physics and the methods of differential calculus and integral calculus originated and were developed. However, this notion of a strict proportionality between cause and effect is wrong, in general, even for many other problems of celestial mechanics.

One of the fundamental questions that is posed in celestial mechanics, where this vision fails, is the study of the consequences of the interaction between the Sun and the various planets of the solar system, not in pairs, but *taken as a whole*.

The roots of the problem can be found in astronomical studies in the eighteenth and nineteenth centuries: it regarded the problem of the n bodies, which we mentioned in Chapter 2, a problem that was tackled by Lagrange and Laplace himself, among others. The movements of the planets appear regular in human time-scales, i.e. from the existence of historic documents to date; this, on the surface, would appear to confirm the idea that the small disturbances in the dynamics of the planets that the planets themselves, taken as a whole, reciprocally cause to each other do not intensify and do not end up by compromising the general stability of the solar system. However, even for the solar system, like any other system, we are faced with the problem of whether the apparent stability maintains itself on a time-scale that is much longer than the human one.

The question has various aspects that complicate it further. One particular problem is emblematic and is worth remembering: the problem of the resonance between the revolution movements of Jupiter and Saturn. It has been known from ancient times that the ratio between the revolution periods of these two planets was *circa* 2/5: for each two orbits that Saturn makes, Jupiter makes five. This means that Saturn and Jupiter, periodically, find themselves in a position of maximum reciprocal vicinity always at the same point in their orbits and in the same position with respect to the Sun. We would therefore expect the following type of resonance phenomenon: the mutual attraction between Jupiter and Saturn causes periodic disturbances to the movement of the two planets, which are always equal and always occur at the same position in each of the two orbits. These individual periodical disturbances caused by the Jupiter–Saturn attraction, even if the latter is small with respect to the attractive force of the Sun–Jupiter and the Sun–Saturn, and if the two periods are exactly in the ratio 2/5, should give rise to a bona fide resonance phenomenon and cause increasingly significant effects over time, in line with the typical behaviour of chaotic dynamics: the amplification of small disturbances. The orbits of Jupiter and Saturn, if this schema is correct, would not be stable.

The issue is that the precision of the available measurements does not enable us to determine whether the said 2/5 ratio is 'exactly' the rational number that is written in decimal form 0.4000... This 2/5, therefore, should be understood as a number 'close' to 2/5: not necessarily 0.4000... Perhaps it could be an irrational number, for instance, 0.400010010001... in which the first figure that is not 0 does not appear because the precision of the measurements is not able to provide it. If the periods were incommensurable, there would not be long-term resonance and there would not be an accumulation of disturbances causing instability. The mathematician Karl Weierstrass, in the 1870s, however, had observed how difficult it was to accept the idea that the answer to the question of the solar system's stability had to depend only on the precision of a measurement of periods.[61]

---

[61] Weierstrass linked this point to the problem of the non-convergence of certain trigometric series that were obtained when trying to write the equation of the orbit of the planets around the Sun, taking into account the mutual interactions between the bodies of the solar system. In particular, the problem originated from the use of the Fourier series development of functions, a technique of linear mathematics commonly used to describe periodic phenomena, whose convergence, however, in that case, presented grave difficulties. The issue extended and then ended up touching on the problem of the continuity of the set of real numbers and the foundations of mathematics itself. The problem is very delicate, and is at the origin of Poincaré's studies, which first led to the 'discovery' of chaos. Weierstrass convinced himself that the problem of the stability of the planetary orbits and, in general, of a system of n bodies had to be tackled in a completely new way with respect to the ordinary techniques of mathematical analysis of that era. The prize that the King of Sweden, Oscar II, who wished to provide proof of his interest in scientific progress, had decided to award an important discovery in the field of mathematical analysis, gave Weierstrass the opportunity to invite the mathematicians of that period to investigate whether there was a new way to solve the problem. The prize, as we have already mentioned (p. 24, note 12), was awarded to Poincaré the following year, for his fundamental study that appeared, more than a century later, as the first real description of chaos in mathematical terms (Bottazzini, 1999).

An important element in the solution of the problem of the stability of the orbits was proposed at a later date, when, in 1954, the Russian mathematician Andrei Nikolaevich Kolmogorov formulated a renowned theorem, which was then demonstrated in 1962 by Vladimir Igorevich Arnold and Jürgen Moser, known as the KAM theorem, taken from the initials of the three scholars (Arnold, 1987; Peterson, 1993; Burke Hubbard and Hubbard, 1994). According to the KAM theorem, in certain conditions, stable orbits, in reality, are much more numerous than the unstable ones, because, for sufficiently small perturbations, such as the Jupiter–Saturn attraction with respect to the Sun–Jupiter and Sun–Saturn attractions, they can have orbits that are not periodic but so-called quasiperiodic, the possibility of which had already been foreseen by Lagrange and Laplace; orbits that remain, so as to speak, in a region of stability in the neighbourhood of the unperturbed orbit, even for irrational values of the ratio between the two periods, as long as they are close to 2/5. However, it has never been proven that the solar system satisfies the conditions established by the KAM theorem.

The stability properties of the model and therefore of the solar system remained unclear. The problem lay not so much in the technical limitations of astronomy, as rather in the lack of clarity regarding mathematical concepts of the stability of dynamical systems that was encountered and in the technical failings resulting from this.

In more recent years, Jacques Laskar (1989) conducted numerical simulations of the evolution of the solar system. According to the results, there are numerous resonances, the solutions of the equations of motion do not converge in stable, or periodic or quasiperiodic orbits. The dynamics of the solar system are such that they would appear to exponentially amplify a disturbance Δx_0 according to a law such as the following:

$$\Delta x(t) = \Delta x_0 e^{t/5}$$

expressing time t in millions of years.

This means, for instance, that an error in the determination of the initial conditions of the motion of a planet to the order of a ten billionth would become a 100 per cent error in approximately 100 000 000 years. In other words, considering the solar system in its entirety, the conclusion of Laskar's simulations is that the movements of the planets are chaotic and that, basically, in an interval of time in the order of 100 000 000 years, they become unpredictable. As Laskar himself observed, however, in reality, even though the idea of a chaotic solar system may seem surprising, and may even turn our conception of the world, built on centuries of stability, upside down, it is rather the opposite that should seem extraordinary, knowing, after Poincaré, that the majority of dynamical systems cannot be integrated and are therefore not subject to this type of behaviour.

The problem of whether the solar system is stable or not, in any event, is far from being solved for good. On one hand, in fact, there is the intrinsic stability of the evolution of the dynamical system that we use as a model of the solar system (Newton's renowned law); but on the other hand, assuming that we possess the complete law of motion of the solar system is asking too much. In reality, what we possess here, as in all natural and social systems, is merely a set that is very large but not unlimited, of observations regarding how the system has behaved in the past; based on these, we have attempted to define a valid law for future evolution without limitations of time (Casti, 2000), forgetting that the model constructed is inevitably an approximate model, as are all models, due to the limited nature of the data on which the dynamic law is based.

16 Strange and chaotic attractors

Some preliminary concepts

In the early 1970s, mathematicians David Ruelle and Floris Takens introduced the concept of the strange attractor to describe the phenomenon of turbulence (Ruelle and Takens, 1971; see also Ruelle, 1991). A strange attractor is a subset of points in the phase space that is fundamentally different to that of the objects belonging to ordinary Euclidean geometry; it is a geometric object characterized by a dimension that is not an integer number. The French mathematician of Polish origin Benoît Mandelbrot, also in the 1970s, coined the expression 'fractal' to indicate geometric objects whose dimension is not integer[62] (Mandelbrot, 1975): strange attractors are fractal objects.

An ordinary attractor and a strange one are different objects from a topological point of view,[63] as we cannot obtain a strange attractor simply by deforming an ordinary attractor, without generating the fragmentation of such.

Chaotic attractor and *strange attractor* are different concepts. By indicating an attractor as 'chaotic', we are referring to the chaotic dynamics of

---

[62] To discuss the non-integer dimension of fractal objects necessitates the introduction of a new (and more general) definition of the concept of dimension, with respect to the one commonly used in Euclidean geometry, within which we speak only of spaces with dimension one, two, three or, possibly, any integer number greater than three. A redefinition, that includes the ordinary Euclidean dimension as a particular case, can be given in various ways; the one most frequently used was provided before the birth of fractal geometry by the work of Felix Hausdorff (1919) and Abram Samoilovich Besicovich (1934, 1935) and this is why it is referred to as the Hausdorff–Besicovich dimension (we will return to this in chapter 19). The concepts of dimension and fractal objects are illustrated in numerous texts, for example: Mandelbrot (1975); Peitgen and Richter (1986); Falconer (1985, 1990); Pickover (1990, 1994); Crilly, Earnshaw and Jones (1991); Jones (1991); Schroeder (1991); Peitgen, Saupe and Jürgens (1992); Czyz (1994); Frankhauser (1994); Vulpiani (1994); Abarbanel (1996).

[63] Two geometric objects are equivalent from a topological point of view if there is a transformation that enables us to pass from one to the other without breaking them. For example, a circumference and a square are topologically equivalent, just as a sphere and a cube, while an annulus and a rectangle are not, just as a torus (see p. 130, note 58) and a cube are not.

the system in which the attractor originates, not to the geometric shape of the latter: we are therefore referring to an attractor generated by a dynamic which largely depends on the initial conditions, but whose orbits are confined to a limited region in the phase space. By indicating an attractor as 'strange', on the other hand, we are referring to its geometric properties, because we say that it is a fractal object. The question we ask ourselves is whether the strangeness, i.e. a characteristic of the geometric form, and chaoticity, i.e. a characteristic of the underlying dynamics, are two different perspectives, on the basis of which we can define the same object, or whether strange attractors and chaotic attractors are, on the contrary, fundamentally different objects.

In most cases chaotic attractors are strange and strange attractors are chaotic, which forms the common premise to distinguish between the two concepts (this was true for the first non-ordinary attractors encountered). However, cases of strange attractors that are not chaotic have been described, both in a theoretical ambit (Grebogi et al., 1984) and in experimentally observed phenomena (Ditto et al., 1990), as well as cases of chaotic attractors which are not strange (Holden and Muhamad, 1986; Grebogi, Ott and Yorke, 1987).

In this regard, we find the renowned example of the two-dimensional map, known as *Arnold's cat map* (Schroeder, 1991), proposed by the Russian mathematician Vladimir Arnold, who demonstrated it using the image of a cat's head (Figure 16.1).

We assume a square with side l whose points are identified by a system of x and y coordinates, centred at the bottom left vertex. Each iteration of Arnold's map involves three steps:

(1) extension along the x axis, according to the following map:

$$x_{n+1} = x_n + y_n \qquad (16.1)$$

(2) extension along the y axis, according to the following map:

$$y_{n+1} = x_n + 2y_n \qquad (16.1')$$

(3) recomposition of the square with side l, starting from the figure obtained from the previous deformations, subtracting l from the x_{n+1} and the y_{n+1} that exceed l.[64]

---

[64] The operation described in step (3) as a geometric operation can be seen, from an arithmetic point of view, as the result of a calculation that attributes the values of the

Some preliminary concepts   143

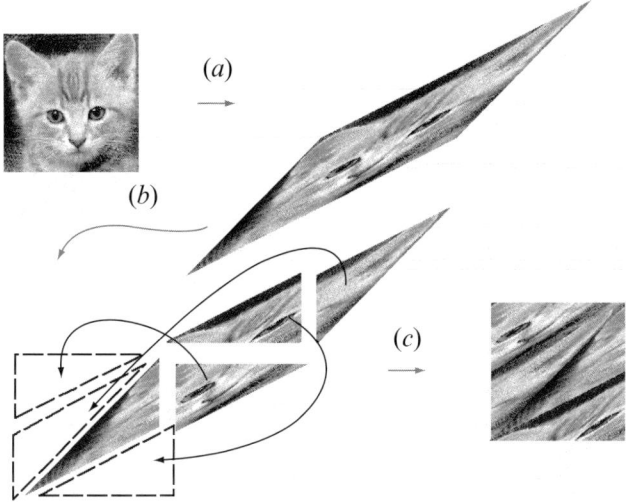

Figure 16.1   The first iteration of Arnold's cat map.

The steps of Arnold's cat map are shown in Figure 16.1: steps (1) and (2) in the transformation are indicated by (*a*), and step (3) is indicated by (*b*) and (*c*).

In the subsequent iterations, repeating steps (1), (2), and (3) in sequence, the points increasingly distance themselves from one another, while remaining within the surface of the square that contained the original image, according to a typical chaotic evolution. However, at a certain point, and this is the interesting aspect, the original image recomposes itself as it was at the start (this is known as the Poincaré recurrence[65] phenomenon; see Crutchfield et al., 1986). The image is therefore a chaotic attractor, due to the progressive distancing of the points from one other, which indicates the instability of the trajectories, but it is not strange, because the points, moving in a limited space, end up in their initial positions, re-forming the

remainders of the divisions x_{n+1}/l and y_{n+1}/l to x_{n+1} and y_{n+1}, respectively; the operation is called 'modulus' (mod):

$$\begin{aligned} x_{n+1} &= x_{n+1}(\text{mod } l) \\ y_{n+1} &= y_{n+1}(\text{mod } l) \end{aligned} \qquad (16.2)$$

[65] It can be proved (Poincaré recurrence theorem) that systems of limited size that contain only discrete elements repeat their evolution after a sufficiently long time; more generally, if the system has a fixed total energy that restricts its dynamics to bounded subsets of its phase space, then the system will eventually return closer to any given initial state than any arbitrary distance.

initial image. This is a simple and elegant demonstration and illustration of one of the principles of chaos: the underlying order of an apparently random evolution of a system.

In all the cases of models of oscillating systems described in Part 1, with one exception, the chaotic pendulum (Model 4), we used a two-dimensional phase space in which we observed ordinary attractors made up of one point or a closed line (the limit cycle). It is important to note that systems of differential equations with two variables cannot give rise to strange attractors. In fact the latter cannot exist in two-dimensional phase spaces, but can only exist in phase spaces with more than two dimensions; on a two-dimensional surface, in fact, the orbit generated by a system of differential equations is a closed line, as with the harmonic oscillator (Chapter 9), or gives rise to an ordinary attractor, i.e. it aims towards an attractor point or a limit cycle, as occurs, for example, with the orbit of a damped pendulum (Chapter 8). An orbit different to the two preceding ones would be a line of infinite length, which, because it is contained in a limited region of a surface, necessarily has to intersect itself in one or more points. This latter case, however, cannot occur, because, if it did, for a point in the phase space in which the orbit intersects itself, we would have two possible directions of the dynamics, which would go against the theorem of existence and uniqueness of the solutions of a differential equation.[66] In other words, this means that models in the form of differential equations (i.e. the dynamical systems) can be chaotic only if characterized by at least *three* independent state variables. Only in this case can an orbit be a line of infinite length which, although confined to a limited region, never intersects itself, because it has at least a third dimension available for 'lifting itself' above the branch of the encountered curve and thus, we could say, 'climbing over' the intersection.

---

[66] The theorem of existence and uniqueness, as it is sometimes called for the sake of brevity, can be applied to the case of a dynamical system because the hypotheses on which theorem is based are satisfied due to the fact that a dynamical system, basically, describes a movement in the phase space, which is necessarily characterized by continuity, and instantaneous, univocally defined and non-infinite velocities. By virtue of the theorem of existence and uniqueness, an orbit cannot intersect itself at any point, neither by having two different directions (i.e. two separate tangents), nor by passing through the same point again in the same direction (i.e. with the same tangent); given a dynamical system and a point, the orbit passing through that point is unique. This is the foundation of determinism: given the evolutive law and the initial state, the evolution is univocally determined. The impossibility of the existence of different attractors in a two-dimensional space, other than a point or a limit cycle, is substantially the concept of the renowned and fundamental theorem of Poincaré–Bendixon (for a discussion of this theorem, see for example Hirsch and Smale, 1974; Beltrami, 1987; Bellomo and Preziosi, 1995).

If the dynamical system is controlled by forces that depend on time (this can occur only if the forces considered are external to the system in question, such as the forced pendulum, Model 4, described in Chapter 9), namely if the system is not autonomous, then time also comes into play, directly or indirectly, as a third coordinate needed to describe the dynamics. In this case chaos can arise even with two coordinates only, because time is added in what is called an 'essential' way, i.e. in such a way that it cannot be ignored without losing a fundamental element that characterizes the system's evolution.

The forced pendulum, in certain particular situations, can demonstrate chaotic evolution. To describe its dynamics we need to consider, more than for the other non-chaotic models of the pendulum, a variable dependent on time, imposed by the fact that a force that varies with time acts externally to the system. This is, in actual fact, a non-autonomous system, in which the total resulting force depends explicitly on time; therefore, the differential equation that describes the system is also non-autonomous. Basically we also need to take into account the fact that the force acting on the system is not only determined by the state of the pendulum during oscillation.

In order to be able to 'contain' the system's orbit, the phase space must now have an extra dimension with respect to the other cases of oscillation, that regarding the dependence of the force on time, or even simply time. Therefore, there must be a three-dimensional phase space. It is precisely the presence of the third dimension that enables the orbit of the dynamical system in question to pass through the same pair of coordinates ϑ and $\dot{\vartheta}$ again, while continuing to satisfy the conditions of the theorem of existence and uniqueness; in actual fact it is the third coordinate, time t, or a third state variable in addition to the previous ones, that now assumes different values at the two moments in time when the orbit passes through the same values ϑ and $\dot{\vartheta}$.

While a one- or two-dimensional dynamical system cannot be chaotic if it is in continuous time, it can be, however, if it is in discrete time, i.e. if it is a map. For dynamical systems described in discrete time by finite difference equations, such as the logistic map that we introduced in Chapter 15 and to which we will return in Chapters 21 and 22, we saw, in fact, that it is also possible to have chaotic dynamics in a single dimension, i.e. even if the system is described by a sole variable.[67]

---

[67] More generally, it can be demonstrated that this can occur only if the map is not invertible, i.e. if one can obtain the value x_{n+1} starting from more than one value of x_n. This is precisely the case of the logistic map, which can be easily verified by using map (15.2): $x_{n+1} = 3.7x_n(1 - x_n)$. For example, we have the same x_{n+1} in two cases:

Still on the subject of the different behaviour of differential equations and finite difference equations, it is important to note that the use of numerical methods to search for an approximate solution necessarily involves, in some way, the transformation of a differential equation into a finite difference equation, i.e. into a map. This occurs both because any numerical method, in itself, is not able to conceive infinitesimals, and because any computer, used to obtain the numerical solution of an equation, works on a discrete set of numbers. Therefore, it *might* occur that the numeric algorithm also gives rise to chaos for a one-dimensional dynamical system, not because the model intrinsically envisages chaos, but because chaos can have been generated by the approximate integration method used in practice.

Two examples: Lorenz and Rössler attractors

A chaotic system evolves towards a chaotic attractor that is almost always also a strange attractor: a curve (orbit) of infinite length, because the evolution of the dynamical system may occur in infinite time, at least in principle, that remains enclosed in a finite volume, without ever passing through the same point (Crutchfield et al., 1986; Sprott, 1993). An open curve of infinite length contained in a finite volume is possible only if the structure of the attractor is that of a curve that folds back on itself, that continually passes close to points that it has already passed, but never through the same points.

Curves of infinite length enclosed in a finite volume are not a new concept to geometry; they had already been studied towards the end of the nineteenth century, well before the study of strange attractors, in particular by the Italian mathematician Giuseppe Peano. One example that is particularly famous is that of a continuous curve of infinite length (one of the so-called monster curves), known as 'Peano's curve', that 'fills' the surface of a square (its description appears in numerous texts, for example: Simmons, 1963; Schroeder, 1991; Vulpiani, 1994; Bertuglia and Vaio, 1997b).

With the sole aim of presenting some examples of curves of infinite length enclosed in a limited volume, without entering into the theoretical details of

$$x_n = 0.25 \Rightarrow x_{n+1} = 3.7x_n(1 - x_n) = 0.69375$$
$$x_n = 0.75 \Rightarrow x_{n+1} = 3.7x_n(1 - x_n) = 0.69375$$

The linear map (15.1), $x_{n+1} = 2x_n$, on the other hand, is an invertible function, i.e. we can obtain x_{n+1} from a single x_n (in this case this is called a bijective map) and does not give rise to chaos.

the models, Figures 16.2–16.4 show the two-dimensional projections of two cases, from among the numerous renowned ones, of strange attractors in a three-dimensional space that have nowadays become classical.

The first case, probably also the most famous example of a strange attractor, as well as being the first to be discovered, is the Lorenz attractor (1963). It originates from a model that aims to describe the stability and the onset of convective or turbulent motion in a fluid heated from the bottom and cooled from the top. The system's equations are as follows:

$$\begin{cases} \dot{x} = \sigma(y - x) \\ \dot{y} = -xz + rx - y \\ \dot{z} = -bz + xy \end{cases} \quad (16.3)$$

in which parameters σ, r, and b have a specific meaning within the physical phenomenon that the model is aiming to describe, atmospheric dynamics, which we will not be examining here. Variables $x(t)$, $y(t)$, and $z(t)$ are not spatial variables, but represent magnitudes referring, respectively, to the remixing of the fluid and to the horizontal and vertical variations in temperature.

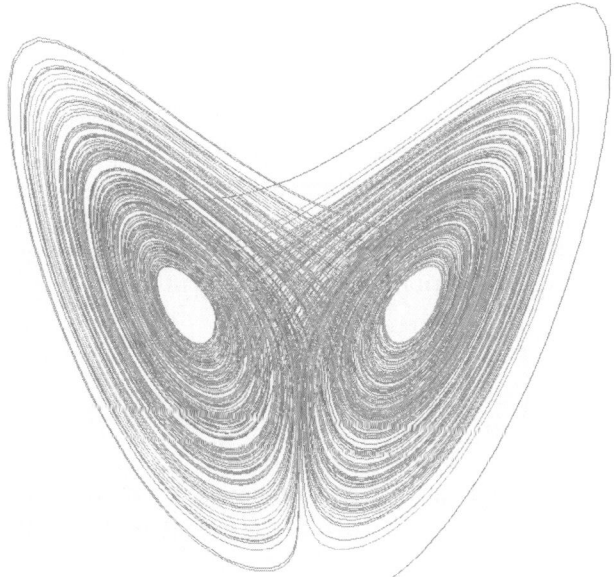

Figure 16.2 Lorenz attractor calculated for the following parameter values: $\sigma = 5$, $r = 15$, $b = 1$ (two-dimensional projection on the xy plane).

16: Strange and chaotic attractors

Figure 16.3 Rössler attractor calculated with the following parameter values: $a = 0.2$, $b = 0.2$, $c = 5.7$ (two-dimensional projection on the xy plane).

A two-dimensional projection of the Lorenz attractor is shown in Figure 16.2.[68] The Lorenz attractor was generated by transforming the system into discrete time (dt is now the finite step from which the system's variables have evolved at each iteration), i.e. in the form of the following map:

$$\begin{cases} x_{n+1} = x_n + \sigma(y_n - x_n)dt \\ y_{n+1} = y_n + (-x_n z_n + r x_n - y_n)dt \\ z_{n+1} = z_n + (-b z_n + x_n y_n)dt \end{cases} \quad (16.3')$$

The phase space in Figure 16.2 is obviously an abstract space and the attractor shown has nothing to do with the actual motion of the fluid or of

---

[68] The aspect of the broken line that the curves in Figures 16.2–16.4 show, particularly in certain parts of them, is due to the numerical generation of the attractor, which is conceived as a continuous and differentiable curve, but is drawn in finite steps.

Figure 16.4 Enlarged view of Figure 16.3 (two-dimensional projection onto the *xy* plane).

one of its molecules. The curve shown, obtained by integrating the equations of the model numerically, is the strange attractor of the Lorenz model, a curve of infinite length, if the steps of integration are infinite, which never intersects itself, while remaining confined to a limited region in the phase space (in Figure 16.2, the curve appears to intersect itself only due to the effect of the projection of the depiction of the three-dimensional attractor on the *xy* plane).

The second example that we are going to illustrate is also very famous, and is the attractor obtained by Otto Rössler (1976ab; see also Eubank and Farmer, 1989), who attempted to write the simplest example of a dynamical system featuring a chaotic attractor. The equations, which have a purely abstract meaning and are not intended for the modelling of any phenomenon whatsoever, are as follows:

$$\begin{cases} \dot{x} = -y - z \\ \dot{y} = x + ay \\ \dot{z} = b + z(x - c) \end{cases} \qquad (16.4)$$

Figure 16.3 shows a two-dimensional projection of the Rössler attractor.

To generate the attractor, we wrote the system in the form of a map, using the version of the equations in discrete time:

$$\begin{cases} x_{n+1} = x_n - (y_n + z_n)dt \\ y_{n+1} = y_n + (x_n + ay_n)dt \\ z_{n+1} = z_n + (b + x_n z_n - c z_n)dt \end{cases} \quad (16.4')$$

Figure 16.4 shows an enlarged view of a detail of Figure 16.3, the region in which the line (which is always the same one) *seems* to cross over itself numerous times (but the coordinates of the z variable are different, as the latter has been 'squashed' by the projection).

Sensitivity to initial conditions is reflected in the fact that for these attractors, as for any strange or chaotic attractor (sacrificing a little rigour for the sake of simplicity, from now on we will consider the two concepts to be identical), two initially close points, corresponding to two initial sets of values of the state variables close to one another, evolve distancing themselves from one another, but tend to move along the attractor independently. The result is that the initial set of points is dispersed as the evolution proceeds, scattering themselves on the attractor.

A two-dimensional chaotic map: the baker's map

To better understand how a curve of infinite length can be enclosed in a finite volume, it is useful to refer to some of the laws of transformation of a surface in itself and to trace the movements of a point on it, specified at the start of the series of transformations (we encountered an example of these transformations in Arnold's cat map: equations (16.1), (16.1') and (16.2)). For certain transformations, all points travel along open trajectories, i.e. they never return to a position that they have already occupied (this, however, is *not* true for Arnold's cat map, which, after the transformation has been repeated a certain number of times, re-forms the original image; in fact we used this map as an example of an attractor that was *not* strange).

We can imagine how a curve of infinite length contained in a closed surface originates if we think of the movement of the points of that surface according to a determined law. The trajectory travelled by each point of that surface is an infinite line, at least in principle, if the motion lasts for an infinite time; the trajectory of each point remains confined to the surface, but in general no point passes the same position that it has previously occupied (Arnold's cat map is an exception to this).

A renowned example of a deterministic law with two variables, the two coordinates of the points of a surface, that 'remixes' the points of the surface itself is the so-called *baker's map* (Farmer, Ott and Yorke, 1983; see also Nicolis and Prigogine, 1987; Schroeder, 1991; Moon, 1992). It can be expressed in mathematical terms and depicted in a fairly simple way. Each point, identified by a pair of coordinates x, y, belonging to the surface of a unit square, undergoes a change of coordinates according to the following law:

$$\begin{aligned} \text{for } 0 \le x < 1/2 \quad & x_{n+1} = 2x_n \quad & y_{n+1} = y_n/2 \\ \text{for } 1/2 \le x \le 1 \quad & x_{n+1} = 2x_n - 1 \quad & y_{n+1} = y_n/2 + 1/2 \end{aligned} \quad (16.5)$$

The transformation represented by the map (16.5), even if it occurs in a single step, can be visibly understood very simply if we identify two subsequent phases for them, as shown in Figure 16.5, in which the circle and the cross approximately demonstrate the dynamics of the two points of the surface: first of all, for each point, the x coordinate is doubled and the y coordinate is halved; therefore a square with a side of 1 is transformed into a rectangle with a side of 2, along the x axis and with a side of 1/2, along the y axis; subsequently, the right half of the rectangle is superposed over the left half.

The result of the transformation is a square again, but the points on its surface are arranged differently with respect to the previous step (with the exception of the point on the vertex of the coordinates (0,0), which is a fixed point of the transformation).

The transformations that the points of the surface progressively undergo recall those that a baker performs when, during the remixing of the dough, stretches the dough in one direction, squashing it and flattening it with his hands, cuts it and brings a part back upon itself, returning the dough to its original shape, but with the points of the dough in new positions. In map (16.5), excluding the cited fixed point, as the transformation is repeated, no point will ever return to the same coordinates that it had in a previous transformation, and will travel an open trajectory of infinite length (if

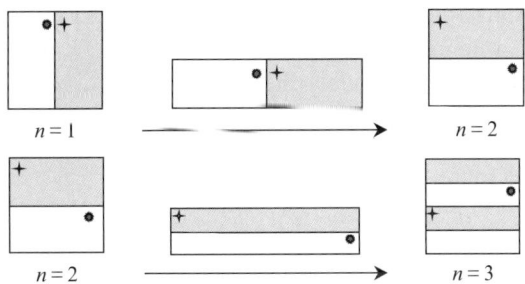

Figure 16.5 Transformation generated by two applications of the baker's map.

subsequent transformations are infinite), confined to the surface of the square: a curve with fractal properties.

One aspect needs to be clearly understood: the remixing that the map (16.5) causes is not due to the fact that each point goes 'all over the place' jumping here and there inside the square, something which is not really true,[69] as rather the fact that close points travel along diverging trajectories. The trajectories are obviously confined to the unit square, whose area remains constant (i.e. the system is conservative); therefore the trajectories have to fold back on themselves, with the result that a cloud of points initially gathered in one part of the surface disperses over the course of the evolution to all the regions of the square (see the examples shown in Figure 16.6).

We are faced with a dynamic that shares several important aspects with that of the logistic map illustrated in (16.2): for (16.5), as for (16.2), the dynamics are described by known, deterministic laws of motion, that can be calculated recursively. Furthermore, the distances travelled by a point in an iteration do not observe laws of linear proportionality, as occurs instead with the linear map (16.1). Finally, the distances between two points can increase or decrease during an iteration (16.5), as for map (16.2); it is easy to recognize this, even without calculations, simply by looking at Figure 16.6. The fact that two points, that are close to each other after an iteration, can considerably distance themselves from one another after the next iteration, as occurs for example with points A and B, is an index of the sensitivity to initial conditions, one of the elements that characterizes the chaotic situation. Both the logistic map (16.2) and the baker's map (16.5) are nonlinear maps, the first one-dimensional, the second two-dimensional, that both give rise to chaos (with the logistic map, in reality, chaos is generated only in specific circumstances, as we will see in Chapters 21 and 22).

It is evident that the analogy between transformation (16.5) and the transformation that the baker actually makes on the dough is only partial and approximate. There are various reasons for this. First of all, because in the case of the remixing of dough, it is *extremely unlikely* (but in principle not impossible) that the particles of dough, at random, return fully or partly to the same initial coordinates; in map (16.5), which, on the other hand, is a mathematical model, the evolution is deterministic, and not probabilistic, and therefore the points *definitely* never return to their initial coordinates.

---

[69] The remixing of the points caused by (16.5) in reality is not complete, as there are some invariants; for example, the points of rational coordinates remain confined to the square's subset made up of the pairs of rational coordinates; in the same way, the points of irrational coordinates will continue to have irrational coordinates.

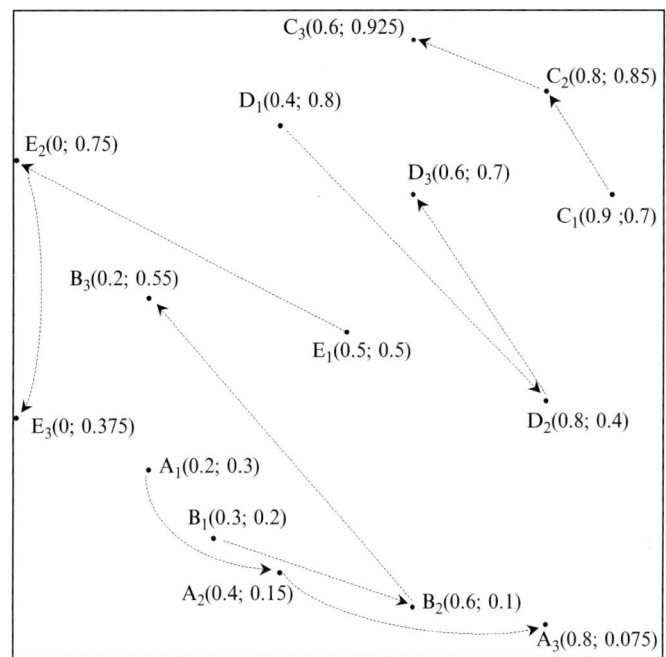

Figure 16.6 First two iterations of the baker's map (16.5) for some points.

A similar phenomenon in physics, the mathematical formulation of which is however much more complicated, if not impossible without the use of drastic approximations, is, for example, the remixing of a liquid, or the mixing of two different liquids (Crutchfield et al., 1986). By remixing a liquid, molecules that were initially close to one another, distance themselves from one another, then get closer again and then distance themselves and so on, but the orbit of each molecule remains confined to the container. In principle, the orbits are of infinite length, because the action of remixing, i.e. the succession of transformations of the coordinates of each molecule, can be infinite.

Abstracting a little, we can talk of points in the liquid that, under the action of successive transformations, follow an orbit of infinite length that takes them 'close' to any other point in the liquid, where 'close' means a distance that is less than any arbitrary pre-established distance. In a model of this nature, each point in the liquid follows a trajectory with fractal properties.[70]

---

[70] A dynamics of this type, i.e. with fractal properties, also occurs in the irregular motion of solid particles of small dimensions suspended in a fluid, called Brownian motion (Nicolis and Prigogine, 1987; Moon, 1992).

17 Chaos in real systems and in mathematical models

Phase space, orbit, basin of attraction and attractor are all mathematical concepts and as such are abstract; this means that they relate more to the abstract models that we construct of the processes observed, rather than to the processes themselves. In a certain sense, we imagine attractors and dynamical orbits on the basis of our mathematization of phenomena (when we succeed in doing so). But the passage from the real system, that we *assume* to be deterministic, despite the fact that its evolution might seem unpredictable to us, to the mathematical model that describes it, from a certain viewpoint, is an arbitrary passage, because a real system and its mathematical model are conceptually found in two completely different spheres. The correspondence between the two is based only on the fact that very often the model created 'works well' in predicting the behaviour of the real system, as we have already seen. The differential equations of the model are nonlinear (in general) and deterministic, so much so that often we call them *the law* that controls the real corresponding process; they give rise to deterministic chaos that seems to do a good job of reproducing the phenomenology observed. But it is still the chaos of a model, i.e. a chaos of a mathematical nature; the problem is linking this with the chaos observed. This passage, however, is not always easy to make.

The fact that chaos is possible starting from even very simple deterministic laws, and that the chaos encountered in models corresponds to the evolutive processes of real systems, was an important discovery, and has led us to hope that the phenomena considered up until now to be random, i.e. impossible to describe without the use of statistical laws, are in reality chaotic, i.e. dependent on deterministic laws. If this is true, once the laws have been identified, the systems would always remain unpredictable, but in some way would be more comprehensible. Attractors (mathematical concept) are characterized by a basin of attraction that provides the attractors themselves with the stability that distinguishes them. Other phenomena, whose manifestations are (apparently) characterized by unpredictability, such as fluctuations in the stock market, heartbeats, the metabolism of a cell, life itself in biological terms, etc., according to this version, could (perhaps) be 'controlled' by chaotic attractors which, despite their unpredictable nature, would nevertheless intrinsically maintain them in a condition of relative stability, as long as variations in external conditions are not too large. In other words, chaotic phenomena, unlike stochastic ones, are in a certain sense 'robust' phenomena, because they are driven by a strange attractor that 'sucks them back' into a state of dynamic stability, until such time as a too large a perturbation removes them from the basin of attraction of that attractor, transferring them to the basin of attraction of another attractor. Basically, if this is true, it would be sufficient to identify the attractor (or attractors) that an effective model of the process envisages, to 'understand' the dynamics of an apparently unpredictable real process.

Chaos in models is also linked to another matter that becomes important only when a dynamical system is made to evolve. This is a concept fundamental to mathematics in general, that of real numbers, and the fact that the onset of instability can in some way be linked to the numerical approximation that we inevitably have to make when a real number has to be replaced by a rational number 'close' to it, in order to be used in practice, namely in measurements.

To better clarify this point, let us return to the question raised by Weierstrass regarding the assumption of a rational ratio equal to 2/5 between the periods of revolution of Jupiter and Saturn, which we mentioned in Chapter 15. Weierstrass found it unacceptable that the stability of the orbits could depend on the rationality or otherwise of a certain ratio that, above all, could only be measured approximately. Stating that the ratio between the periods of the two planets is *exactly* 2/5 is the same as stating that *all* the infinite decimal places of that ratio are known and equal to 0, and that this number is 0.40000... This only makes sense from a mathematical, i.e. abstract, point of view, whereas physics, like all the experimental sciences, is founded on experimental measurements. We will never be capable of formally establishing, on a purely mathematical level, whether the value obtained from measurements is an approximate value of a rational or an irrational number. Therefore, if this is true, how can the stability of an observed orbit be established? In other words, does an orbit that becomes perturbed in some way behave like a limit orbit, towards which all orbits 'very close' to it tend, or not?

In chaotic systems, like in chaotic models, small initial differences become increasingly less negligible over time, and, so to speak, emerge from the shadows and in the long term give rise to evolutions that are macroscopically different. This is deterministic chaos: the substantial and intrinsic unpredictability of the future evolution of both a real system and an abstract model.

The result that Poincaré was the first to unexpectedly stumble upon in his study of the stability of planetary orbits, i.e. the emergence of chaotic phenomena, was disconcerting even on a philosophical level. The fact that a small perturbation that he introduced into the equation (namely the model) of a periodic, stable orbit, through a parameter with a very small value, gave rise to an orbit which behaved in a way that we now call chaotic, clashed with a deep-rooted conviction. As Poincaré himself wrote in his book *Science et méthode*: 'A very small cause which escapes our notice determines a considerable effect that we cannot fail to see, and then we say that the effect is due to chance. If we knew exactly the laws of nature and the situation of the universe at the initial moment, we could predict exactly the situation of the same universe at a succeeding moment. But even if it were the case that the natural laws had no longer any secret for us, we could still know the situation approximately. If that enabled us to predict the succeeding situation with the same approximation, that is all we require, and we should say that the phenomenon had been predicted, that it is governed by the laws. But it is not always so; it may happen that small differences in the initial conditions produce very great ones in the final phenomena. A small error in the former will produce an enormous error in the latter. Prediction becomes impossible, and we have the fortuitous phenomenon' (1908, Part I, Chapter 4).

As we said in Chapter 3, the vision developed by classical determinism fails on this point. According to the latter we assume that the same cause, in similar circumstances, produces similar effects and, vice versa, similar effects correspond to similar causes, according to a rigid conception of the relations between the phenomena. In the evolution of real systems, this is not always true, and therefore we talk about the sensitivity of a system to the initial conditions, which, however, can *never* be 'completely' known and, therefore, different evolutions can originate 'within' the area of approximation with which we define the initial state. In the model that we construct of the system, on the other hand, we talk about a deterministic law also characterized by sensitivity to the initial conditions; in this case, this means that in the numbers that quantify the values of the state functions it makes small differences or small approximations, relegated to decimal places far from the point, emerge from the 'negligible' zone.

The problem, as we have said, is strictly linked to the concept of a real number and to the fact that we generally associate a real number to a coordinate on a straight line, i.e. to a point on that same line. We *imagine* a geometric object made up of points like the line to be continuous, and at the same time, we *conceive* the set of real numbers to be continuous (whereas neither the infinite set of natural numbers 1, 2, 3, 4, ..., nor that of rational umbers is continuous).[71] The real number concept is a cornerstone of mathematical analysis and, therefore, of the very idea of the differential equation, the fundamental tool used to construct dynamical systems. Any measurement, however, provides a number that cannot have infinite decimal places; this means that when we measure and operate on measured data, we do not use the set of real numbers, but a much more limited subset of the same, that of rational numbers, finite or periodic decimals.[72]

We associate the idea of 'infinite' precision to real numbers, but in practice we necessarily 'confuse' them with their subset made up of rational numbers. Rational numbers have a mechanism, known as a 'period' that enables any decimal place to be obtained, without having to explicitly calculate all the previous numbers: in fact the result of the division between two whole numbers (by definition a rational number) always gives a periodic result, this property allows us to know any decimal place by just iterating the period, once this has been identified, without going on computing the division explicitly. For real, non-rational numbers, on the other hand, this is no longer true and each of the infinite decimal places must be calculated step by step in order to determine that number with 'absolute' precision.

---

[71] A renowned theorem of Georg Cantor proves that between the set of natural numbers and that of rational numbers there exists a one-to-one correspondence (i.e. they are bijective sets): these sets, so as to speak, are 'infinite to the same order' or 'equipotent', which, in practice means that they are interchangeable. Another of Cantor's theorems demonstrates that it is not possible to establish any one-to-one correspondence whatsoever between the set of rational numbers and that of real numbers; they are sets of different cardinality, so to speak; the infinity that characterizes the set of real numbers, for example, is of an order higher than that of the set of rational numbers (and therefore also of the set of natural numbers). The first, in a manner of speaking, is an infinite 'more infinite' than the second. More precisely, we say that the rational numbers are countably infinite, i.e. it is *possible* to enumerate all the rational numbers by means of an infinite sequence of ordinal numbers. By contrast, the real numbers are uncountable, i.e. it is *impossible* to enumerate them by means of an infinite sequence of ordinal numbers, as the latter are not 'numerous' enough. These discoveries underlie the idea of cardinality, which is expressed by saying that two sets have the same cardinality if there is a bijective correspondence between them.

[72] The standard division of rational numbers into finite decimals and periodic decimals, in reality, is merely a scholastic habit that doesn't consider any intrinsic property of rational numbers. Any rational number, in fact, can be written in an infinite number of different ways, some with a period, others without, depending on the numerical base adopted to represent the number in question. However, apart from this, if we remain on the standard numerical base of 10, any number usually called an 'finite decimal' can be written as a periodic number in two different ways: (1) by adding a period made up of the single figure 0, or (2) by writing the number with the last decimal place decreased by one unit and followed by the period 9. For example, the number 2.3 (non-periodic) can also be written as 2.3000000... (period 0) or as 2.299999... (period 9). It is not wrong, therefore, to state that the property that characterizes *any* rational number is in fact the presence of a period; whether this is actually written or not is simply a matter of representation.

In this profound difference between real numbers and rational numbers that we use to approximate them, we have the seed that generates chaos in a dynamical system (we will return to this point in Chapter 33). Let us imagine that we are using a set of real numbers, as we usually do when we apply mathematical analysis: for each particular 'collection' of real numbers that characterizes a state of a system, we have, in the evolution of the system, a particular sequence of 'collections' of real numbers that form the orbit of the system in question. As the use of any measurement tool and of any calculator necessarily imposes the use of finite decimal numbers, the result is approximate with respect to the ideal case already after the first iteration. This means that, instead of continuing along the starting orbit, the system necessarily 'jumps' to a nearby orbit: if the system is unstable, the new orbit will increasingly distance itself from the initial theoretical, abandoned one. The chaos actually lies in the fact that, taking a certain number of close points in the phase space, these will disperse over time to a region that is a *chaotic* (strange) attractor and that, as time passes, the dispersion into the chaotic attractor continues to increase, without the appearance of any *ordinary* attractor (point or closed line).

We will return to chaos in real systems in Chapter 19, after we have clarified some technical aspects regarding the problem of the stability of dynamical systems. In Chapter 18, therefore, we will be discussing the concept of stability with regard to models, with the primary aim of being able to establish the sensitivity of a dynamical system to the initial conditions, and therefore, of being able to characterize the degree of stability of the orbit and obtain an indication on the predictability of the same.

18 Stability in dynamical systems

The concept of stability

As a mathematical model in the form of a dynamical system evolves, it generates a sequence (sets) of numbers that represent states in a phase space, i.e. an orbit; the meaning that is attributed to these (sets of) numbers depends on different factors, as we have discussed on several previous occasions. Of these, we have the problem of sensitivity to (small) changes in the initial conditions, that is to say the study of 'by how much' said changes influence ('propagate') the numerical values calculated for future states, provoking large or small variations of the same. The stability characteristics of the orbit have to be analysed, which requires, first and foremost, a definition of stability and a method to quantify it. We will be examining this in this section.

A dynamical system appears unstable if its evolution is unpredictable, because, as time passes, a small variation of the orbit can become increasingly larger. This means that a small perturbation can cause large alterations of the orbit that follows and that the evolution displays very particular phenomenologies. From this point of view, even very simple models can become unstable in certain conditions, a phenomenon overlooked by eighteenth century mathematicians.

A dynamical system is unstable, in general, if the orbits established from points near the phase space diverge exponentially, over time, i.e. if the distance between the trajectories grows proportionally by a function $e^{t/\tau}$ where the τ, called the Lyapunov time, is the time-scale according to which two trajectories diverge from one another. The Lyapunov time therefore characterises the actual 'time horizon' of the dynamical system; the smaller it is, the more the system is unpredictable, chaotic, and the linear approximation becomes even less acceptable. We need to examine, therefore, to what extent the solution of a differential equation (or of a map in discrete time) is effectively a function that is sensitive to small variations of the parameter values or of the boundary conditions.

The stability of a model means 'robustness' of the function which solves a system of differential equations, that is to say its lack of sensitivity to small changes in the initial conditions. We can refer to stability at particular points of the evolutive trajectory (local stability), but we can also refer to stability of the entire trajectory, as in asymptotic stability. We could also discuss how an entire family of trajectories converges towards a set of stable points, an attractor, as well as the existence and characteristics of the latter.

Furthermore, we can distinguish between two types of stability, depending on the type of perturbation that is applied to the system. We talk, in fact, about the *dynamic stability* of the dynamical system (or of the model) if we are referring to the behaviour of the system in the face of a perturbation of the states at a certain point in its evolution. In this case, the stability of the system following the perturbation originating from the environment external to the system is under discussion. On the other hand, we talk about *structural stability* if we are referring to the behaviour of the dynamical system (or of the model) following perturbations that affect the values of the dynamical system's parameters. In this second case, our attention is focused on the operating mechanism of the system, overlooking any alteration in environmental conditions.

The point we are most interested in is the study of how a dynamical system reacts to perturbations to the initial conditions, namely the study of dynamic stability. This area has two main aspects. The first regards the way in which the real system that the model reflects responds to actions originating from outside the system itself, due to interference of some environmental element. The second aspect regards how the inevitable imprecision of measurements of the real system affect the determination of the initial conditions, in other words, how sensitive the dynamical system is to an improvement in the precision of the measured data.

In this regard, we saw in Chapters 16 and 17, how the very discovery of the existence of a chaotic phenomenology by Poincaré towards the end of the nineteenth century originated from the problem of the stability of planetary orbits subjected to small perturbations caused by the presence of other planets, stability that was explored using mathematical techniques on a model obviously derived from that of Newton, and referred to the first of the two cited aspects: how a planet reacts to its interaction with all of the others. More recently, however, Lorenz (1963), who we cited in Chapter 16 with regard to the attractor that bears his name, returned to the question of chaos in a model, but this time concerning a problem that regards the second aspect, i.e. the different approximations with which he entered the numerical data describing the state of the atmosphere in different simulations in a program. In fact, James Gleick (1987) recounts how Lorenz realized that by iterating the system (16.3) starting from numerical values of the state variables approximated in different ways, even simply by cutting off numbers with different decimal places, different orbits were obtained and, what is fundamental, the difference did not appear to disappear as the system evolved, but became increasingly larger. The result was that the points in the phase plane, during the course of their evolution, dispersed following different orbits, which did not tend to gather in a *single* orbit, as would have been the case, on the other hand, for a stable system with an ordinary attractor. The different orbits that originated from the different truncations, on the contrary, all tended towards the same region in the phase space, the chaotic attractor shown in Figure 16.2.[73]

Structural stability, as we have said, refers to the system's behaviour following changes that affect its parameters; the study of structural stability mainly concerns, therefore, the creation and validation stages of a model. A mathematical model, as we discussed in Chapter 2, is never a loyal and complete reflection of any objective situation, but provides a picture of the situation that is able to include only some aspects, which can be represented by a great number of very different, and at times contradictory, models. The modelling perspective implies, therefore, abandoning the vision of the relationship between mathematics and reality as a one-to-one correspondence (Israel, 1994, 1996, 1999; Boncinelli and Bottazzini, 2000). The problem of choosing which elements to consider as the main ones of the model and which as its corrections, and therefore which parameters to insert in the model and in what form, is therefore fundamental. A first point that concerns structural stability, therefore, refers to the study of how a change made to certain elements of the form of the equations influences the dynamics generated by the model, while a second point refers, on the other hand, to the effects that originate from the presence of conceptual errors in the model itself or even just to the consequences of imprecision and approximations in the values of the constants and the parameters (Bellomo and Preziosi, 1995).

A basic case: the stability of a linear dynamical system

Let us further explore the stability of a linear dynamical system; to this end we will actually be referring to the differential equation that defines its evolution in the simplest case of all: a first order equation with constant

---

[73] Only later in the 1970s were the fractal properties of the Lorenz attractor and of geometric figures in general defined; when Lorenz described the attractor that bears his name, the concept of fractal geometry had not yet emerged.

18: Stability in dynamical systems

coefficients, focusing our attention on the stability of the solution function. We will reconsider Malthus' law (4.2):

$$\dot{x}(t) = k\, x(t) \qquad (4.2)$$

where k is a constant parameter and t is time, an independent variable. If the initial condition is known, $x(t_0) = x_0$, then the general solution of (4.2) can be written as:

$$x(t) = x_0 e^{k(t-t_0)} \qquad (18.1)$$

When $x_0 = 0$, the solution becomes simply $x = 0$ which we will call the unperturbed solution.

Let us now suppose that the initial value x_0 is not 0, but close to 0. In other words we are changing the previously established initial condition by a small value; we will call the new solution perturbed. How will the perturbed solution now act with the passing of time? Will it tend to return to the unperturbed solution $x = 0$, thus progressively reducing the initial perturbation with a form of exponential decay, or will the small alteration intensify over time, giving rise to the increasing divergence of the function $x(t)$ of the unperturbed solution? This depends on the sign of the parameter k.

If $k > 0$, solution (18.1) is a function that increases (in absolute values) over time t. In this case the unperturbed solution $x = 0$ is unstable, because small perturbations generate an $x(t)$ that increasingly distances itself from $x = 0$ as time passes; the initial perturbation x_0 intensifies and the system is unstable (Figure 18.1).

If $k < 0$, (18.1) immediately shows that, as t increase, the $x(t)$ solutions tend to return to $x = 0$. In this case, the unperturbed solution $x = 0$ is asymptotically stable with respect to the perturbation of the initial conditions (Figure 18.2).

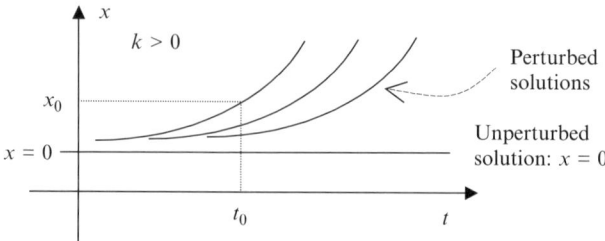

Figure 18.1 Perturbed solutions of the linear model (18.1): $x = 0$ is the unstable solution.

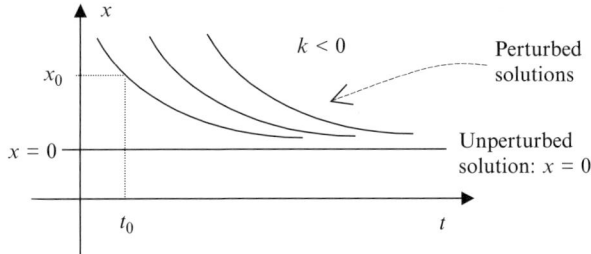

Figure 18.2 Perturbed solutions of the linear model (18.1): $x = 0$ is the stable solution.

If $k = 0$, then (18.1) simply becomes $x = 0$. If the initial state $x_0 = 0$ is altered, the perturbed solution does not tend towards zero, nor does it distance itself from zero, but it remains constant; we are in a condition of indifferent equilibrium.

The above can be generalized to the case of a linear equation of any order. If the order is above the first, however, the graphic representation is not as simple and obvious as the case of the first order differential equation described above, because the geometric representation of the successive derivatives is not as immediate as for the first derivative.[74]

Poincaré and Lyapunov stability criteria

Let us now look more closely at what we mean by the dynamic stability of a dynamical system such as (8.1):

$$\frac{dx_i}{dt} = F_i(x_1, \ldots, x_n, t) \quad i = 1, 2 \ldots, n \qquad (8.1)$$

with t extended to arbitrarily large intervals and with initial conditions $x_i(t_0)$. We want to define criteria that can be adopted to establish, beyond

---

[74] The case of a second order linear system like those discussed in Chapter 11 is particularly interesting because in the latter a perturbation of the solution $x = 0$ produces oscillations that amplify and dampen, according to the values of the equation's coefficients. A real unstable system that can be described effectively by a second order linear model occurs, for example, in the case of noise captured by a microphone situated in front of the speaker that diffuses, amplified, the sound captured by the microphone. In this situation, a small noise (perturbation of the silence corresponding to the initial unperturbed state), passing from the microphone to the amplifier and to the speakers, and from these back to the microphone, amplifier and speakers again, and so on continually, again and again, intensifies, generating a increasingly louder whistle, a typical phenomenon of positive feedback, commonly called the Larsen effect.

the simple approximate idea of instability based on intuition, when we can say that a (small) change in initial conditions generates a new orbit that, after a certain period of time, 'has distanced itself' from the unperturbed one beyond a pre-established distance. Obviously the term 'change' can be understood as a perturbation or as an effect of the improvement in the precision of measurements, which, inevitably, are affected by approximations (errors).

There are a number of criteria for dynamic stability.

A first criterion is that of stability according to Poincaré. Poincaré's problem, as we have mentioned on several occasions, concerned the study of the stability of the planetary orbits. The definition of stability adopted in Poincaré's era dated back to Siméon Denis Poisson and was the following: the trajectory of a moving point is stable when, tracing a circle or a sphere with radius r around an initial point, the moving point, after having exited this circle, or this sphere, re-enters it an infinite number of times, regardless of how small r is (*Sur les inégalités séculaires des moyens mouvements des planètes*, Journal of the École Polytechnique, 1809, see Bottazzini, 2000, p. 67). In his book *Les méthodes nouvelles de la mécanique céleste* (1892–99) Poincaré added the conditions that no body may distance itself to infinity and then return, and that the distance between any two bodies of the (planetary) system considered cannot fall below a certain limit.

The stability criteria adopted by Poincaré, basically, introduced a band, or better still, a sort of 'tube' of stability around the unperturbed orbit, within which a planet can move 'jumping' here and there, following perturbations, but without ever distancing itself 'too far' from the unperturbed orbit. If this is so, the orbit of the planet is stable, even if subjected to perturbations due to the effect of its interaction with other planets[75] (Figure 18.3).

In this criterion, all of the aspects of the system's dynamics are not taken into consideration, only the system's orbit in the phase plane; it must be stationary, i.e. it shouldn't depend on time, which, in fact, never directly comes into play. This is a relatively undemanding criterion of so-called orbital stability that can only be applied to autonomous models, i.e. to dynamical systems governed by laws that do not expressly depend on time, made up of autonomous equations.

---

[75] More precisely, a system is stable according to Poincaré, if by taking a positive number ε, one can determine, as a consequence, another number δ, with $0 < \delta \leq \varepsilon$, such that all the (stationary) trajectories that originated within the distance δ from the initial point of the unperturbed orbit permanently remain within the maximum distance ε from the unperturbed orbit (Jordan and Smith, 1977).

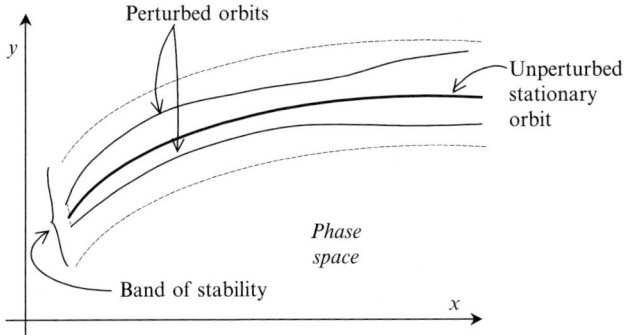

Figure 18.3 Orbital stability according to Poincaré in the phase space xy.

This criterion, apart from its limitation of only being applicable to stationary states generated by autonomous dynamical systems, is not effective in highlighting certain situations where divergences evolve over time. For example, let us consider the oscillation of a linear pendulum again (Model 1) and let us perturb the stationary movement by increasing the 'small' initial amplitude. As we discussed in chapter 6, in a situation of this nature, linear approximation is no longer acceptable; in this case the dynamic law of the pendulum is system (7.1), that of the nonlinear pendulum (Model 3). This means that the orbit in the phase plane (Figure 18.4) deforms with respect to that of the elliptic form of the linear pendulum, remaining, in any event, in the vicinity ('close', in the Poincaré sense) of the elliptic orbit of the linear case.

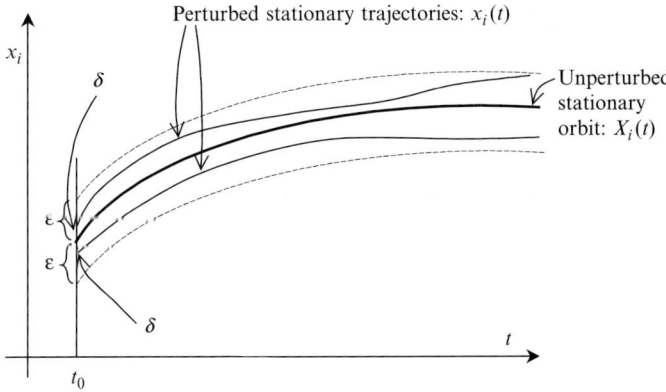

Figure 18.4 Lyapunov stability in the space tx.

As the oscillation periods are now different, the ellipse of the linear case and the orbit of the perturbed case are travelled in different times; in particular, the pendulum that follows the perturbed orbit accumulates an increasing delay with respect to the one that follows the unperturbed orbit. This different behaviour over time escapes Poincaré's criterion, which is only a criterion of orbital stability and doesn't consider the increasing distance that is created between successive states, due no so much to the geometrical form of the orbits, but rather to the speed with which the latter are travelled.

The criterion of stability introduced by Lyapunov,[76] is more specific to the subject we are interested in, as the latter is also applicable to non-autonomous systems and is more restrictive than that of Poincaré [there is a vast amount of literature on the subject; see for example Elsgolts (1977); Jordan and Smith (1977); Beltrami (1987); Anosov and Arnold (1988); Arnold (1992); Bellomo and Preziosi (1995); Puu (2000); Riganti (2000)]. In Lyapunov's definition, the fact that the perturbed orbit stays 'close' to the unperturbed orbit is no longer satisfactory; now we ask that the evolution over time maintains the system close to that which evolves unperturbed.

More specifically, the solution $X_i(t)$ with $i = 1, 2\ldots, n$ of the system (8.1) is Lyapunov stable on $t \geq t_0$ if, given a positive arbitrary number ε, a number δ can be determined such that, for any other solution $x_i(t)$ of the system, whose initial value $x_i(t_0)$ is such that:

$$|x_i(t_0) - X_i(t_0)| < \delta \quad \text{with } i = 1, 2\ldots, n \quad (18.2)$$

the following inequality is satisfied for each $t \geq t_0$:

$$|x_i(t) - X_i(t)| < \varepsilon \quad \text{with } i = 1, 2\ldots, n \quad (18.3)$$

(compare Figure 18.4 with Figure 18.3, concerning stability according to Poincaré, in which time does not appear explicitly; note that in Figure 18.4,

---

[76] The Russian mathematician Alexandr Mikhailovich Lyapunov, towards the end of the nineteenth century, an era that marked a fundamental stage in the long process of establishing the fundamentals of infinitesimal calculus, which had started at the end of the seventeenth century, conducted a series of studies in which he investigated, on a very general level, the problem of the stability of the solutions of a differential equation (republished in Lyapunov, 1954; see Parks, 1992). It is worth remembering that his studies on instability led to an open debate, starting from 1905, with the astronomer George Darwin, son of Charles Darwin, who sustained that the Moon had separated from the Earth when the latter was still a liquid mass. According to George Darwin, the process of detachment had given rise to a stable configuration: the Moon. According to Lyapunov, on the contrary, the solution of the process was unstable and the liquid mass would not have been able to break and cause the detachment of the satellite. Only in 1915 was George Darwin's idea demonstrated to be wrong by the astronomer James Jeans (Bottazzini, 1999).

the independent variable is time, while Figure 18.3 displays the phase space in which there is no real independent variable).[77]

Systems (18.2) and (18.3) state that the perturbed orbit and the unperturbed orbit never distance themselves from one another, *at any time*, while, according to Poincaré's orbital stability criterion, they should not distance themselves from one another, *at any place*. Moreover, if solution $X_i(t)$, as well as being stable according to Lyapunov, is such that, as time passes, the perturbed orbit $x_i(t)$ tends to return towards the unperturbed $X_i(t)$, this means it also satisfies the following condition:

$$\lim_{t \to \infty} |x_i(t) - X_i(t)| = 0 \quad \text{with } i = 1, 2 \ldots, n$$

then $X_i(t)$ is said to be asymptotically stable (Figure 18.5).

Lastly, we observe, without going into detail, that for an autonomous system, it can be shown how stability in the Lyapunov sense also implies stability in the Poincaré sense.[78]

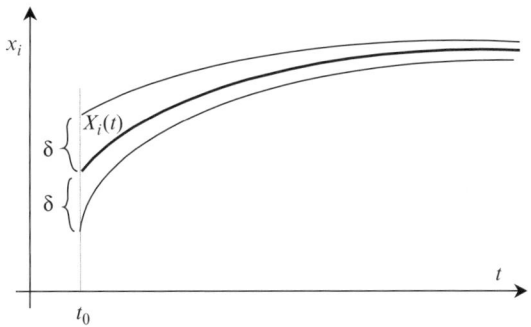

Figure 18.5 Asymptotic Lyapunov stability in the space tx.

---

[77] To be thorough, the criterion defined in the text is of *uniform stability*. As a matter of fact, we should consider δ, in principle, as a function not only of ε but also of t_0, that is different values of δ for the same ε might be chosen along an orbit $X_i(t)$ if a perturbation arises at different instants t which therefore become the different initial values t_0 of a perturbed orbit $x_i(t)$. The stability is said to be non-uniform in this case. It is clear that any stable solution of an autonomous system must be uniformly stable, since the system is invariant with respect to time translation.

[78] The Lyapunov stability criterion described in the text is also called the first method of Lyapunov. In fact there is also a second, very generalized method of Lyapunov, for the study of the system's stability, which claims that a system is stable at point $x_i = 0$ in the phase space, if a differentiable function $v(x_i)$ exists, known as the Lyapunov function, that satisfies the two following conditions in a neighbourhood of $x_i = 0$:

Application of Lyapunov's criterion to Malthus' exponential law of growth

In the light of the previous discussion, let us apply the definition of Lyapunov stability, for example, to the basic case of a population model (or a capital model with compound interest) that grows according to Malthus' law (4.2), that we have already used in this section:

$$\dot{x}(t) = k\, x(t)$$

Let us study the stability of the solutions of (4.2) for two values of k: first Case A, when $k = -1$; then Case B, when $k = +1$.

Case A: $k = -1$.

Equation (15.1) is therefore written:

$$\dot{x}(t) = -x(t) \tag{18.4}$$

and its general integral is a decreasing exponential:

$$x(t) = x_0 e^{-t} \tag{18.5}$$

assuming, for the sake of simplicity, that time starts from instant $t_0 = 0$.

Imagine, for example, the process of the extinction of a population or a resource or of any other magnitude that varies over time, according to a decreasing exponential function given by (18.5). Even altering the initial value x_0, the different dynamics represented by (18.5) all lead to zero, for any value of x_0. The only magnitude in play that defines the system's state is $x(t)$ (for example, the number of individuals in a population). Regardless of the initial value assumed, therefore, the evolution of the dynamical system (18.4) is such that, as time tends towards infinity, the number of individuals in the population tends towards 0. Small variations in the initial value of $x(t)$, as time passes, become increasingly smaller, until they are negligible. The population system described by model (18.4) ends up, therefore, losing all memory of the initial state of its evolution.

(1) $v(x_i)$ has a local minimum in $x_i = 0$;
(2) $\dfrac{dv(x_i)}{dt} \leq 0$ for $t \geq t_0$.

We will not go into discussions on the demonstration or the illustrations of applications here. Interested readers should consult the references cited in this section.

In plane tx, (15.1) represents a set of decreasing exponentials; as all the exponentials of that type tend to flatten out towards zero, the reciprocal distance between all of the exponentials tends to diminish as time passes. This means that, if instead of the solution $X(t)$ passing through the point of coordinates $t_0 = 0$ and X_0, that we will call unperturbed (Figure 18.6), we considered a perturbed solution $x(t)$, passing through the point of coordinates $t_0 = 0$ and $x_0 = X_0 + \delta$, the difference between the two would tend to cancel itself out. The definition (18.2)–(18.3) is satisfied and therefore solution $X(t)$ is stable according to Lyapunov (it can be noted that it is also asymptotically stable).

A numerical example of evolution of the type (18.4) is shown in Table 18.1, where the values of the two state variables at different instants are calculated, starting from initial values $X_0 = 1$ and $x_0 = 1.1$. As we can see, the differences Δ between the states calculated for the same values of time decrease as the latter increases.

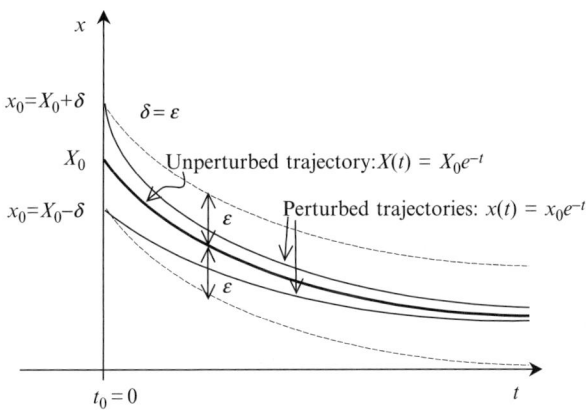

Figure 18.6 Stability of the solutions of equation $\dot{x} = -x$.

Table 18.1 Two examples of orbits of system (18.4), starting from initial values that are close to one another (approximate numerical values)

t	$X(t)$	$x(t)$	$\Delta = x(t) - X(t)$
0	1.00000	1.10000	0.10000
1	0.36788	0.40467	0.03679
2	0.13533	0.14887	0.01354
3	0.04979	0.05476	0.00497
4	0.01831	0.00201	0.00184

Case B: $k = +1$.

Equation (15.1) is now written:

$$\dot{x}(t) = + x(t) \tag{18.6}$$

whose general integral is an increasing exponential:

$$x(t) = x_0 e^{+t} \tag{18.7}$$

assuming also here that time starts from instant $t_0 = 0$. In the plane tx, (18.7) represents a set of increasing exponentials, the distance between which, unlike the previous case, increases as time increases. This means that, if instead of the unperturbed solution $X(t)$, passing through the point of coordinates $t_0 = 0$ and $x_0 = X_0$ (Figure 18.7), we considered the perturbed solution $x(t)$, passing through $t_0 = 0$ and $x_0 = X_0 + \delta$, the difference between the two would tend to increase. Now the definition (18.2)–(18.3) is no longer satisfied, the perturbed trajectory and the unperturbed one diverge, increasingly distancing themselves from one another; the given solution $X(t)$ of the equation is Lyapunov unstable.

Let us once again take the same example presented for Case A, and let us calculate the values of the two state variables $X(t)$ and $x(t)$, for the same values of time, starting from the initial values $X(0) = 1$ and $x(0) = 1.1$, but now according to (18.7); the results are shown in Table 18.2. As we can see, the differences Δ between the states calculated for the same values of time increase as the latter increases.

In Case B, the dynamical system evolves amplifying the changes in the initial state and preserving the memory of such, in the sense that, unlike

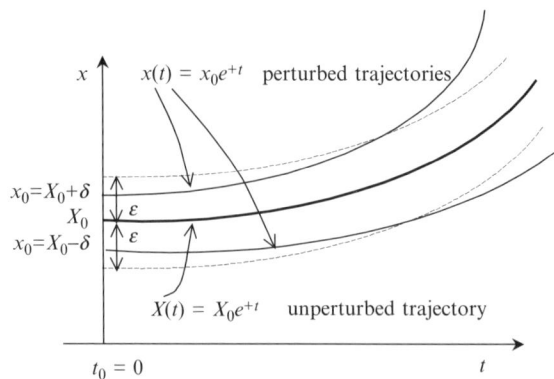

Figure 18.7 Instability of the solutions of equation $\dot{x} = +x$.

Table 18.2 Two examples of orbits of system (18.6), starting from initial values that are close to one another (approximated numerical values)

t	$X(t)$	$x(t)$	$\Delta = x(t) - X(t)$
0	1.00000	1.10000	0.10000
1	2.78183	2.99011	0.20828
2	7.38906	8.12796	0.73890
3	20.0855	22.0941	2.00860
4	54.5981	60.0580	5.45986

what occurred in Case A above, at any instant, the value of $x(t)$ depends on the initial value of x_0, without ever tending towards a single final asymptotic value. As shown in the numerical example illustrated in Table 18.2, the differences Δ between the values of the state variable in the two evolutions become increasingly higher as time passes. We find ourselves faced with a situation of instability. Just as with the example of the map (15.1) shown in Table 15.1, there is no chaos here either, because the divergence that originates between close points amplifies and extends to increasingly large areas in the phase space; the dynamics do not generate any attractor.

We are now faced with the problem of identifying a method that allows us to quantify the degree of instability of an orbit. In order to gain a greater insight into the evolution of a dynamical system, we therefore need a method that enables us to establish not only if the amplification of a small initial perturbation takes place over time, but also *how fast* it grows.

Quantifying a system's instability: the Lyapunov exponents

Before concluding our reflection on the stability of a system, we need to define a parameter that enables said stability, which we have so far discussed only in qualitative terms, to be quantified. To this end, we will introduce the so-called Lyapunov exponents: numbers that provide a measurement of the stability of a dynamical system and that, in a certain sense, quantify its level of chaos.

Let us again consider a dynamical system (a model) made up of a system of differential equations such as (8.1):

$$\frac{dx_i}{dt} = F_i(x_1, \ldots, x_n, t) \quad i = 1, 2 \ldots, n$$

18: Stability in dynamical systems

As previously, we will indicate the stable unperturbed solution of (8.1) with $X_i(t)$. We now wish to establish 'how' unstable orbit $X(t)$ is, i.e., 'how quickly' any perturbed solution $x_i(t)$ that starts (or passes) close to $X_i(t)$ in the phase space, diverges from $X_i(t)$ over time.

We have often mentioned the concept of sensitivity to initial conditions, basically meaning the intensification of small divergences from the initial values over time. The reference that we assume, so as to speak, as 'unit of measurement' to quantify the speed of growth of the perturbations is the law of exponential growth. This assumption is justified by the fact that, if we 'test' the instability of an orbit in its immediate neighbourhood, we can therefore simplify the matter assuming that the instability follows, at least in a first approximation, a linear law, which, as we have observed (see equations (4.2) and (18.1)) envisages orbits expressed by exponentials.

The situation that we are studying can be schematized as in Figure 18.8, where we assume, for the sake of simplicity, that the system of equations (8.1) comprises only three variables $x_i(t)$, $x_j(t)$ and $x_k(t)$. Figure 18.8 represents an unstable orbit, in all points of which small perturbations give rise to new orbits that exponentially diverge from the unperturbed orbit.

In the case of an unstable orbit, the new perturbed orbit does not tend towards the unperturbed one, but distances itself according to a trend, the initial state of which can be expressed in an exponential form:

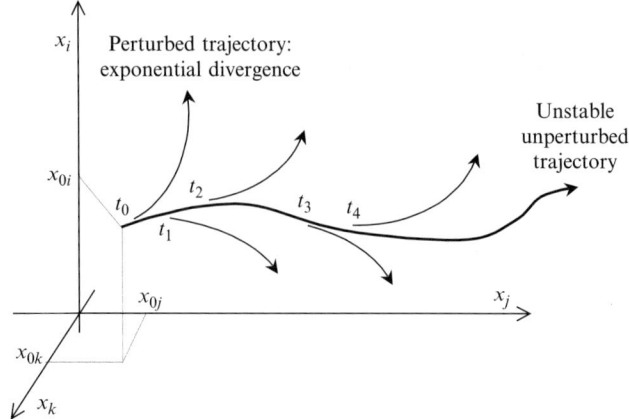

Figure 18.8 Dynamics of an unstable system in which we have exponential divergence. The dashed arrows represent trajectories that diverge from the unstable trajectory, exponentially amplifying the initial perturbation that has generated them.

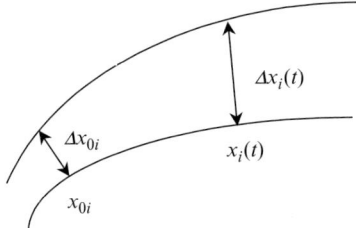

Figure 18.9 Divergent orbits.

$$\Delta x_i(t) = \Delta x_{0i} e^{\Lambda_i t} \tag{18.8}$$

In (18.8) $\Delta x_i(t)$ indicates the difference, expressed as a function of time, between the values of the i-th state variable x_i on the perturbed orbit and the corresponding value on the unperturbed one, starting from a difference between the two initial values of the same state variable (two initial states), indicated by Δx_{0i} (Figure 18.9).

The parameter Λ_i is the *Lyapunov exponent*[79] for variable x_i. Its value indicates the speed of growth (instability) or reduction (stability) of the 'small' perturbation made to x_i, that we mentioned earlier.

For the sake of simplicity, let us examine the variation of just one of the state variables, $x_i(t)$. Isolating Λ_i in (18.8), we have:

$$\Lambda_i = \frac{1}{t} \ln \left| \frac{\Delta x_i(t)}{\Delta x_{0i}} \right| \tag{18.9}$$

in which we introduce the modulus, as we do not distinguish between a positive or a negative direction for the distancing between the perturbed and unperturbed orbits.[80]

---

[79] In reality, more than a speed in the strict sense, Λ_i represents the rate at which the orbits diverge, namely the ratio between the speed at which they diverge and the initial perturbation. The fact that Λ_i is a rate can be obtained by dividing the derivative of (18.8) by (18.8) itself:

$$\frac{\Delta \dot{x}_i}{\Delta x_i} = \Lambda_i \tag{18.8'}$$

[80] As we have said, (18.8) implies that the evolution over time of the difference $\Delta x_i(t)$ is evaluated through the linear approximation of the dynamical system's equations for variable x_i; the said approximation, therefore, only makes sense if we remain in the immediate neighbourhood of x_{0i}. In other words, Lyapunov exponents thus defined express to what extent an infinitesimal variation of the unperturbed orbit influences the system's evolution for an infinitesimal time. A stricter definition of the Lyapunov exponent than the one illustrated in the text,

The stability of the dynamics of a system with respect to the state variable x_i, therefore, depends on the sign of the corresponding Lyapunov exponent Λ_i:

- $\Lambda_i < 0 \rightarrow$ Convergent orbits with respect to the state variable x_i. In this case we have asymptotic stability.
- $\Lambda_i = 0 \rightarrow$ Constant distance between the orbits with respect to the state variable x_i. In this case we have stability.
- $\Lambda_i > 0 \rightarrow$ Divergent orbits with respect to the state variable x_i: the orbits are unstable and diverge exponentially from one another. In this case we have chaos.

The general stability of the dynamical system, therefore, is guaranteed only if each of the state variables $x_i(t)$, individually, is characterized by a negative Lyapunov exponent. On the contrary, a measurement of the degree of chaoticity of the dynamical system can be given by the sum of the Λ_i exponents of the x_i variables for which they are positive. This magnitude is known as the Kolmogorov–Sinai entropy (Eckmann and Ruelle, 1985) and can be assumed as an indicator of the extent to which the system is unpredictable: the higher the Kolmogorov–Sinai entropy, the more unpredictable the system.

As we have said, the trajectories of a dynamic defined by n state variables are characterized by n different Lyapunov exponents: this set is called the Lyapunov spectrum.

The values of the Lyapunov exponents enable us to draw some important conclusions on dynamics:

1. If all the Lyapunov exponents are negative, then we have a fixed point that behaves like an attractor. Each negative Lyapunov exponent provides a measurement of the speed with which the corresponding state variable directs itself towards the point attractor.

however, must add two further elements to (18.9): (1) the fact that initial perturbations $|\Delta x_o|$ are considered infinitesimal, which leads to the introduction of a limit for $|\Delta x_o| \to 0$; (2) the fact that, to obtain a measurement of the divergence of the orbits on the whole unperturbed orbit, an average of the exponential divergences at the different points along the orbit (in theory at all points) needs to be calculated. This involves the introduction of the limit for $t \to \infty$, which basically extends the calculation of instability to an average on N points, with N becoming increasingly larger. The formal definition of the Lyapunov exponent for a state variable x_i, is:

$$\Lambda_i = \lim_{\substack{t \to \infty \\ |\Delta x_{0i}| \to 0}} \frac{1}{t} \ln \left| \frac{\Delta x_i(x_{0i},t)}{\Delta x_{0i}} \right| \quad i = 1, \ldots, n \qquad (18.10)$$

2. If some exponents are nil and others are negative, then we have an ordinary attractor that is less simple than a fixed point, such as for example a limit cycle.
3. If at least one of the Lyapunov exponents, let us suppose the i-indexed one, is positive, then we are in the presence of evolutive dynamics which, in that particular dimension i, diverge exponentially, and give rise to chaos. Vice versa, we can also state that in a chaotic system at least one Lyapunov exponent is positive.

In the case of one-dimensional maps such as:

$$x_{n+1} = f(x_n)$$

in which the analytical expression of the map f is known, the Lyapunov exponent assumes the following form (obtained from (18.10), on p. 000, note 78) that is particularly important, as we will see when we examine the logistic map (Moon, 1992; Çambel, 1993; Puu, 2000):

$$\Lambda = \lim_{N \to \infty} \frac{1}{N} \sum_{n=1}^{N} \ln \left| \frac{\partial f(x_n)}{\partial x_n} \right| \qquad (18.11)$$

where the role of time t is played by the index n which counts the iterations, the total number of which is N, and the speed of divergence is simply the derivative of the evolution function.

A final consideration regards the fact that the phase space can also be of an infinite dimensionality. This happens, for example, if the evolutive dynamic depends on one or more parameters that vary with continuity, such as with the logistic map (Chapter 22). In this case, the Lyapunov exponent becomes a continuous function of the parameter (or of the parameters), and the values of the parameter (or parameters) in correspondence of which the Lyapunov exponent has a positive value are only those for which the system is chaotic. (For further details on the subject of Lyapunov exponents and chaos in dynamical systems see Wolf, 1984; Wolf et al., 1985; Eubank and Farmer, 1989; Rasband, 1990; Medio, 1992; Moon, 1992; Katok and Hasselblatt, 1995; Abarbanel, 1996; Robinson, 1999, Puu, 2000; Riganti, 2000; Medio and Lines, 2001).

Exponential growth of the perturbations and the predictability horizon of a model

Lyapunov exponents, as we have said, provide important quantitative information on the chaotic properties of the evolution of a dynamical system. Mathematical models are more often than not written in the form of dynamical systems; therefore, the concepts that we have illustrated are applicable to models, but transferring them to real systems requires some adjustments. To conclude this section, we would like to add some considerations on the meaning of Lyapunov exponents in models, if the latter are in the form of dynamical systems; in Chapter 19, on the other hand, we will see how concepts regarding a system's stability can be applied to real systems.

We ask ourselves, in particular, what is the time t^* beyond which the uncertainty of the prediction of a system's future state exceeds a pre-established tolerance that we will indicate with δ. The answer can be obtained by isolating time in (18.9) (ignoring, for the sake of simplicity, index i and the symbol of the modulus, assuming all magnitudes to be positive):

$$t^* = \frac{1}{\Lambda} \ln \frac{\Delta x(t)}{\Delta x_0} \qquad (18.12)$$

Now, Δx_0 in (18.8) was the perturbation of an initial state, but can also be interpreted as the approximation, or uncertainty, of the measurement of the initial state of the evolution. In this sense, we assign Δx_0 the meaning of an interval of values within which 'the' initial value, that we are not able to know with any degree of precision, of the state variable $x(t)$ lies; assuming an initial value x_0 with an approximation Δx_0 means that the initial value could be very close to x_0, but not x_0. We ask ourselves the usual question: to what extent does a 'small' difference of the initial state, hidden due to our imperfect knowledge of the same, influence its future evolution? If the evolutive trajectory is stable, then it has little or no influence; if it is unstable, however, the effect of the initial uncertainty gets increasingly larger over time. It is exactly this initial uncertainty that is intensified over the course of the evolution, if the dynamics of the model are chaotic, until it exceeds the pre-established threshold δ. The latter occurs in correspondence to the time t^* obtained by assuming $\Delta x(t) = \delta$ in (18.12). This value of the time t^*, that depends on δ, thus assumes a particular role that we can define as a sort of predictability horizon of the model's future.

In principle, it should be sufficient to take the initial uncertainty Δx_0 that is sufficiently small to generate an arbitrarily large value of t^* for that given value of $\Delta x(t) = \delta$. In practice, however, the presence of the logarithm in (18.12) makes the dependence of t^* on Δx_0 so weak that the term that is by far the most dominant in determining the predictability horizon is actually the Lyapunov coefficient Λ. In other words, to extend predictability, i.e. the stability of the evolution over time, it would be better, where possible, to succeed in improving the model, rather than the precision of the initial values from which we make the model evolve. For the purposes of the stability of the model, and therefore of the extension over time of the validity of predictions, basically it is important that Lyapunov exponents have small values. Lyapunov exponents, in fact, are inversely proportional to the predictability horizon, while the predictability horizon depends on the precision of the initial values only in logarithm terms.

By way of example, let us once more consider the very simple case in which we employ an exponential model to describe the growth of a population (of any nature), using Malthus' linear model (4.2):

$$\dot{x}(t) = k\, x(t)$$

As we know, the solution of (4.2) is $x(t) = x_0 e^{kt}$. A variation (or uncertainty) of the initial value x_0 will therefore give rise to a variation (or an uncertainty) of the subsequent states $x(t)$ that increases exponentially over time we thus have $\Delta x(t) = \Delta x_0 e^{kt}$ which, if we identify the parameter k with the Lyapunov coefficient Λ, is exactly (18.8). For the sake of simplicity, let us now suppose that the pre-established threshold of the uncertainty of the prediction is, for example, $\delta = 1$; we now obtain that the predictability horizon t^* of the model is:

$$t^* = \frac{1}{k} \ln\left(\frac{1}{\Delta x_0}\right)$$

Assuming $\delta = 1$ means that the uncertainty Δx_0 on the initial data must be $\Delta x_0 < 1$, because this must be at least less than the threshold of uncertainty of the future evolution δ. If we wanted, for example, to double the horizon of predictability t^*, wishing to improve *only* the approximation of the initial data Δx_0, then we should square the argument of the logarithm, i.e. replace $1/\Delta x_0$ with $(1/\Delta x_0)^2$; this operation, in fact, would double the value of the logarithm. As $\Delta x_0 < 1$, squaring Δx_0 means, in practice, replacing the number that expresses the uncertainty of the data with another much smaller one, i.e. considerably reducing the initial uncertainty.

The same doubling of the horizon t^* could be obtained, on the other hand, maintaining the same uncertainty of initial data, but improving the model, so that the Lyapunov exponent Λ is halved. In order to extend the timeframe of the 'glance into the future' that a model enables us to do, based on the hypotheses that comprise its foundation, what counts, in practice, is almost exclusively the value of the Lyapunov exponent.

This last observation on the exponential growth of a 'small' initial approximation can have fundamental practical consequences when a model (even if written in continuous time) is made to evolve using, as always happens, an approximation in discrete terms and therefore an iteration of a map calculated by a computer.[81] Remember that any computer *always* works on a discrete series of numbers, because each number is stored in the computer's memory (i.e. 'exists' for the computer) taking only a finite number of figures into account; this can mean that a number is subjected to an approximation that leads to the elimination of all of the figures that exceed the memory capacity envisaged to represent it. Working in this way, however, we end up *effectively* examining a trajectory (an orbit) that is different to the *theoretical* one along which we had intended to follow the system's evolution, if the latter is unstable; the theoretical evolution predicted by the model remains, in practice, unknown because it is numerically unachievable. In other words, if the model is a dynamical unstable system, the evolution that is obtained by materially making the calculations, in reality, is different to what would be obtained by solving the equations in a formal way, without using numerical approximation. Because, with the exception of a few simple cases, we do not have the formal methods to solve a system of equations (i.e. to make a model evolve) in *exact* terms, the formally 'exact' theoretical orbit, basically, is nothing more than a completely abstract concept.

---

[81] As we already mentioned at the beginning of this section, it was precisely the comparison between two different evolutions starting from different numerical values as a consequence of a simple numerical approximation, that drew Lorenz's attention to the chaotic characteristics of the dynamical system that he was studying (Gleick, 1987).

19 The problem of measuring chaos in real systems

Chaotic dynamics and stochastic dynamics

In the previous sections we have mostly talked about the stability and chaos of dynamical processes in mathematical models. Now we would like to look at the problem of the stability and chaos (if any) in *observed* dynamical processes, i.e. referring to the analysis of historic data series observed in real systems.

Definition (18.9) of the Lyapunov exponents does not have only a theoretical meaning, but can be used in important applications to real cases. The problem we are faced with is how to use a series of data in order to be able to calculate quantities such as Lyapunov exponents and to identify the presence, if any, of an attractor and its dimension (fractional or integer). This would enable us to establish, in particular, if the historic series of data is produced by a system that 'functions' according to a deterministic law, regardless of whether we have chaos or not, or if the data observed are stochastic, i.e. if they are the result not of a law, but of that which is commonly called 'chance', and follow a probability distribution. The problem is anything but easy, in particular for systems in whose dynamics it is difficult, if not impossible, to observe regularities, as usually happens in social sciences.

A historic series of data (i.e. a set of successive measurements of a certain magnitude that varies over time), in fact, can be identified by a set of states in an appropriate phase space, i.e. referring to appropriate state variables. If with some type of measurement we manage to obtain a set of coefficients that, with respect to the presumed state variables, plays the role of Lyapunov exponents, then the quantitative analysis of these coefficients can lead to the empirical verification of the system's stability and the observed evolution of the real system. A method of this kind can represent an effective test, valid both in the case of conservative systems and in the case of non-conservative systems, to diagnose the presence of chaotic situations (Wolf et al., 1985; Abarbanel, 1996).

19: The problem of measuring chaos in real systems

In practice, no temporal series can be assumed to consist of data that result *only* from a (hypothetical) deterministic law, i.e. from a so-called 'pure signal', that has no stochastic components. Any series of experimental data, in fact, is always mixed with a stochastic process that superposes on it, like a background noise, a so-called 'white noise', that reduces the quality of the information. This is due to a number of reasons. On one hand, for example, there are the inevitable practical difficulties of measuring data, which, in any event, never constitute a continuous sequence over time; on the other hand, any measurement is always affected by approximations of various types and of various origin, even those due simply to the truncation of the numbers. A deterministic law, precisely because it is identified (or rather, 'constructed') starting from an incomplete set of approximated data, can never be *completely* predictable, precisely because *in principle* all that we have available is a set of measurements that may be more or less numerous and more or less approximated, but is never an infinite quantity and are never absolutely precise.

The aim of the analysis of a temporal series of data, therefore, is to answer questions such as the following:

1. Is it possible to identify an attractor underlying the temporal series? In other words can the data of the series be considered as the expression of a deterministic dynamics and not of a stochastic process?
2. Assuming that the attractor exists, what is its dimension? Is it a chaotic dynamics (non-integer dimension of the attractor) or not (integer dimension of the attractor)? And what different types of dynamics (limit cycle, quasiperiodic orbits, chaotic orbits, etc.) does its dimension allow?
3. How many state variables are needed to constitute the phase space (the dimension of the phase space) to contain the attractor (if such exists)? In other words, what is the minimum number of state variables needed to identify the system and to define its dynamics? Which variables are they?

The methods used to distinguish deterministic processes from stochastic ones are based on the fact that a deterministic system always evolves in the same way, starting from given conditions. At least in principle, given a temporal series of data, we can proceed as follows:

1. Choose a reference state A in the historic series.
2. Search for a state B in the same historic series that is as close as possible, in the phase space, to the reference state A.
3. Compare the evolution that state A follows with that of state B and call the difference between the corresponding values of the same state vari-

ables in the two successive evolutions of A and B 'deviations'. There are two alternatives: (a) with a stochastic system, the values of the deviation are distributed in such a way that we cannot identify any relationship between them, i.e. the data appear to be distributed at random; (b) with a deterministic system, the deviations remain small, if the evolution is stable, or some of them grow exponentially over time, at least initially, if the system is chaotic. In case (b), we can *measure* (and not calculate, as we do in models) Lyapunov exponents by following the two temporal series that start from A and B (close to one another in the phase space), and simply measuring the trend of the deviations as a function of time.

The above is from a theoretical standpoint. In practice, we are faced with the problem of how to determine two states 'close' to one another from 'all' of the state variables. The systems we encounter in the real world are defined by very large sets of state variables. In particular, as we have observed on several occasions, this is true for social systems, where it is often difficult not only to follow the evolution of the state variables, but also to identify the most significant state variables, or even to establish how many of them there are.

One approach to solving this problem entails initially taking a very limited set of variables that we try to identify as fundamental and analysing the behaviour over time of the deviations between the successive evolutions of states A and B, that now, with relatively few state variables, are relatively easy to identify. If said deviation appears to behave in a random way, this means that we have not considered a sufficient number of state variables to construct an effective deterministic law. Therefore the number of state variables, i.e. the dimensionality of the system's phase space, needs to be increased. A higher number of variables could now be sufficient to depict that deterministic law which previously appeared hidden, camouflaged, so to speak, by the apparent random nature of the data.

The first difficulty that we encounter, therefore, regards the construction of the phase space in which to place the system's dynamics and an attractor, if such exists. In principle, it should be sufficient to continually increase the number of state variables to be certain of obtaining a deterministic law sooner or later. This would occur, again in principle, when we come to closing the system, isolating it completely from any relation with the outside. A system of this nature would certainly be deterministic, because all of its dynamics would be endogenous, i.e. caused by laws internal to the system, that do not assign a role to the interaction with the external environment, which is subject to change and out of the system's control (i.e. what we call 'chance'). Probably, however, the number of variables needed to consider the system closed would be so high that it would render the calculation time unacceptably long, in practice, to effectively identify the deterministic law at the origin of the observed dynamics, rather than limiting ourselves to imagining its presence.

On the other hand, a system that contains the entire universe, therefore with a very high number of state variables, to describe the *entire* contents of the universe, would certainly be closed and deterministic in theory, but absolutely impossible to treat from a practical point of view.

Basically, if as the number of variables considered increases, at a certain point we observe that the deviation between the trajectories, i.e. the *apparent* random dispersion of data, continues to grow more or less proportionally to the number of variables, then we can say that the system is stochastic, because the state variables that we are considering are not sufficient to identify any underlying dynamic. If, on the other hand, as the number of variables considered increases, at a certain point, once a certain number n of variables has been exceeded, we observe that the distance between the trajectories stops increasing and tends to assume an (almost) stable value, which doesn't increase even if additional new variables are introduced, then we can claim to have identified the 'real' number of variables n needed to 'close' the system and to define its state and dynamics. If this is so, we can state that there is a deterministic process at the origin of the historic series of data, whose law involves exactly n state variables, and that the system's dynamics is chaotic and non-stochastic, and can be described by a deterministic law (Nicolis and Prigogine, 1987; Vulpiani, 1994).

The simplest case of system dynamics is that in which there is a point attractor that acts as the system's stable equilibrium, regardless of the size of the phase space. On the other hand, if we consider n state variables, there are, on the contrary, systems whose successive states disperse over the *whole* n-dimensional phase space in question, without the configuration of any regularity or any attractor; in this second case, we say that the evolution appears to be random. Chaotic systems, with their strange attractors, fall between these two extremes. The problem, therefore, is determining a dimension d (that will not be integer) of the chaotic attractor.[82]

---

[82] We will repeat what has already been stated in Chapter 16: the concept of strange attractor regards the fact that the attractor is a fractal and is therefore a geometric concept that regards the shape of the attractor; on the other hand, the concept of chaotic attractor refers to the instability of the orbit and is, therefore, a concept that regards the system's dynamics. In practice the two concepts are often identified together even though in certain cases this is incorrect. In this context, for the sake of simplicity, we will also identify them together, as this simplification will not effect our discussion.

If the (assumed) deterministic dynamics of the real system envisages an attractor whose dimension d is larger than n, then in the n-dimensional phase space the attractor cannot appear (or at least not entirely) and the system appears to be stochastic. For example, a set of three state variables x, y, and z is sufficient to define a three-dimensional space in which a strange attractor, whose dimension is less than 3, can be found as in the case of the Lorenz attractor (Figure 16.2) that is generated by a system of three equations with three variables [the system of equations (16.3)]; with just two state variables, the system would not give rise to an attractor. The problem of recognizing the attractor is strictly linked, therefore, to the problem of the definition and of the choice of the minimum number of state variables needed to contain it.

A method to obtain the dimension of attractors

To attribute a dimension to strange attractors, for which ordinary topological dimension loses meaning, the dimension known as Hausdorff–Besicovich is often used (sometimes called *capacity dimension*), which we mentioned in Chapter 16 and which we will indicate with d_H[83] (Grassberger, 1981). From a practical point of view, however, the application of the Hausdorff–Besicovich definition of dimension requires very long calculation times, because the number $N(l)$ of the cells to be considered (see note 83 at foot of page) increases very quickly as their side l decreases, and the calculations become too onerous if the number of state variables of the

---

[83] Imagine dividing the space containing the geometric object (in our case the attractor), the dimension of which we want to define, into 'cells' of side l: if the space in question is one-dimensional (a line), the cells are segments of length l; if it is two-dimensional (a plane), the cells are square of side l; if it is three-dimensional, the cells are cubes of edge l, and so on. The total number $N(l)$ of cells of side l that are crossed by the object in question is counted: $N(l)$ depends on the length l assumed and will be a rather small fraction of the total number of cells into which the space has been divided. In general $N(l)$ is related to l by a power law such as:

$$N(l) = \frac{1}{l^{d_H}}$$

In the case of ordinary Euclidean geometry objects, d_H is a natural number and coincides with the (topological) dimension usually used. If d_H is not an integer, then we refer to a fractal figure that cannot be described using the usual techniques of Euclidean geometry, of which d_H is the Hausdorff–Besicovich dimension:

$$d_H = -\frac{\log(N(l))}{\log l}$$

system (i.e. the dimensionality of the phase space) is higher than 3 or 4. Furthermore, the calculation of d_H, is substantially inefficient as the majority of the $N(l)$ cells in reality are empty.

James Farmer, Edward Ott and James Yorke (1983) showed that, even from a conceptual point of view, the use of d_H to measure the fractality of a strange attractor is not completely adequate as it does not properly take into account the *frequency* with which an orbit enters the cell (see note 82 on p. 00) used to measure d_H. In other words, d_H does not consider the dynamic aspects of a chaotic attractor, only the geometric ones, i.e. those relative to the attractor's shape. For these reasons, in more recent times other definitions of dimension have been developed that are more pertinent to the problem in question with respect to the Hausdorff–Besicovich dimension, which, incidentally, had been proposed for other purposes, long before the concepts of fractal and strange attractor were introduced (Hausdorff, 1919; Besicovich, 1934, 1935). In particular (Swinney, 1985) from the 1980s onwards, the so-called correlation dimension d_c has often been used.[84]

Peter Grassberger and Itamar Procaccia (1983ab) proposed a method to obtain an estimation of the value of the correlation dimension d_c of the strange attractor, if any, that we are attempting to identify in a historic series of data, and therefore also to determine the number n of state variables needed to obtain the deterministic law that describes the system's evolution and generates the attractor (Rasband, 1990; Schroeder, 1991; Moon, 1992;

---

[84] There have been a number of working definitions of the correlation dimension. One of the most popular involves the following operations (Grassberger and Procaccia, 1983ab): (1) we take K points of the geometric object (in the phase space) the dimension of which we wish to define, and we consider all of the pairs that can be formed with said points; (2) we calculate the distance δ_{ij} between all of the points of the pairs formed and we use $M(r)$ to indicate the number of pairs whose distance δ_{ij} is less that a pre-established distance r; (3) we calculate, starting from the numbers obtained, a function $C(r)$, called the correlation function, defined by:

$$C(r) = \lim_{K \to \infty} \frac{M(r)}{K^2}$$

In general, we note that $C(r)$ is related to the distance r by a power law:

$$C(r) \sim r^{d_c}$$

The exponent d_c is the correlation dimension of the set of points considered.

It should be noted, however, that for many strange attractors, the different definitions of dimension lead approximately to the same results (see Moon, 1992). For Lorenz attractor, for example, the Hausdorff–Besicovich dimension is $d_H = 2.06 \pm 0.01$, while the correlation dimension is $d_c = 2.05 \pm 0.01$ (Grassberger and Procaccia, 1983a).

Peitgen, Saupe and Jürgens, 1992; Çambel, 1993; Casti, 2000; Riganti, 2000).

The method originates from the fact that the *calculated* value of the dimension d_c of a *known* strange attractor increases as the number of state variables n being considered increases, if the latter are not in sufficient numbers to generate the appearance of an attractor. If, on the other hand, the number of state variables n that is being considered exceeds the dimension d_c of the attractor, then the phase space has the dimensionality needed to contain the attractor, and further increases of n do not influence the attractor: d_c no longer depends on n. For example, the fact that the Lorenz attractor requires at least a three-dimensional space to appear, while in a one or two-dimensional space it would not be seen, means that the equations with one or two variables would not succeed in generating it. In a four or more dimensional space, the attractor would not be altered: additional variables would simply be introduced that do not have any effect on the system's dynamics.

Basically, Grassberger and Procaccia's model assumes that the contrary is true.[85] They assume, that is, that if we observe that the dimension d_c of a set of points that is *presumed* to be part of an *unknown* attractor increases as the number of state variables considered n increases until it reaches a maximum value d_c^*, then this can be taken as an indication of the existence of a strange attractor. In this case, d_c^* is the dimension of the strange attractor, and the value of n that we have reached is the minimum number of state variables needed for a phase space to contain the presumed attractor.

In practical terms, the method comprises the following steps (Figure 19.1):

1. A function $f(n)$ is constructed by points that relate the correlation dimension d_c, calculated on the basis of the historic series to different values of the number of state variables n that are considered.
2. Initially, for a small n, d_c increases as n increases; the state variables considered are too few and the system *appears* to be stochastic.
3. Subsequently, as n increases, if $f(n)$ tends to flatten out and to stabilize at a maximum value (i.e. if the slope of $f(n)$ tends to zero) beyond a value of n, then n is the minimum number of state variables needed to contain the attractor. We conclude that the system represented by the temporal series possesses an attractor whose dimension d_c^* is given by the same maximum value to which $f(n)$ tends. If this is not an integer, then the attractor is strange and indicates a chaotic dynamic.

---

[85] Their assumption, however, is not strictly justified; on the contrary, it has been demonstrated that there are cases that contradict it (Osborne and Provenzale, 1989).

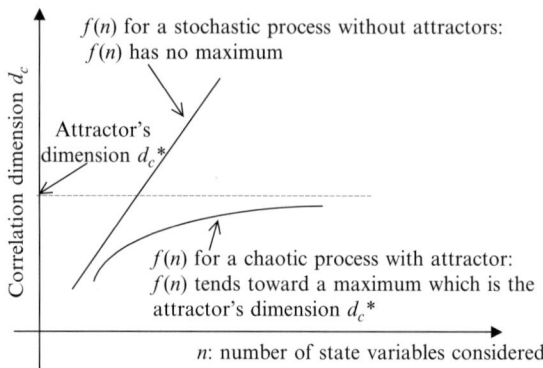

Figure 19.1 Relation between the number n of state variables considered and the correlation dimension d_c, for the dynamics of a stochastic system and those of a chaotic system.

Obviously, the most interesting situation is that in which we observe an attractor with a low dimensionality, i.e. in a phase space with an n that is not too high.

A method like the one described has been applied, amongst other things, to the historic series of (indirect) measurements of the total volume of glaciers on Earth that have emerged over the course of the past million years, interpreted as an indicator of the average temperature of the Earth's surface. The presence of a chaotic attractor with a dimension of 3.1 was observed (the so-called climatic attractor), contained in a phase space of $n = 4$ state variables (Nicolis and Prigogine, 1987). This shows that the climatic evolution of the past one million years can be seen as the manifestation of a deterministic dynamics characterized by a chaotic (strange) attractor with a low dimensionality. This result, obviously, is still too generic to be able to claim that we have identified the entire dynamics of the climate and from which we can make predictions. The fact that the presence of a chaotic deterministic mechanism has been identified, in any event, represents a fundamental step in the description of dynamics that on the surface appear to be incomprehensible.

An observation on determinism in economics

In dynamical system theory, as we know, chaos means irregular fluctuations in a deterministic system; the system behaves irregularly due to its own internal logic, not due to the effect of forces acting upon it from the outside. In a closed (and therefore also autonomous) system nothing acts externally;

a closed economic system can be assimilated to a dynamical system but its dimensions (i.e. the number of state variables) are usually so high that it is impossible to identify the determinism to which the chaotic dynamics observed is subjected (provided that we are dealing with chaos). So that the chaos can be usefully 'managed', i.e. identified as the outcome of a deterministic mechanism that can be given a description, in order to gain some form of advantage from it, we at least need to identify the deterministic law that governs (or more correctly that 'defines') the system. A law of this type, however, can be identified only if the number of state variables required to describe the observed dynamics, and in a certain sense to account for them, is low enough to make the construction of a model possible.

The basic problem, then, is precisely that of recognizing which state variables are necessary to characterize the collective dynamics observed, ignoring the others. These others, however, even if ignored, do not actually cease to exist and act on the system's evolution, even if as state variables of a lesser importance. Ignoring the less important state variables means, in practice, conferring a certain degree of openness to the system, because the ignored variables act as if they were external to the system, which, therefore, ends up operating in an environment that is continually changing, due to the changing of the values of the variables not taken into consideration (Brock, Hsieh and Le Baron, 1991; Benhabib, ed., 1992).

Deterministic processes in economics have been studied for almost three centuries, basically from the foundation of modern economic science. This has occurred according to very different viewpoints, approaches and methods, applying both principles and techniques of classical science, and using techniques developed more recently, above all in the last few decades[86] (for example from Mandelbrot, 1975, 1997ab), but the results have been poor and partial, especially with regard to the identification and the definition of general laws and effective models.

If these deterministic schema exist in economics, then either they are more difficult to recognize, as they do not clearly emerge from the background noise created by individual actions, or they last too short a time to be able to be used, drawing some form of advantage (or even a profit in the strict sense), or both of the above hypotheses are true.

---

[86] With regard to the evaluation of Lyapunov exponents, starting from a temporal series of economic data, we cite the calculation method proposed by Eckmann et al. (1986) and its application to the analysis of the trends of financial market indices (Eckmann et al., 1988), which provides an assessment of market unpredictability. The study of economic systems in general, and of the financial markets, in particular, conducted using techniques taken from the dynamics of physical systems, has led to the development, over the course of the past 30 years, of a new discipline called econophysics (Mantegna and Stanley, 2000; Mantegna, 2001).

Nevertheless, it should be observed that one of the main difficulties that obstructs the identification of deterministic schema in economics is linked to the fact that the numerous actors operating in economic systems are each seeking maximum individual utility (profit). A fundamental aspect of the matter is that, if in financial markets we actually succeeded in identifying deterministic laws with low dimensionality, laws that therefore use only a few state variables, that were relatively simple to recognize and use to effectively account for the small and large fluctuations that we observe and to formulate predictions on the future trends of values, then all market operators would become aware of such and would use such as rapidly as possible, organizing financial operations aimed at achieving maximum profit.

The consequence of this, if a 'deterministic law of economics' really existed, would be that the knowledge of this law, unfortunately, would end up making the market inactive; this law, therefore, would be completely useless, not to mention lacking in any meaning. Let us suppose, in fact, that a law was discovered that described the dynamics of the economy in a deterministic way, and that all the operators in a stock market became aware of such and used it to make predictions (deterministic and therefore equal for everyone) on future price trends. The result of this would be that all the operators would act on the market not only with the same objective (profit), but also in the same way. All of the operators would want to buy, for example, a given stock *now*, if the deterministic law guaranteed that the price will not fall any further. For the same reason, however, none of the operators that owned said stock would want to sell it, preferring to wait to sell it at a future date, at a higher price. In reality, therefore, transactions aimed at seeking profit based on the deterministic law valid for all operators would not take place, because there would be no possibility for individual operators to make different evaluations of the stock value, nor different opinions on future price trends. The rigid 'deterministic law of economics' tool, equal for all operators, if used systematically, would unfortunately end up destroying economic trends, as soon as the latter were recognized and used in practice, and blocking the market.

In reality, in economics and in finance, as in many social sciences, the role of the individual opinions of the actors, which conflict with one another and which change over time, appears to be too important to be able to identify deterministic laws applicable to models that predict (or account for) dynamics that are more varied and complex than simple oscillations (cycles). Simple oscillations (of prices, of production, of the level of employment, ...) in fact, are the forms of a supply–demand dynamics based on elementary rules, the result of subjective evaluations such as: 'sell when you think that the current price will not rise, and buy when you think that the current price will not fall'. They are subjective rules of empirical origin, in which only the interactions between a few variables of very general significance are taken into account (this is all one can do), which consider average values calculated according to some definition. These rules can be used without destroying the market, precisely because they are subjective and therefore lead the various operators to personal evaluations and therefore to differing decisions.

Roughly, an oscillatory dynamics similar to that cited also originates from the linear model in the case of a predation relationship between two populations (Chapter 11, Cases 2A and 2B), or from the Volterra–Lotka model (Chapter 2). These models, however, are ineffective in economics because they do not take a number of aspects into account, for example, the fact that the cycles never repeat themselves in exactly the same way, insofar as there are general trends that superpose and render situations new each time. An effective model should at least take the variability of the individual actions of operators into account, considering them not as statistical fluctuations with respect to average values, but as real state variables that follow their own deterministic dynamics. This, however, would lead to a huge increase in the number of state variables to be taken into consideration, and thus the problem of identifying which variables are necessary to 'close' the system would resurface. On the other hand, even if we actually managed to close the economic system, in accordance with our observations above, it would cease to work, which would impose a substantial review of the deterministic law, as soon as such was identified.

Precisely due to the effect of the continual individual action of the operators and the variability of their decisions, in the absence of a (based on the above, probably useless) 'deterministic law of economics', it does not appear very plausible that the markets would remain in conditions of stationarity for long enough to be able to identify its chaotic attractors. We could say that anyone who claims to have recognized the mechanisms of deterministic chaos in financial markets has not learnt anything of any use; and anyone that really knew something useful would not say so, otherwise that something would cease to be useful.

20 Logistic growth as a population development model

Introduction: modelling the growth of a population

In this section we would like to return to the description of the dynamics of a single population interacting with the external environment, a subject that we have mentioned on several occasions in previous sections, however without ever going into any further detail.

A population can consist of bacteria, insects, people, or any other type of entity able to reproduce itself. The growth model of a population describes how the number x of the members of the latter evolve, i.e. increase or decrease over time.

Let us attempt to formulate a model that reproduces the evolution of a population over time. We can make various types of simplifying assumption: from those, for example, that are extremely schematic and non-realistic used as the basis of the model proposed by Fibonacci[87] in the *Liber Abaci* dated 1202 (a modern English translation is in Siegler, 2002), regarding the problem of describing the growth of a population of rabbits in a closed space, which gave rise to the well-known sequence of whole numbers now called the Fibonacci sequence. To those on which Malthus' model (1798) is based, which establishes the growth of a population according to a geometric law [or, expressed in a continuous form in equation (4.3), according to an exponential] faced with the growth of wealth according to a arithmetic law.[88] We have repeatedly used Malthus' law (4.3) in previous sections every time we needed an example of a linear differential equation.

---

[87] Leonardo Pisano, mathematician from Pisa (Tuscany) better known as Fibonacci, short for *fillio Bonacci* (or *Bonaccii*), which in middle-age Italian means Bonaccio's son; he introduced in Europe the Hindu-Arabic decimal number system in the XIII century and is considered the greatest European mathematician of the middle ages (see, Smith, 1958; Boyer, 1968).

[88] In his renowned *Essay on the Principle of Population*, Thomas Robert Malthus attributed, albeit in a rather a prioristic way, the cause of social poverty to an imbalance between

Growth in the presence of limited resources: Verhulst equation

Just before the mid-nineteenth century, the Belgian mathematician Pierre-François Verhulst (1838, 1850) proposed a correction to Malthus' growth model to take into account the limited nature of environmental resources, by

population growth and the development of resources. Malthus' theory was based on two basic hypotheses: (1) there are different laws for population growth, that develops at a constant rate according to a geometric progression which leads it to double every 25 years, and for resources, which grow at a constant speed only, according to an arithmetic progression; (2) the tendency of the working classes to react to an improvement in wages, and therefore of the level of well-being, by procreating. Malthus sustained the need for forms of demographic control as a solution to the problem of poverty, indicating some so-called preventive checks, such as celibacy or delaying marriage until an older age for fear of poverty, as a way to avoid that other causes, called positive checks, i.e. the checks that repress an increase which is already begun (Malthus indicated 11 of these, amongst which: war, famine, infant mortality, etc.), bring humanity to levels of survival able to avoid a destiny of poverty and hunger that the entire human race would face due to uncontrolled demographic development. Unfortunately, those that were most willing to understand this line of reasoning belonged to the middle and upper classes, while the poor are less disposed to demographic control, therefore the preventive check ends up acting on the upper classes, while the positive one acts on the lower ones. Malthus saw the precipitation of mankind towards hunger as inevitable: 'To prevent the recurrence of misery, is, alas! beyond the power of man' (Chapter 5); 'Famine seems to be the last, the most dreadful resource of nature. The power of population is so superior to the power in the earth to produce subsistence for man, that premature death must in some shape or other visit the human race. The vices of mankind are active and able ministers of depopulation. They are the precursors in the great army of destruction; and often finish the dreadful work themselves. But should they fail in this war of extermination, sickly seasons, epidemics, pestilence, and plague, advance in terrific array, and sweep off their thousands and ten thousands. Should success be still incomplete, gigantic inevitable famine stalks in the rear, and with one mighty blow levels the population with the food of the world. Must it not then be acknowledged by an attentive examiner of the histories of mankind, that in every age and in every state in which man has existed, or does now exist. That the increase of population is necessarily limited by the means of subsistence. That population does invariably increase when the means of subsistence increase. And that the superior power of population it repressed, and the actual population kept equal to the means of subsistence, by misery and vice?' (Chapter 7). (Malthus, 1798). Malthus' prediction that population growth would outstrip agricultural production and lead to mass starvation proved fallacious, as he didn't take into account the consequences of technological innovation, e.g. the introduction of mechanical aids, chemical fertilizers, hybridisation techniques and so on that, over time, would change for the best the dramatically insufficient resources' growth law. The first version of the *Essay*, conceived as a somewhat polemic response to the optimistic vision of society proposed by William Godwin, Jean-Antoine de Caritat de Condorcet, Adam Smith, and other economists, was published anonymously in 1798, in the form of a very brief compendium (around 50,000 words), and enjoyed immediate success; a second version followed in 1803, with the addition of a considerable amount of material, a real treatise of demographics, no longer anonymous, that made Malthus extremely famous.

introducing a constant: the carrying capacity. This, as we know (Chapter 12), is the maximum value that the number of individuals in a population can reach, compatible with the available environmental resources. In this and the next section we will be discussing some of the characteristics of Verhulst's growth model.[89]

Let us go back to Malthus' exponential growth model (4.3), which we write as follows:

$$n(t) = n(0)e^{rt} \tag{20.1}$$

in which $n(t)$ is the variable that indicates the value of the population, a function of time, $n(0)$ is the initial value of the population, and r is the constant growth rate of the population.

Differentiating (20.1), we have:

$$\dot{n}(t) = rn(t) \tag{20.2}$$

which, as is common practice, can be put into discrete form:[90]

$$n(t+1) = n(t) + rn(t) \tag{20.2'}$$

In the above case, $n(t)$ indicates the population at the beginning of the t-th interval of amplitude $\Delta t = 1$, a population that remains constant for the entire duration of the interval. A differential equation that describes a dynamical system in continuous time, such as (15.2), is called a flow, while the version in discrete time, such as (20.2') (finite difference equation) is known as a map.

At this point, let us introduce Verhulst's carrying capacity, a constant that we will indicate with k. When the number n of individuals in the population reaches the value k, growth itself has to become nil. However, this is not enough; k must, so to speak, display its influence even before n reaches the maximum permitted amount. It therefore has to show its presence by curbing the population's speed of growth to an extent that is proportional to the size of n.

---

[89] Verhulst's model was resubmitted to the attention of scholars and examined in depth only relatively recently, in a famous work of the biologist Robert May (1976).

[90] Very often, in economics, in demographics, and in the majority of other social sciences, it is common to put models into the form of finite difference equations, in contrast to physics, in which models are almost always formulated in continuous time. This is due to the fact that, as a norm, the model attempts to simulate historic series of data that are made up of observations (measurements) made at discrete time intervals.

This can be obtained by summing an appropriately selected negative term that is a function of the value of population n to the second member of the equation (20.2), and to its discrete version (20.2′). In this way we introduce a real feedback to the system: population growth is now controlled, not only by a mechanism of liberal growth (namely unperturbed), à la Malthus, but also by a regulation mechanism that puts itself in competition with the liberal growth, the action of which depends on environmental circumstances, i.e. on the relation between the population at a given time and the carrying capacity. The presence of this second term ends up destroying the linearity of the law of growth (20.2).

According to Verhulst's assumption, one of the many possible ones, Malthus' model in discrete form (20.2′) can be adapted with the introduction of a negative corrective term, proportional to the product of the growth rate r and the square of the value of the population n^2, as follows:

$$n(t+1) = n(t) + rn(t)\left(1 - \frac{n(t)}{k}\right) \tag{20.3}$$

Equation (20.3) can be usefully rewritten in a more concise and schematic form, that lends itself better to subsequent analyses:

$$x(t+1) = \lambda x(t)(1 - x(t)) \tag{20.4}$$

Map (20.4) is called a *logistic map*,[91] let us take a brief look at it.

We rewrite (20.4) in the form $x(t+1) = \lambda x(t) - \lambda x^2(t)$. The second member has two addends that act in opposition to one another to generate the value of the population at a later time: the first, $\lambda x(t)$, acts, in a certain sense, as an 'engine' that provokes the growth of the population according to a linear law; the second, $-\lambda x^2(t)$, being negative, acts, on the other hand, like a term that provokes the diminution of the population. It is the effect of the feedback, that in (20.3) and (20.4) was postulated in quadratic form, caused by the interaction between the state of the system, in this case the population, and the environment, in this case the available resources (Figure 20.1).

We also observe that, if we interpret $x(t)$ as a population, even if only in an abstract sense, this can never be negative; this means that (20.4) makes sense only when $0 \leq x(t) \leq 1$; in fact, if $x(t) > 1$, at a subsequent time we would have a negative $x(t+1)$. The usefulness of the transformation of

---

[91] Equation (20.4) can be obtained by replacing the variable in (20.3). In fact, if we assume $n(t) = k(1+r)x(t)/r$, we obtain: $x(t+1) = x(t)(1 + r - (1+r)x(t))$; if we also assume $\lambda = 1 + r$, the equation obtained can be simplified into (20.4).

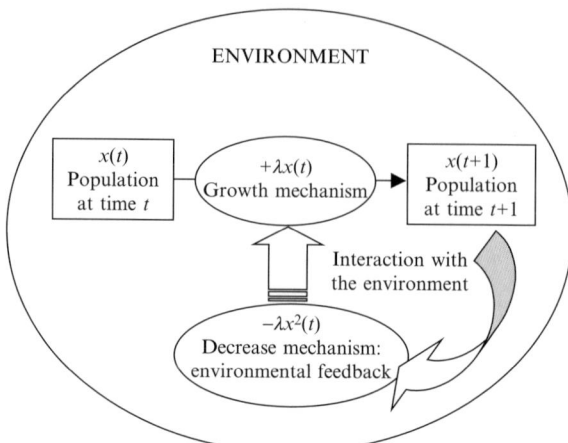

Figure 20.1 The logistic map as a representation of the interaction between a growth mechanism and a control mechanism exercised by environmental feedback.

(20.3) into the form of (20.4) lies precisely in the fact that with it we normalize, so to speak, the variable $x(t)$ between 0, the minimum value, and 1, the maximum value. This facilitates the analysis of the model, which can be conducted at a general level, without considering the particular value of the carrying capacity, but expressing, on the other hand, the population in percentage terms with respect to the permitted maximum number.

The logistic function

Let us return to Mathus's model in continuous time (15.1). By introducing the term k following the same line of reasoning used for the discrete form, we obtain the following first-order nonlinear differential equation with separable variables, a continuous correspondent of (20.3):

$$\dot{n}(t) = rn(t)\left(1 - \frac{n(t)}{k}\right) \tag{20.5}$$

Equation (20.5) satisfies the conditions of the theorem of existence and uniqueness and therefore provides a single analytic solution, if the initial conditions are given. Integrating (20.5), we obtain:

$$n(t) = \frac{k}{1 + Ce^{-rt}} \tag{20.6}$$

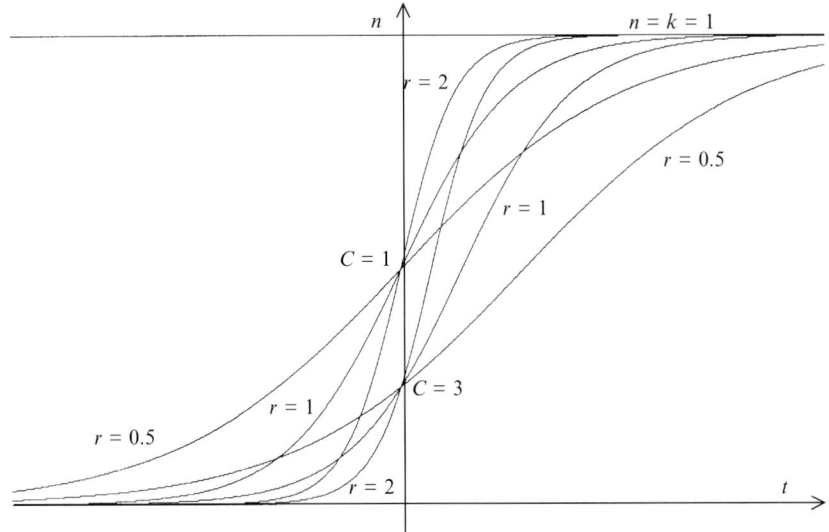

Figure 20.2 The logistic function (20.6) portrayed for several values of C and r.

in which the integration constant C depends on the initial value of the population $n(0)$ and is given by $C = (k - n(0))/n(0)$. Figure 20.2 shows several cases of the general integral (20.6) for three values of parameter r, combined with two values of constant C (parameter k is just a scale factor on the axis $n(t)$; it does not significantly affect the trend of the function and, therefore, we can overlook it without losing any important elements).

Equation (20.6) is known as the logistic function, or curve. As Figure 20.2 shows, the typical trend of the logistic curve is an S form: slow initial growth, followed by an acceleration in growth, the intensity of which depends on the value of parameter r, and then by a subsequent slowing down when it is close to the maximum permitted value $n(t) = k$, which thus constitutes an asymptotic limit of function (20.6).

It is interesting to observe how Verhulst, when he introduced the carrying capacity and the logistic function, or law which derives from the former in continuous form, conceived the idea on the basis of an analogy with physics, in perfect physicalist style (see Chapter 1), taking inspiration from the resistance that friction has on motion (Le Bras, 1999). The analogy was based on two replacements: the speed of a body with the value of the population, and the acceleration of the body with the rate of growth of the population, i.e. with the increase per unit of time divided by the value of the population: $r = \dot{n}/n$. Just as, in the presence of friction, the acceleration of a body subjected to a force is equal to the constant value that it would have in the absence of friction, diminished by a term proportional to the speed of the body, caused by the resistance to its progress opposed by the friction, in the same way, proposed Verhulst, the growth rate of a population, $\dot{n}/n$, is given by a constant term a, diminished by a term proportional to the value of the population by a factor b, i.e.

$$\frac{\dot{n}}{n} = a - bn$$

which is precisely the logistic function in continuous form (20.5), of which the logistic map (20.4) is the version in discrete time.

In the 1920s, the American biologist and demographist Raymond Pearl, one of the first scholars along with Alfred Lotka to use mathematical methods in biology, made extensive use of the law of logistic growth in a series of demographic surveys and showed that the law of logistic growth effectively described the evolution of numerous limited groups of appropriately defined human populations[92] (Pearl and Reed, 1920, Pearl, 1925). Pearl believed that this law could be used to calculate the future values of the entire world population, even going as far as predicting a number of individuals equal to 1.9 billion for 2000, which was a far cry from the values actually reached (Le Bras, 1999). The human population, therefore, if considered in its entirety, does not appear to follow Verhulst's logistic law.

According to Hervé Le Bras (1999), it is possible to construct basic demographic models that clearly lead to a type of logistic growth law, by considering individuals in a limited territory and making drastic, but fairly simple assumptions: that the individuals do not die in the time period considered, that they randomly choose where to live, and that, once established, they no longer migrate.

The following is a very simple example of a model of this nature. We divide a square territory with side k into k^2 unit side cells, and we presume that: (1) a certain number n of said cells is occupied by a single individual, (2) each individual has a probability a of generating another individual, (3) the latter individual randomly chooses one of the existing k^2 cells and moves there, if it is empty, or perishes if he finds it occupied. The model assumes that, at each moment in time, the (instantaneous) speed of growth of the population is interpreted as a probability of growth and is proportional to three factors: (1) the probability a that a new individual is born, (2) the percentage of free cells, (3) the value of the population, which is equal to the number n of occupied cells. The percentage of free cells is equal to the difference between the total number k^2 of cells and the number n of occupied cells, all divided by

---

[92] For this reason, the logistic law is sometimes also referred to as Pearl's law.

k^2. Using the compound probability theorem, the probability of growth, indicated by H, can be expressed as follows:

$$H = an(k^2 - n)/k^2$$

The growth rate $\dot{n}/n$ of the population, which is expressed as the growth rate of the occupied cells, thus becomes equal to the probability of growth H divided by the total population n, i.e.:

$$\frac{\dot{n}}{n} = \frac{H}{n} = a\left(1 - \frac{n}{k^2}\right)$$

The equation obtained has the same form as (20.5) and therefore represents a law of logistic growth, the law according to which the population throughout the territory considered in the model grows.

The logistic curve constitutes a relatively simple growth pattern that can be applied to a number of different contexts: in any situation in which we wish to model the nonlinear growth of a variable in the presence of a factor that acts, in the same way as the carrying capacity, establishing itself as the maximum value of the variable and limiting the speed of growth.

In times more recent than to those in which Pearl made his analyses, the logistic function has found application, amongst others, in models that describe the evolution of territorial systems (Allen et al., 1978; Wilson, 1981; Allen, 1997a). However, it has been noted that, in applications of this nature, the hypothesis of a constant limiting factor k is rather implausible, because it is difficult to exclude a priori effects due to the interaction between the value x of the population and the limiting factor k. In fact, a human population adapts itself to the environment and, in this way, moves the limits that have been imposed on growth by using new means of production, by improving productivity and, in general, through technological innovation. For reasons of this nature, therefore, the maximum number of individuals that can live together in a unit of surface area is very low in a rural area, providing that only agricultural activities are practised therein, in comparison to an urban area, in which the availability of food resources for each individual is much higher than what would be supplied by the primary sector in the same urban surface area.

Several variants of the model of logistic growth have been repeatedly applied to the description of the processes of diffusion, in particular in social sciences, such as for example, the diffusion of epidemics in a population or the diffusion of a commercial product in a market. In general they are processes in which there is an initial period of slow growth of the diffusion, followed by an increase in the rapidity of said diffusion and by a final slowing down, when the diffusion is close to a saturation level (almost the entire population is ill, in the example of the epidemic, or almost everyone has purchased that particular product, in the case of the diffusion of a product).

Different types of approach, in terms of assumptions and viewpoints, have been proposed, particularly to describe the diffusion of technological innovation in a territory (Fischer, 1989; Sonis, 1990b, 1995; Frenkel and Shefer, 1997). In this context, the diffusion of innovation is defined as a process, which, over time and in various ways, causes an innovation to be transferred within a group of individuals (businesses), in a given social system. It can be compared to a type of communication that transfers information and ideas.

The use of the logistic model in its basic form to model the diffusion of innovation calls for several limitative assumptions that are at the basis of several criticisms (Davies, 1979). They contest, for example, that the influence of a new technology can occur only in the long term: the potential user adopts it, in fact, only when the return satisfies his expectations, i.e. only after the superiority of the new technology with respect to the old has been proven. This means that for a certain length of time there is a sort of technological pluralism, which does not guarantee that the replacement of the old with the new technology is merely a question of time, as is assumed in the basic logistic model. Furthermore, the latter assumes that the potential users only have the option of adopting or not adopting the innovation, without considering intermediate stages such as its study, analysis and assessment, the acquisition of know-how, etc., and also assumes that the innovation doesn't change during the diffusion process and that it is not affected by the diffusion of other innovations.

Despite the criticisms of its intrinsic limitations, however, some variants of the logistic model have been used repeatedly, usually expressing the diffusion of the innovation in equations such as the following (Allen, 1997a):

$$\dot{N}(t) = (a + bN(t))(\bar{N} - N(t))$$

in which $N(t)$ is the number of individuals (or businesses) that adopt the innovative technological element at time t, $\bar{N}$ is the total number of individuals that can adopt a given technological innovation and $(a + bN(t))$ is the diffusion coefficient of the given technological innovation.

In the following section we will be looking at the version of Verhulst's law in its discrete form. The evolution that it provides, as we will see, is considerably different to that of the version in its continuous form, and in certain conditions, becomes chaotic.

21 A nonlinear discrete model: the logistic map

Introduction

In this section we want to continue our discussion of the growth model presented in Chapter 20, the logistic law, illustrating the dynamics of the version in discrete time (20.4). We are particularly interested in showing how a simple nonlinear growth model such as (20.4), that determines the logistic growth of a population, can in some cases generate chaos. In Chapter 22 we will be illustrating various examples produced by means of numerical simulations, as well as some applications of the logistic map in regional sciences. Readers interested in a complete description of the properties of the dynamics of the logistics map can find detailed and in-depth discussions of the mathematical aspects, which in this text are only briefly mentioned, in two fundamental articles. These are works that have reproposed the logistic map to scholars in recent decades, more than a century after Verhulst introduced it: the work of Robert May (1976), the first, in recent years to examine the properties of the map, and that of Mitchell Feigenbaum (1978), whose content is more technical, in which specific aspects are looked at in more depth and generalized.

Let us reconsider the logistic map in the form of (20.4). For the sake of simplicity we will rewrite it, without altering its substance, replacing the dependence on time with an index k that 'counts' the successive iterations:

$$x_{k+1} = \lambda x_k (1 - x_k) \qquad (21.1)$$

In this way, the value of variable x at iteration (time) $k+1$ depends, according to (21.1), on the value that the same variable x has at the previous iteration k. Note that, as discussed in Chapter 20, the values of x_k in (21.1) fall between 0 and 1.

Equation (21.1), like (20.4), is a quadratic map, whose sequence of x_k values, for a given value of λ, provides the evolution of the population at discrete time intervals. The phase space of the dynamical system (21.1) is made up of only one variable, x_k, and the set of the successive values of x_k is the population's orbit.

As is immediately apparent in (20.4) and (21.1), if during the course of the evolution, the value $x_k = 0$ is reached, then all the subsequent values of x are still 0. On the other hand, we note that the other extreme of the permitted interval for the x_k, i.e. $x_k = 1$, is also a particular value: if $x_k = 1$ is reached, the subsequent value is $x_{k+1} = 0$, and from then on all subsequent x_k are equal to 0.

The fact that the population x falls between 0 and 1 also imposes a limitation on the values of the growth parameter λ, which, on one hand, must not be negative, so that the population never becomes negative, and on the other hand, must not exceed 4, as we will see very soon, so that the population doesn't exceed 1, the maximum permitted value. More specifically, depending on the value of λ, the following cases can arise:

(a) When $\lambda < 0$, the population becomes immediately negative, which, obviously, is nonsensical, thus negative values of λ are not acceptable.
(b) When $\lambda = 0$, the minimum acceptable value, the population becomes nil at the first iteration and then remains so.
(c) When $0 < \lambda < 4$, the population evolves, always maintaining a value in the range between 0 and 1. More precisely, as we will see at the end of this paragraph, an attractor exists which, depending on the value of λ, acts in three different ways: (1) it drives the population towards one asymptotical final value only, (2) it makes the population oscillate on a set of asymptotical values or (3) it drives the population to take only values within a certain interval, whose limits, lying between 0 and 1, depend on the value of λ. One can show that, in the latter case, the attractor is chaotic and that the set of the values that the population can assume in that interval is dense, i.e. given any two values (states), the population, in its evolution, will sooner or later assume a value that falls between them. $0 < \lambda < 4$ is the most interesting case which we will be discussing in the next paragraph.
(d) When $\lambda = 4$, we have another limit case, which is still acceptable, whose particularity lies in the fact that, for any initial state of the population between 0 and 1, the attractor is also the unit interval $(0,1)$; in other words, the population evolving from any initial state can take any value between 0 and 1.[93]

---

[93] If the population, during the course of its evolution, reaches the value $x_k = 0.5$, then at the next step the map (21.1) gives $x_{k+1} = 4 \times 0.5 \times 0.5 = 1$. But if x_{k+1} becomes 1, the map (21.1) means that the value of the population at the next step will be 0, and from then on, will always be 0.

(e) When $\lambda > 4$, the speed of growth of the population is too fast. This means that when x_k reaches values included in a certain neighbourhood of 0.5, the amplitude of which depend on λ, at the next iteration, x_{k+1} will exceed 1, the maximum permitted value. Let us clarify using an example, assuming that $\lambda = 4.1$. If, for example, during the course of its evolution, the population reaches the value $x_k = 0.49$, then at the next step, the value $x_{k+1} = 4.1 \times 0.49 \times 0.51 = 1.02459$ will be obtained, which is outside the interval permitted for x. Therefore values of λ higher than 4 are not acceptable.

Before proceeding with the analysis of the dynamics of (21.1), let us remind ourselves of the concept of fixed point of a function: we will use it immediately afterwards, applying it to (21.1).

The iteration method and the fixed points of a function

One of the simplest methods to identify the approximate roots of an equation is that known as the iteration method. It involves writing the equation for which a root is being sought in the following form, where an unknown x has been isolated as the only term in the first member:

$$x = f(x) \qquad (21.2)$$

assigning a value x_1 to the x value, then calculating $f(x_1)$, and then assuming the following values of $x_2 = f(x_1)$, $x_3 = f(x_2)$, and so on, according to the iterative map, which is just a simple discrete dynamical system:

$$x_{n+1} = f(x_n) \qquad (21.2')$$

For example, using a simple pocket calculator, an approximate root for the equation $x = \cos x$, that cannot be solved in exact terms, can be identified by proceeding as follows. Assume $x_1 = 1$ (radians), calculate its cosine, then the cosine of the cosine, then the cosine of the cosine of the cosine and so on. As the iterations progress, the values of x_n tend to stabilize, defining a real number with increasing precision. The following values are subsequently obtained:

$$x_1 = 1$$
$$x_2 = \cos x_1 \cong 0.54030$$
$$x_3 = \cos x_2 \cong 0.85755$$
$$x_4 = \cos x_3 \cong 0.65429$$
$$x_5 = \cos x_4 \cong 0.79348$$
$$x_6 = \cos x_5 \cong 0.70137$$
$$x_6 = \cos x_5 \cong 0.70137$$
$$x_7 = \cos x_6 \cong 0.76396$$
$$\ldots$$
$$x_{20} = \cos x_{19} \cong 0.73894$$
$$x_{21} = \cos x_{20} \cong 0.73918$$
$$x_{22} = \cos x_{21} \cong 0.73902$$
$$x_{23} = \cos x_{22} \cong 0.73913$$

As we can see, from the twenty-first iteration onwards, the sequence provides values whose first three decimal places no longer change. The limit that the sequence of numbers generated by iterating (21.2') tend towards, if it exists and is finite as in the example of the cosine, is an approximation of the root X we were seeking:

$$X = \lim_{n \to \infty} \cos x_n \qquad (21.3)$$

The value X defined by the limit (21.3), the root of equation (21.2), is called a fixed point attractor of $f(x)$.

It can be shown that the sequence $x_{n+1} = f(x_n)$ converges towards a number X, a fixed point attractor, only if in a neighbourhood of X we have $|f'(X)| < 1$.

The above method has an effective geometric interpretation. The graphs of the two members of (21.2) are traced separately: $y = x$, the bisector of the first quadrant, for the first member, and $y = f(x)$, for the second member (Figure 21.1). The fixed point attractors (like point A in Figure 21.1) can be identified graphically in some of the points of intersection of the function $y = f(x)$ with the bisecting line of the first quadrant $y = x$.

In addition to the fixed point attractors of the abscissa X, for which $|f'(X)| < 1$, points of a different nature can also be found between the points of intersection of the function $y = f(x)$ with the straight line $y = x$. These are fixed point repellers: those points of abscissa X, common to the $f(x)$ and the straight line $y = x$, in a neighbourhood of which we have

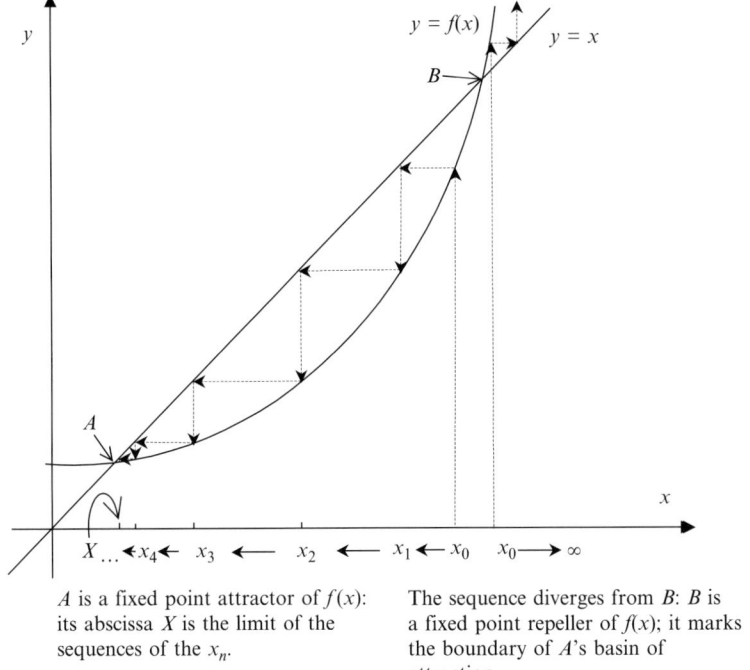

A is a fixed point attractor of $f(x)$: its abscissa X is the limit of the sequences of the x_n.

The sequence diverges from B: B is a fixed point repeller of $f(x)$; it marks the boundary of A's basin of attraction.

Figure 21.1 The fixed points of a function $f(x)$.

$|f'(X)| > 1$ (like point B in Figure 21.1). From the fixed point repellers, the $x_{n+1} = f(x_n)$ sequence, initiated by assuming an x_1 value close to X, increasingly distances itself as n increases. The X abscissa of a fixed point repeller, therefore, cannot be reached by seeking the limit of the sequence of the x_{n+1}; it can be obtained by using particular devices, which we are not going to discuss here, which enable us to go backwards in a numerical sequence. In general, fixed point repellers act like a border between the basins of attraction of attractor points.

Basically, fixed point attractors are the points at which the function $f(x)$ meets the bisector $y = x$ with an inclination (the derivative's value at that point) less than that of the bisector itself, which is 1; fixed point repellers, on the other hand, are the points at which $f(x)$ meets the bisector, with an inclination higher than that of the bisector itself.

The search for the fixed point attractors or repellers of a function is only one of the aspects of the eternal problem of algebra: the search for the roots of any equation. With this technique, in practice, the search for the roots leads to the analysis of a particular sequence generated by a function or, in some cases to the examination of a graphic representation of the same function, while the analysis of the derivative of the function enables the nature of the fixed points to be analysed.

Let us close this short diversion into the fixed points of a function[94] and return to our discussion of the logistic map, now paying particular attention to its fixed points, which we will be applying as mentioned previously.

The dynamics of the logistic map

As we mentioned at the beginning of this section, the successive values of the x_k variable, i.e. the values of the population at successive intervals obtained by iterating the logistic map (21.1), define a set of points in a one-dimensional phase space, that we have called the orbit of the population. Iterating the map means following the progress of subsequent x_k, identifying, if such exist, the fixed point attractors or repellers of the second member function of (21.1), which in the rest of the section we will often indicate simply as $f(x) = \lambda x(1 - x)$.

A fundamental characteristic of the logistic map is that different values of the growth parameter, which fall between 0 and 4, not only determine different orbits, but can give rise either to different configurations of fixed point attractors or to a chaotic evolution that does not have any fixed point attractor. In the remainder of this section, we will be looking at the qualitative aspects of the different situations that arise as the parameter λ increases from 0 to 4.

It is immediately apparent that there is a fixed point that does not depend on the value of λ: $x = 0$. Let us indicate the abscissa of this (first) fixed point with X_1. It is an attractor or repeller, as we have seen, depending on whether the derivative of $f(x)$:

$$f'(x) = \lambda - 2\lambda x = \lambda(1 - 2x) \tag{21.4}$$

---

[94] In the text we have provided a very succinct and rather approximate version of the iteration method and of the theory of fixed points, both at the theoretical and applicative level, limiting ourselves to aspects that are relevant to our subject area only. For example, the case in which we have $|f'(X)| = 1$, i.e. the case in which the straight line $y = x$ is tangent to the function $y = f(x)$, would require further specific considerations that we do not present here. For a more complete and exhaustive discussion the reader should consult texts regarding dynamical systems and numerical calculus (for example, Tricomi, 1965; Collatz, 1966; Householder, 1974; Zeldovich and Myškis, 1976; Von Haeseler and Peitgen, 1988/1989; Devaney, 1992).

calculated when $x = 0$, is, respectively, less than or more than 1. However, (21.4) tells us that $f'(x)$, calculated when $x = 0$, is exactly equal to the growth parameter λ; thus we can simply conclude that $X_1 = 0$ is the fixed point attractor, if $\lambda < 1$, or fixed point repeller if $\lambda > 1$.

If $\lambda < 1$, therefore, $X_1 = 0$ represents a state of stable equilibrium. In this case model (21.1) represents a population that becomes increasingly less numerous and tends asymptotically to extinguish itself, a situation towards which the said population is inexorably led by too low a growth parameter. On the other hand, if $\lambda < 1$, it is evident that the limit of the sequence of the x_k generated by (21.1) can be none other than 0. In fact, the product $x_k(1 - x_k)$ is always less than 1 (remember that $0 < x_k < 1$) and, at each iteration, it decreases further because it is multiplied by λ.

When $\lambda = 1$, the derivative (21.4) does not help us understand the nature of the fixed point, because we obtain $f'(0) = 1$, which does not tell us whether the iterations converge or not. Nevertheless, by iterating (21.1) with $\lambda = 1$, it is immediately apparent that the sequence of the x_k converges towards 0 starting from any initial value that falls between 0 and 1. Thus also when $\lambda = 1$, $X_1 = 0$ is a fixed point attractor. As there are no other roots of equation (20.4), we conclude that when $\lambda \leq 1$, model (21.1) always leads to the extinction of the population.

If, on the other hand, $\lambda > 1$ (21.4) always gives $f'(0) > 1$. This assures us that $X_1 = 0$ is a fixed point repeller, i.e. a point of unstable equilibrium. When $\lambda > 1$, the model represents the case of a population that, starting from a low initial value, close to 0, tends to grow.

However, now the growth is maintained only until when, for sufficiently large x_k, other elements become important that slow it down and invert it. For x_1 close to 0 and $\lambda > 1$, in fact, *the first* terms of the sequence of the x_k are increasing ones, because in $x_k(1 - x_k) = x_k - x_k^2$ the positive term x_k is dominant with respect to the negative one $-x_k^2$. The situation is different, however, when, as the population grows, the value of the second order term grows, or when the iteration starts from a high value of x_1 that is therefore not close to 0. In the latter two cases, the subsequent progress of the population depends on the particular value of λ. In fact *whether and where there is* equilibrium between the growth mechanism, the positive first order term x_k, and the decremental one, the negative second order term $-x_k^2$ depend on the said value of λ. The situations that are generated are different and need to be examined further. Let us therefore illustrate the different cases that arise as the value of λ passes from 1 to 4; for the sake of simplicity, the discussion will be mainly qualitative. Interested readers can find useful mathematical elements to support the descriptions below in the bibliographical references cited at the end of this section.

Case 1: $1 < \lambda \leq 3$. As we have seen, point $X_1 = 0$ is now a repeller. However there is now another fixed point that we can obtain by writing (21.1) in the form $x = f(x)$, obtaining a simple second degree equation:

$$x = \lambda x(1-x) \qquad (21.5)$$

It has a second root X_2, that can be obtained using basic algebra, the value of which depends on λ. When $1 < \lambda \leq 3$, it can be seen after several operations that (21.4), calculated in X_2, now gives $f'(X_2) < 1$, therefore we can conclude that X_2 represents a fixed point attractor, regardless of its specific value.

In Case 1, therefore, the logistic map (21.1) describes the evolution of a population which, as the iterations progress, tends asymptotically towards a value X_2. It is important to note that X_2 does not depend on the initial value from where the evolution started, but only on the value of the growth parameter λ. As the iterations progress, in a certain sense, the system loses its memory of the initial state; for any initial value of the population, the final outcome (asymptotic) is always the same: X_2.

As λ increases from 1 to 3, two effects are observed: on one hand, the fixed point X_2 increases with continuity until the value $X_2 = 2/3$, which is reached when $\lambda = 3$; on the other hand, the rapidity with which the population tends towards the attractor X_2 decreases, because, as λ gets closer to 3, the value that (21.4) generates for $|f'(X_2)|$ increases and gets increasingly closer to the critical value 1. When $\lambda = 3$, we have $X_2 = 2/3$, but in this situation X_2 loses its characteristic of attracting the value of the population x_k, because (21.4) gives $|f'(2/3)| = 1$. The value $\lambda = 3$ is therefore a critical value that we will call λ_1: when $\lambda = \lambda_1$, the sequence of the x_k ceases to converge at an asymptotic value and we have the transition to a dynamics different to that of Case 1.[95]

---

[95] To be more precise, for $\lambda = 3$ the sequence (21.1) tends towards the limit $x_\infty = 2/3 = 0.66666\ldots$, but this occurs very slowly right from the start of the iterations, and at a speed that continues to diminish as the iterations progress; for example, after 10000 iterations the sequence of the x_k has still not stabilized at the third decimal place: $x_{9999} = 0.669014$; $x_{10000} = 0.664303$.

Note, incidentally, that the value $\lambda_1 = 3$, after which more than one fixed point exists, can find a justification on an algebraic basis. Consider for period 2, in place of (21.5), the second iterate of the logistic map, i.e. the following equation, where the composed function $f(f(x))$ is introduced:

$$x = f(f(x)) = \lambda \times \lambda x(1-x) \times (1 - \lambda x(1-x))$$

and look for the conditions that the parameter λ has to satisfy in order to let real roots of the fourth degree polynomial in x, so obtained, exist. It can be easily seen that the roots are only real for $\lambda \geq 3$ (see Strogatz, 1994).

Case 2: $3 < \lambda < 3.56994\ldots$ When λ enters this third interval, the fixed point X_2, whose value still depends on λ, becomes a repeller (unstable equilibrium), and two new attractor points are generated on opposite sides with respect to X_2. The x_k variable now tends to asymptotically oscillate between two new values, one higher than X_2, the other lower, that alternate periodically, iteration after iteration, one for even iterations and one for odd ones. The two new states that are formed (fixed point attractors) represent, as a whole, a single attractor, so to say 'split' into two values.

After several operations, however, it is apparent that the splitting of the attractor into two new attractor points is maintained only until λ, which increases, reaches a second critical value, given by:[96]

$$\lambda_2 = 1 + \sqrt{6} \cong 3.44949$$

When $\lambda > \lambda_2$, each of the two fixed point attractors becomes a repeller and gives rise to ('splits into') two new fixed point attractors, as occurred for X_2 when λ has exceeded the critical value $\lambda_1 = 3$. The population now tends to oscillate between four stable states (naturally we are still talking about asymptotic stability). The specific values of these four states depend only on the growth parameter λ and not on the initial value of the population. As soon as we have $\lambda > \lambda_2$, we therefore have four fixed point attractors and four fixed point repellers alternated with the fixed point attractors: the four repeller points mark the basins of attraction of the four fixed point attractors. For example, for $\lambda = 3.5$ and initial value $x_1 = 0.2$, after 25 iterations, the population system oscillates between the following four values: $0.5009\ldots, 0.8875\ldots, 0.3828\ldots$ and $0.8269\ldots$

When λ exceeds λ_2, therefore, the number of final states between which the population oscillates doubles, and consequently, the oscillation period doubles. On further growth of λ, we observe that the splitting of the final states occurs again, in the same way, in correspondence with a new critical value λ_3 and then again, in correspondence with another critical value λ_4. To be more precise, what we are observing is that the splitting of the states

---

[96] Here again, as for $\lambda_1 = 3$ (see footnote 92), the critical value λ_2 can be computed algebraically as the minimum value that the parameter λ has to reach in order to let the roots of the 16th degree polynomial equation that we get using the fourth iterate (the composed map):

$$x = f(f(f(f(x))))$$

be real (hereinafter we will write irrational numbers either showing the first figures followed by three dots or preceded by the symbol $\cong$ as shown in the text).

repeats in correspondence with an infinite sequence of critical values λ_n of the growth parameter λ. In this way we have oscillations first between four states (2^2), then between eight states (2^3), then between sixteen states (2^4), etc., which means that the oscillation period of the population between the final states (asymptotic), as λ increases, first quadruples, then is multiplied by eight, then by sixteen, and so on. For each $\lambda_n < \lambda < \lambda_{n+1}$, therefore, the population evolves over time, tending to bring itself to a condition in which it oscillates between 2^n values that follow each other cyclically, with a period represented by 2^n values of x_k.

The successive doubling of the oscillation period takes place according to the same mechanism (a fixed point attractor becomes a repeller, generating two fixed point attractors, one smaller and one larger than the former, which in turn split, and so on), for values of λ_n, that continue to increase but that are increasingly closer to one another. Following several operations that are not shown here, the first values of λ_n can be obtained as shown in Table 21.1.

The limit of the sequence of the λ_n is an irrational number λ_∞, an approximated value of which is $\lambda_\infty \cong 3.56994$.

When $3 < \lambda < \lambda_\infty$, the configuration of the attractor, despite the fact that it is split into increasingly numerous states, only depends on the value of λ and not on the initial value of the population. In other words, also in Case 2, it is the value of the growth parameter λ that has full control of the system's evolution, regardless of the initial value of the population. What happens is basically something that we have already met on a number of occasions in

Table 21.1 Logistic map: splittings, periodicities and critical values λ_n of the growth parameter

Splittings n	Periodicity 2^n of cycles	λ_n
1	2	$\lambda_1 = 3.000000$
2	4	$\lambda_2 \cong 3.449490$
3	8	$\lambda_3 \cong 3.544090$
4	16	$\lambda_4 \cong 3.564407$
5	32	$\lambda_5 \cong 3.568759$
6	64	$\lambda_6 \cong 3.569692$
7	128	$\lambda_7 \cong 3.569891$
8	256	$\lambda_8 \cong 3.569934$
9	512	$\lambda_9 \cong 3.569943$
10	1024	$\lambda_{10} \cong 3.569945$
...	...	...
∞	∞	$\lambda_\infty \cong 3.569945672$

this book: the evolution brings the system asymptotically towards a state, whether such is complex or simple, and as the system gets closer to said state it loses its memory of the initial state. It is obvious that the loss of memory of the initial state is contrary to the sensitivity to the initial states that we have mentioned in the previous sections as one of the conditions required for evolution according to a chaotic dynamic. In Cases 1 and 2 we are in a situation of stability in which the dynamics of the population evolves, always tending, for a given value of λ, towards a unique final configuration, regardless of any perturbation (or variation) of the initial state.

As the number n of the states between which the oscillation occurs increases, the period becomes increasingly longer, because the number of iterations after which one of the x_k of the sequence generated by (21.1) repeats a value already taken in a previous iteration is increasingly higher. A very long period, therefore, requires the evolution of the model (21.1) to be followed for numerous iterations before being observed, which makes it more difficult than a short period to recognize, for which the values of the x_k in the sequence repeat after a short interval of time. The growing difficulty in identifying a periodicity in the sequence of x_k, as the parameter λ grows, means that, as λ increases, the evolution of the population according to the model of the logistic map becomes increasingly less predictable, insofar as the periodicity of the values appears to be increasingly 'widened' into a sequence of values in which something similar to a limit or a recurrence is increasingly difficult to recognize.

Recognizing the existence of recurrences, if any, in real systems, when such are characterized by very long periodicities, even in cases in which a model can be constructed according to a relatively simple deterministic law, such as (21.1), can generally prove to be very difficult. Let us assume, for example, that we have a certain number of values x_k of the logistic map, and that we interpret them as a historic series of data observed in a real system. If their number is sufficient to identify a period, then we can, with good approximation, deduce the value of the growth parameter and confirm that the logistic map reproduces its trend. If however the period is not evident, not only can we not establish if an attractor is present, we cannot even be sure that the data come from a deterministic model, and are not, on the contrary, random. Transferred into a situation in which the numbers are generated by a deterministic model (i.e. by a 'law'), this is the most severe difficulty that is encountered when analysing historic series of data, when we are not able to identify any recurrence (or 'law') with any certainty, a difficulty similar to the one we mentioned in Chapter 19.

In Case 2, we are faced with a situation in which the recurrence exists, even if such is not easily recognizable. In Case 3, however, we will encounter a situation in which different initial states give rise to different evolutions and, in particular, initial states that are close to one another can generate, over time, states that are very distant from one another, by amplifying a small initial difference. For now we are in a particular situation: we do not have chaos, but we have an *ordinary* attractor made up of a number of states (isolated fixed point attractors) that increases as λ increases, where, when $\lambda < 3$, there was a single asymptotic state. The population system, in this condition, creates new limit states, as λ grows, but always remains in a condition in which it loses its memory of the initial state of the evolution.

Case 3: $\lambda \geq \lambda_\infty \cong 3.56994$. The situation of successive doubling of the number of states between which the population stably oscillates, and of the consequent successive doubling of the oscillation period as λ grows, described in Case 2, remains until the critical value $\lambda_\infty \cong 3.56994$ is reached. When λ reaches and then exceeds λ_∞, any initial value x_1 gives rise to aperiodic trajectories, regardless of the length of the sequence of the x_k. However, there are still intervals of the values of λ, even in this Case 3, for which the periodicity appears again; one might even show that the values of λ for which the dynamics is non-periodic are dense, in the sense that there is a value of λ which gives a periodic orbit arbitrarily close to each value of λ which gives a non-periodic orbit. We will be discussing this characteristic in Chapter 22.

The fixed points identified in Cases 1 and 2, which constituted the attractor and, in their entirety, defined a set of isolated points, now become, for a given value of $\lambda \geq \lambda_\infty$, so numerous and so close to each other that they constitute an infinite set of isolated points distributed in an interval whose amplitude increases as λ increases, but that remains contained within the interval between 0 and 1 (see Chapter 22).[97] The dissolving of the periodicity, as a consequence of its limitless increase in length, can be interpreted as if, for values of $\lambda \geq \lambda_\infty$, the successive splits have become infinite and have given rise to an infinite set of fixed point attractors and repellers. This means that where, when $\lambda < \lambda_\infty$, an orbit travelled a finite number of states, before passing through a given point again, now, when $\lambda \geq \lambda_\infty$, the system's orbit should pass through an infinite number of points (states) before it passes again through a point (state) that it has already

---

[97] In correspondence of the critical value $\lambda = \lambda_\infty$, beyond which we have chaos, the successive states x_k tend towards a strange attractor, whose structure is similar to that of a fractal known as the 'Cantor set' (Falconer, 1985, 1990; Schroeder, 1991).

touched; in practice it will never pass through it again. In other words, when $\lambda \geq \lambda_\infty$, there is no periodicity in the sequence of the x_k.

When $\lambda \geq \lambda_\infty$ there is no stability at all, not even asymptotic stability. The growth rate is so high that the evolution of the population no longer recognizes any stable set of recurrent states, towards which the orbits that originate from different initial values converge. In this condition, the value of the population, at a given iteration, depends on the initial value from which the map started to evolve. Different initial values with the same λ now lead to different values at the n-th iteration, where, in Cases 1 and 2, the differences between orbits that started from different values tended to disappear as the iterations progressed. In this sense, we can say that the dynamics of the system now preserve the memory of the initial states.

In other words, when $\lambda \geq \lambda_\infty$, the value x_k of the population, which always falls between 0 and 1, appears to evolve in an unpredictable way, in the sense that we discussed in Chapter 15, i.e. without enabling any ordinary attractor to be identified. For values of λ close to 4, the maximum permitted value, x_k can pass repeatedly from a value close to 0, i.e. almost from extinction, to a value close to 1, i.e. to the maximum permitted limit, and vice versa. As there is no periodicity in the sequence of the x_k, each state produced by the iteration is calculated by physically making the calculations, iterating the map cycle after cycle. We cannot now know a future state without calculating *all* of the previous states, one by one. The deterministic law (the model) is still the same, (21.1), but now the visibility that we have of the future starting from a particular state is limited to the state immediately afterwards. If we changed the state, even by a little, the next iteration could give very different values from those that would have followed in the present unperturbed state (see the example illustrated in Table 15.2). In Case 3, the dynamics of the population system are sensitive to initial conditions, in line with our discussions in previous sections regarding chaotic evolution. In this sense, we therefore talk about the unpredictability of the future.

When $\lambda \geq \lambda_\infty$, a situation occurs in which there is sensitivity to the initial conditions, and in which the orbits of the state variable x_k (the value of the population) remain confined to an interval, the extremes of which fall between 0 and 1:[98] this is a chaotic dynamics and the interval allowed for the states is a chaotic attractor. The value of the growth parameter λ_∞ may

---

[98] For $\lambda = 4$, the interval of variability of x_k is the entire closed interval [0, 1]. If $\lambda = 4$, in fact, it can occur that subsequent iterations actually bring the x_k to the value of 1, which is necessarily followed only by 0s, i.e. the extinction of the population. For example, iterating (21.1) from $x_1 = 0.5$ and with $\lambda = 4$, the sequence $x_1 = 0.5$, $x_2 = 1$, $x_3 = 0$, $x_4 = 0$, etc. is obtained.

therefore be seen as the border beyond which unpredictability occurs: the edge of chaos.

Before concluding this discussion on the dynamics of the logistic map, we would like to highlight another fundamental peculiarity of the same. As λ increases beyond the value of 3, we said that we only encounter oscillatory dynamics between states in even numbers: the sequence of powers of 2. When $\lambda \geq \lambda_\infty$, these states, all originating from an infinite sequence of doubling of the oscillation period, represent an infinite set. However, what happens is that from a particular threshold value of the growth parameter $\lambda^* \cong 3.6786$, odd periods also appear; that is to say that the successive x_k values of the population start to oscillate between an *odd* number of states, exiting from the set of the periodicities defined by the powers of 2.

Initially, the odd periods that appear are very long, then, as λ grows, they decrease until when, for $\lambda_{\text{period}-3} = 1 + 2\sqrt{2} \cong 3.8284$, the population oscillation between three states only occurs (in this case, we say that we have a period-3 cycle).[99] According to a well-known theorem of Tien-Yen Li and James Yorke (1975: 'Period three implies chaos'), intermixed among the values of λ for which the trajectories are aperiodic, there are also values of λ that give rise to sequences of x_k that oscillate (obviously we still mean 'tend asymptotically to oscillate') on *any* finite number of states. In other words, when $\lambda > \lambda^*$, we can have periodic evolutions with periods of any length (May, 1976). According to the study by Li and Yorke, one of the first important works dedicated to chaos, prior to the same study by May that brought the logistic map to the attention of intellectuals again, the actual presence of oscillations between three states guarantees the onset of chaos.[100]

---

[99] The exact expression of the $\lambda_{\text{period}-3}$ critical value which gives rise to a period-3 cycle is basically an algebraic problem related to the conditions for the existence of real roots of the 8th degree polynomial equation that we get if we write the third iterate of the logistic map: $x = f(f(f(x)))$, as we already observed for λ_1 and λ_2 in notes 92 and 93, p. 000 (see also Saha and Strogatz, 1995; Bechhoeffer, 1966; Gordon, 1996).

[100] A somewhat similar theorem had been proved 11 years earlier by the Ukrainian mathematician Oleksandr Mikolaiovich Sharkovsky in 1964 (see also, for example, Devaney, 1989, 1992; Tufillaro, Reilly and Abbott, 1992). Although the result did not attract a great deal of interest at the time of its publication, during the 1970s other surprising results were proved which turned out to be special cases of Sharkovsky's theorem (in Western languages often mis-spelled as the Sarkovsky theorem). As Gleick observes, this 'was a typical symptom of the communication gap between Soviet and Western science. Partly because of the language, partly because of restricted travel abroad on the Soviet side, sophisticated Western scientists have often repeated work that already existed in the Soviet Union; on the other hand, it also inspired considerable bewilderment, because much of the new science was not so new in Moscow. Soviet mathematicians and physicists had a strong tradition in chaos

The phenomenon of the progressive doubling of the period when a parameter is changed, which can be traced back to a model such as the logistic map, has been observed in numerous real systems: physical ones (Serra and Zanarini, 1986; Moon, 1992), economic ones (Barkley Rosser, 1991; Nijkamp and Reggiani, 1993; Puu, 2000) and spatial ones (Nijkamp and Reggiani, 1992a, 1993, 1998), characterized by the presence of significant elements of nonlinearity.

We conclude our discussion on the theoretical aspects of the logistic map as one of the simplest examples, and at the same time, most emblematic of the 'first route to chaos'[101] here. In Chapter 22 we will be looking at several numerical examples of the logistic map and some of its applications in regional sciences. The interested reader can find other discussions of a more general nature in the fundamental articles by May (1976), the first scholar to study the dynamics of the logistic map, and by Feigenbaum (1978, 1980), who conducted a very in-depth study on the general properties of the logistic map (some of the results of which will be illustrated in Chapter 22), as well as in the wealth of literature that introduces or brings together other articles on the subject of chaos. For example, amongst the many we cite Schuster (1984); Beltrami (1987); Devaney (1989, 1992); Eubank and Farmer (1989); Tabor (1989); Bellacicco (1990); Falconer (1990); Pickover (1990, 1994); Rasband (1990); Serra and Zanarini (1990); Schroeder (1991); Drazin (1992); Moon (1992); Tufillaro, Reilly and Abbott (1992); Çambel (1993); Ott (1993); Strogatz (1994); Vulpiani (1994); Puu (2000); Riganti (2000).

---

research, dating back to the work of A.N. Kolmogorov in the fifties. Furthermore, they had a tradition of working together that had survived the divergence of mathematics and physics elsewhere.' (1987, p. 76).

[101] The mechanisms that give rise to chaos in dynamical systems are called 'routes to chaos'. Three different routes to chaos are known: (1) that of successive bifurcations, described in the text with reference to the logistic map, (2) that applied to the generation of the turbulence of fluids, of Ruelle and Takens (1971), and (3) that known as the Pomeau and Manneville intermittency mechanism (1980). We will not be looking at the last two, which are less elementary from a conceptual point of view, and furthermore specific to several phenomena related to physics. Interested readers should refer to the sources indicated in this footnote and those cited in the text.

22 The logistic map: some results of numerical simulations and an application

The Feigenbaum tree

In this section we will be dealing with the evolution of the logistic map as a result of the numerical calculation of the states and we will be looking at some of its characteristics on the basis of the numbers supplied by the iterations.

Figure 22.1 illustrates the contents of Chapter 21 regarding the splitting of the fixed points, of the subsequent doubling of the period that are at the origins of chaos and the presence of asymptotic equilibrium on a set of values travelled cyclically.

The ordinate axis shows the value of the growth parameter λ; the abscissa shows the fixed point attractors X_n, between which the population oscillates (they are states of asymptotic equilibrium; the numbers shown are approximated values).

When $\lambda = 2.9$, $\lambda = 3.1$ and $\lambda = 3.460$, the abscissas of the fixed point attractors are indicated. The first four values of λ_n that cause the fixed points to split, and λ_∞, beyond which we have chaos, (see Table 21.1 on p. 208) are shown in the figure by horizontal dashed lines. In the example, several values of λ that are slightly higher than λ_1, λ_2, λ_3 and λ_4 have been chosen and the logistic map (21.1) has been made to evolve in correspondence with these for 500 iterations, enabling the values of the attractor states to be identified to the first six decimal places (in reality one hundred or so iterations are sufficient to stabilize the evolution and identify the first decimal places of the states of equilibrium).

When $\lambda = 3.550$ and $\lambda = 3.566$, on the other hand, the fixed points are only indicated by the number that indicates the order with which they succeed each other periodically, with the numerical values of the corresponding abscissas shown in the legend.

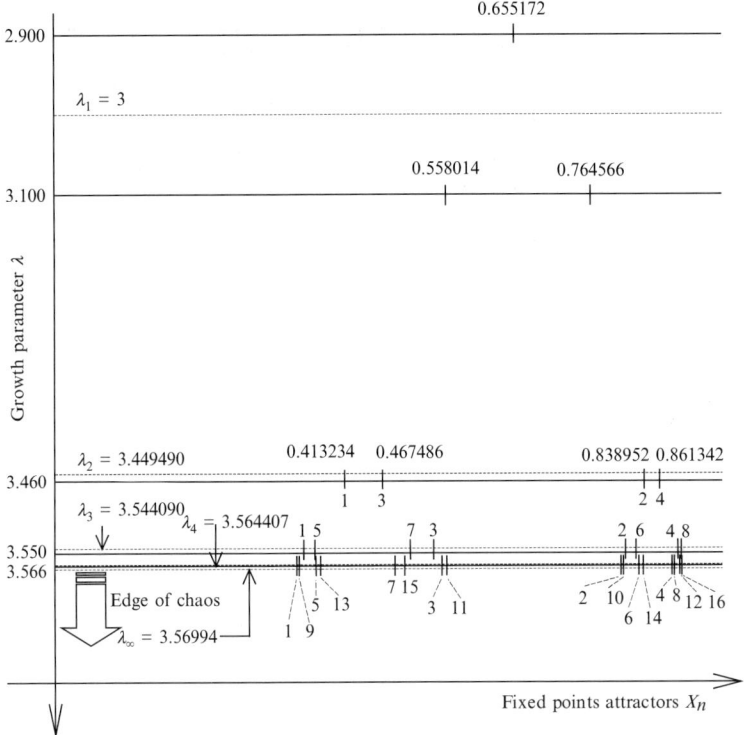

Figure 22.1 The fixed point attractors of the logistic map.

The values of the states of equilibrium that the evolution of the logistic map envisages for the population as a function of the growth parameter λ are shown in Figure 22.2. Figures 22.3 and 22.4 show enlarged details of Figure 22.2.[102]

---

[102] The images in Figures 22.2–22.4 have a granular appearance rather than a continuous curve, because they are all the result of numerical calculations conducted on discrete sets of values of λ and x.

The ordinate of each point of the graph represented in Figure 22.2 indicates the final value x of the population, once this has stabilized, after 15,000 iterations, as a function of the λ parameter shown on the abscissa. The successive splittings (i.e. the successive doublings of the period discussed in the previous pages) generate the successive bifurcations of the graph, which becomes a figure with fractal properties for values of λ higher than $\lambda_\infty \cong 3.56994$, where chaos ensues.

The graph is known as the Feigenbaum tree (Schroeder, 1991). Several regions can be clearly identified:

1. A first region, when $0 < \lambda \leq 1$, in which we have the final extinction of the population.
2. A second region, when $1 < \lambda \leq 3$, which corresponds to Case 1 in Chapter 21, in which the population evolves towards a final stable state, given by the fixed point which, in said Case 1 was called X_2. The value of X_2 increases as λ increases in this second region, describing a hyperbolic arc, as can be shown with a few calculations.
3. A third region, when $\lambda > 3$, in which we observe, as λ increases, successive and increasingly closer doublings of the number of fixed point attractors, each of which travels, as λ grows, a new hyperbolic arc, before splitting.

The Feigenbaum tree enables us to make numerous fundamental observations about the dynamics of the logistic map and about the onset of

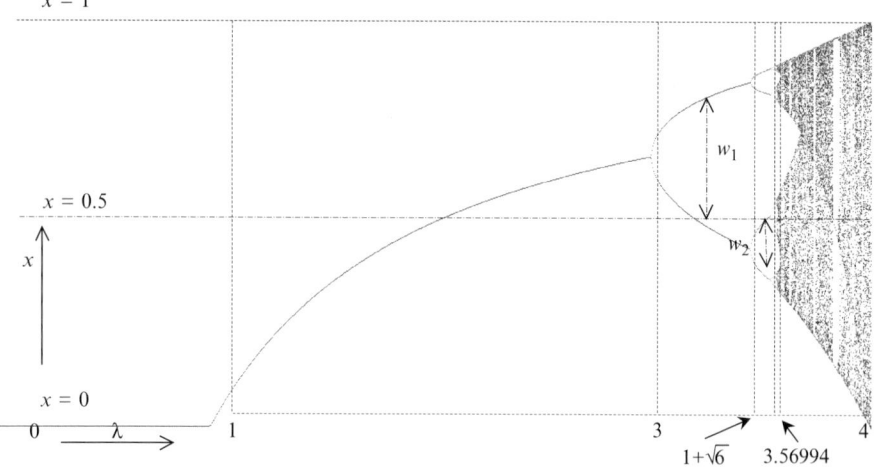

Figure 22.2 The Feigenbaum tree (diagram of the bifurcations) of the logistic map $(0 \leq x \leq 1, 0 \leq \lambda \leq 4)$.

chaos, so to speak visually, observations that escape analyses conducted exclusively on analytical form (21.1).

A first observation is immediately apparent from Figure 22.2 and is the meaning of the accumulation point λ_∞ in the set of values of λ_n: in correspondence of λ_∞ 'grey' areas start to appear, where the states of equilibrium are no longer identifiable individually, because the points have become so dense that they 'cover' intervals of x.

In addition to λ_∞, there is another limit value, indicated with δ, that identifies the speed with which, as n grows, the λ_n become increasingly close to each other (more 'dense') in their tending towards λ_∞, through the limit of the ratio of successive differences between period-doubling bifurcation parameters (Feigenbaum, 1978). More specifically, the following finite limit exists, an approximate value of which has been calculated as follows:

$$\delta = \lim_{n \to \infty} \frac{\lambda_n - \lambda_{n-1}}{\lambda_{n+1} - \lambda_n} \cong 4.66920160910299067$$

A second observation concerns the fact that the splittings of the states, recognizable in the subsequent bifurcations of the diagram, as well as occurring increasingly densely in line with the growth of λ_n, give rise to bifurcations, whose amplitude becomes increasingly smaller. In this regard, there are various ways to quantify the amplitude of the bifurcations. For example, we can observe that, at each successive 2^n-th doubling of the period, that we have in correspondence to λ_n, one of the bifurcations encounters the straight line of equation $x = 1/2$, which identifies the intermediate value of the population x between 0 and 1. Therefore, by using w_n to indicate the amplitude of the bifurcations measured with respect to the intersection with said straight line (see Figure 22.2 for w_1 and w_2, regarding the first two bifurcations), we have the following limit, indicated by α, of the ratio between the amplitudes of two successive bifurcations:

$$\alpha = \lim_{n \to \infty} \frac{w_n}{w_{n+1}} \cong 2.5029078750$$

The constants α (Feigenbaum bifurcation velocity constant) and δ (Feigenbaum reduction parameter) are known as Feigenbaum numbers:[103] they play a fundamental role due to the generality of the characteristics of the

---

[103] In the literature in English, they are occasionally called *Feigenvalues*; they have been calculated many times by various authors (Feigenbaum, 1978; Tabor, 1989; Rasband, 1990; Stephenson and Wang, 1990, 1991ab; Briggs, 1991). Only their approximate values are known, as no formula has ever been found that determines the sequence of the (infinite)

dynamics of the logistic map, common to numerous other functions, in addition to (21.1) (see towards the end of this section).

In the Feigenbaum tree of bifurcations, or at least in parts of it, a real scale invariance (technically known as self-similarity) can be identified, a fundamental characteristic of fractals. We can observe it, for example, in the way in which the bifurcations of the branches repeat in increasingly large numbers on an increasingly smaller scale as λ grows. In fact, by observing Figure 22.2, and then Figures 22.3 and 22.4, which contain enlarged details of Figure 22.2, it is evident how, when $\lambda > \lambda_\infty \cong 3.56994$, i.e. the area where chaos exists, infinite zones reappear in which chaos disappears and in which the phenomenon of successive bifurcations, i.e. the splitting of the stable states that have reappeared after previous areas of chaos, re-presents

Figure 22.3 The Feigenbaum tree of the logistic map; enlarged detail of Figure 22.2, defined by the following intervals: $0.80326629 \leq x \leq 0.89634562$; $3.4252167 \leq \lambda \leq 3.6422535$.

decimal places. Clifford Pickover (1994) cites the values of α and δ with 578 decimal places, calculated by Keith Briggs at the University of Melbourne, as the most accurate that have ever been obtained. At present, no theoretical framework that gives a meaning to these numbers has been identified. It is not known whether the Feigenbaum constant δ is algebraic, and whether it can be expressed in terms of other mathematical constants (Borwein and Bailey, 2004). Feigenbaum numbers have not been proved to be transcendental but are generally believed to be so.

Updated expressions for these numbers, as well as for many of the constants cited in this chapter, are available on 'The On-Line Encyclopedia of Integer Sequences' on the website of the AT&T research laboratories: www.research.att.com/~njas/sequences/index.html.

Figure 22.4 The Feigenbaum tree of the logistic map; enlarged detail of Figure 22.3, defined by the following intervals: $0.88132793 \leq x \leq 0.88510576$; $3.5813515 \leq \lambda \leq 3.5893727$.

itself, on an increasingly smaller scale, as λ grows. The sequence of the bifurcations that ensues brings the condition back to one of chaos. The analogies of structure that repeated changes in scale display, enable us to refer to non-strict self-resemblance in the Feigenbaum tree and to consider it fractal.

Therefore, apart from the inevitable granularity of the figures (remember that we are still dealing with numerical simulations), the chaotic area does not appear as a uniform area, but appears streaked with infinite vertical veins: bands of real order, in which the points appear to be positioned on regular curves. This phenomenon is called intermittency and is a typical feature of a fractal structure. There are infinite, successive waves of intermittency; repeated enlargements of scale highlight this type of order that is generated, so to speak, situated in the middle of the disorder, a phenomenon typical of chaos.

Alongside the splitting of the states that are displayed in the bifurcations of the Feigenbaum tree, we also observe the presence of situations in which the population oscillates on an odd number of states, as we mentioned at the end of chapter 21 (for example, in Figure 16.1, we have the case of oscillation on three states, when $\lambda \cong 3.8284$). As we said in Chapter 21, the triplication cycles are particularly important, because their onset, according to Li and Yorke (1975), guarantees the presence of periodicity of any order and, therefore, the existence of chaos.[104]

Let us now return to what we discussed in Chapter 19 regarding the stability of the trajectories and the Lyapunov exponents. We can identify the presence of chaotic zones in the dynamics provided by the logistic map, according to the value of the growth parameter λ, by using Lyapunov exponents. As we saw previously, if the Lyapunov exponent is negative, the trajectory is stable, whereas if it is positive, the trajectory is unstable and we are in a situation of chaos.

Applying (18.11) from p. 175 to the logistic map, we observe that the Lyapunov exponent depends on the value of the growth parameter λ, on the value of the state variable x_n from which the iterations start, and on the number N of iterations:

$$\Lambda = \frac{1}{N} \sum_{n=1}^{N} \ln |\lambda(1 - 2x_n)| \qquad (22.1)$$

Figure 22.5 shows, at the bottom, the trend of (22.1) as a function of λ in the latter's most significant range, that is from around λ_1 up to 4; the figure also shows, at the top, the part of the Feigenbaum tree of Figure 22.2, for the same values of λ. The onset of chaos when the Lyapunov exponent Λ becomes positive for $\lambda > \lambda_\infty \cong 3.56994$ is apparent, as is the fact that the chaotic zone is streaked by numerous intervals of non-chaos, corresponding to the intervals in which the Lyapunov exponent is negative.[105]

---

[104] More specifically, when $3.8284\ldots < \lambda < 4$, in addition to the bifurcations described, we also observe a cascade of 'trifurcations' in 3^n cycles, in correspondence with particular values λ'_n whose description is similar to that made for the cycles of the bifurcations. If we take only these specific values of the growth parameter into account, we identify the following limits, real generalizations of Feigenbaum numbers:

$$\delta' = \lim_{n \to \infty} \frac{\lambda'_n - \lambda'_{n-1}}{\lambda'_{n+1} - \lambda'_n} \cong 55.247$$

$$\alpha' = \lim_{n \to \infty} \frac{w'_n}{w'_{n+1}} \cong 9.27738$$

The numbers shown have been obtained, again by means of numerical simulations, with a precision lower than that of the constants α and δ of the previously discussed cases (Li and Yorke, 1975).

[105] Note that (22.1) exists only when $x_n \neq 0.5$, a value in correspondence to which the logistic map, naturally, works normally, but for which (22.1) cannot be calculated, because it would give rise to $\ln(0)$ that does not exist in the field of real numbers. The value of the state of

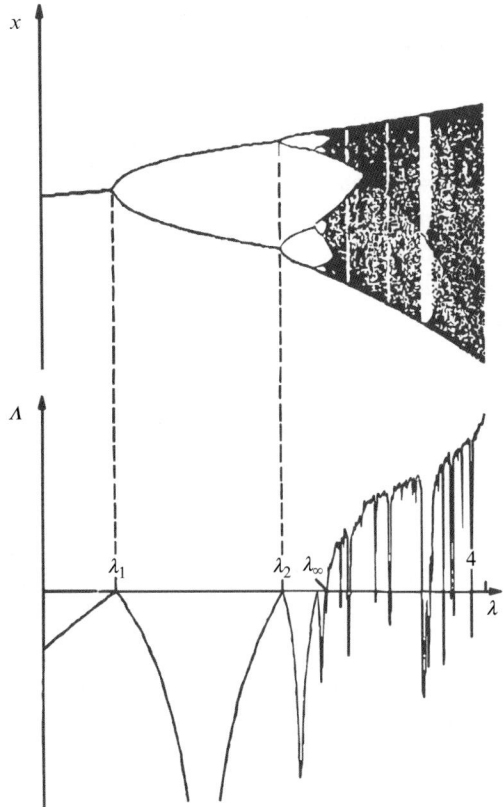

Figure 22.5 Bottom of the figure: trend of the Lyapunov exponent Λ as a function of the growth parameter λ of the logistic map; top of the figure: the Feigenbaum tree for the same values of λ.
Source: Çambel (1993, p. 210).

stable equilibrium $x_n = 0.5$ corresponds to a situation of great stability. The system reaches it very quickly, when $\lambda = 2$, starting from any initial value of the population; for example, starting from $x_0 = 0.1$, from as early on as the seventh iteration $x_n \cong 0.500000$ is stable to the sixth decimal place. In general, the values of the Lyapunov exponent given by (22.1) provide an indication of the speed of convergence of the logistic map towards the value of the stable state, where it exists, a speed that increases the more negative the Lyapunov exponent; when $\lambda = 2$ we have the maximum speed of convergence towards the final value of the population $x = 0.5$.

In the graph in Figure 22.5, the calculation of the curves was also conducted by points. This does not allow us to position the curve traced by the Lyapunov exponent in complete correspondence to the infinites detailed by the Feigenbaum tree, which, incidentally, appear only in part in the figure, as the Feigenbaum tree itself is a fractal.

The instability of the logistic map, and therefore the sensitivity of the latter to the initial conditions, reaches its peak when $\lambda = 4$, a value in correspondence to which also the Lyapunov exponent in the logistic map is at its maximum value. In fact, when $\lambda = 4$, as is apparent in Figure 22.5, the interval of variability of the population in a single iteration is the maximum permitted one, between 0 and 1, and therefore the unpredictability, and basically the level of chaos, are at their highest level.

It can be shown (Schuster, 1984) that, when $\lambda = 4$, the Lyapunov exponent reaches its maximum value, and this can be calculated exactly as follows:

$$\Lambda = \ln 2 \cong 0.693715$$

a result which, incidentally, is also approximated by the numerical calculation of (22.1) after a fairly high number of iterations.[106]

---

[106] Just in this particular case of $\lambda = 4$, one can show that the attractor is chaotic (see for example Kraft, 1999), i.e. the dynamics is unstable and sensitive to initial conditions, but not strange, i.e. the attractor doesn't display a fractal structure. As a matter of fact, if a chaotic dynamical system has an attractor, it is sometimes possible to make precise statements about the *likelihood* of a future state, as a probability measure exists that gives the long-run proportion of time spent by the system in the various regions of the attractor (chaotic unpredictability and randomness are not the same, but in some circumstances they look very much alike). This is the case of the logistic map with $\lambda = 4$ (and of other chaotic systems as well), which maps the interval [0,1] into [0,1]. Even if we know nothing about the initial state of the system's evolution, we can still say something about the distribution of states in the future. More precisely, the future states of the population are distributed between 0 and 1 following a beta probability distribution, according to which the probability density function in the domain $0 \leq x \leq 1$ is:

$$f(x) = \frac{x^{\alpha-1}(1-x)^{\beta-1}}{\int_0^1 x^{\alpha-1}(1-x)^{\beta-1}dx}$$

with parameters $\alpha = 0.5$ and $\beta = 0.5$. At $\lambda = 4$ the sequences are aperiodic and completely chaotic. The dynamics described by the recursive relation are ergodic on the unit interval between 0 and 1, i.e. the sequence x_n will approach every possible point arbitrarily closely or, more formally, for every pair of numbers between 0 and 1, there is an x_n between them. Due to the property of generating an infinite chaotic sequence of numbers between 0 and 1, in the late 1940s John von Neumann suggested using the logistic map with $\lambda = 4$ as a random

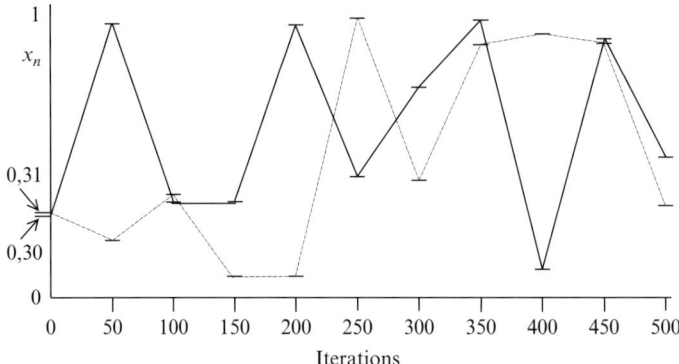

Figure 22.6 Evolutive trajectories of the logistic map (21.1), for two initial population values close to one another. Continuous line: $x_1 = 0.30$; dashed line: $x_1 = 0.31$; in both $\lambda = 3.9$.

With regard to the effects of the sensitivity to the initial conditions on the future evolution of the logistic map and the onset of chaos, Figure 22.6 shows a simple example of two orbits of the logistic map (21.1), calculated for $\lambda = 3.9$, therefore in the interval of λ values in which we have chaos, starting from two initial values of the population that are close to one another: $x_1 = 0.30$ and $x_1 = 0.31$. The subsequent values of the graph in Figure 22.6 are shown at steps of 50 iterations at a time. It is apparent from the figure how the two trajectories develop fully independently of one another, even though they originate from close points.

The numerical (approximate) calculation of Lyapunov exponents for the two values of x_1 used in the examples illustrated in Figure 22.6, after 10,000 iterations and with $\lambda = 3.9$, gives respectively:

$$\Lambda(x_1 = 0.30) \cong 0.493616$$
$$\Lambda(x_1 = 0.31) \cong 0.492634$$

which not only confirms the presence of chaos in the two orbits, but also gives a quantitative measurement of the level that the chaos reaches, highlighting how the two differently shaped orbits are characterized by similar degrees of instability.

---

number generator, but the proposal was unsuccessful, because of strong correlations between the generated sequences of numbers (Andrecut, 1998).

A point of fundamental importance, highlighted by Feigenbaum in his essential study (1978), is that the dynamics of the logistic map that we have illustrated are not specific to the logistic map alone (21.1), but can be extended to a wide category of one-dimensional maps. All the functions which, in the interval considered, are continuous and derivable, display a quadratic maximum,[107] and are unimodal, i.e. this maximum is unique. The function is strictly (monotonically) increasing before the maximum value and strictly (monotonically) decreasing after it.

The importance of this fact is that dynamics extremely similar to that described for the logistic map can exist for numerous other models that appear to be different on the surface, but are not different in their mathematical essence. For example, instead of taking the arc of a parabola for $f(x)$, as in (21.1), we could take, without substantially changing the dynamics described, an arc of a circumference, of an ellipse, of a sinusoid, and so on. As a consequence of this, different assumptions that, in the construction of a model, could generate functions other than (21.1) but that respect the cited conditions of being quadratic and unimodal, in practice, and without fundamental errors, could be replaced by a simple logistic map, the formulation and application of which is simple and easy.[108]

An example of the application of the logistic map to spatial interaction models

We would now like to briefly describe an example in which the logistic map is applied as an element of a more general model of population dynamics. The example, by Peter Nijkamp and Aura Reggiani (1992a) that belongs to a class of models called 'spatial interaction models' is widely applied in the field of urban and regional science (we briefly mentioned this in Chapters 1–3).

The term spatial interaction is generally intended to mean the set of relations within geographic areas that manifest themselves in the form of regularities in flows between the areas considered. Models of spatial interaction attempt to formally represent relations between the various compon-

---

[107] These are functions which, developed in power series in a neighbourhood of the maximum value, display the second order term as the most important one; they are, in other words, functions that, when close to the maximum value, 'behave' in a first approximation, like a parabola.

[108] Other extensions of the logistic map have also been studied; for example in multidimensional systems, such as Hénon's two-dimensional map (1976) or in the field of complex numbers, Mandelbrot's set (1975).

ents (production, services, housing, etc.) located in different zones of the area in question. The aim of this, for example, is to model the movement (the flows) of people between the various zones in which the area in question is divided, usually an urban area.

The remotest forms of the concepts on which spatial interaction is based, initially simple gravitational analogies, date back to the work of Henry Charles Carey (1858).[109] However, gravitational theory does not represent

---

[109] Carey was the first person to attempt to conceptually use Newton's theory of universal gravitation in social sciences, in line with the physicalist trends of that era. An excerpt of his book *Principles of Social Science* is emblematic of the physicalist view: 'The wild man, wherever found, has always proved to be not only destitute of the reasoning faculty, but destitute also of the instinct that in other animals takes the place of reason – and therefore the most helpless of beings. Man tends of necessity to gravitate towards his fellow-man. Of all animals he is the most gregarious, and the greater the number collected in a given space the greater is the attractive force there exerted, as is seen to have been the case with the great cities of the ancient world, Nineveh and Babylon, Athens and Rome, and as is now seen in regard to Paris and London, Vienna and Naples, Philadelphia, New York, and Boston. Gravitation is here, as everywhere else in the material world, in the direct ratio of the mass, and in the inverse one of the distance. Such being the case, why is it that all the members of the human family do not tend to come together on a single spot of earth? Because of the existence of the same simple and universal law by means of which is maintained the beautiful order of the system of which our planet forms a part. We are surrounded by bodies of various sizes, and some of these are themselves provided with satellites, each having its local centre of attraction, by means of which its parts are held together. Were it possible that that attractive power could be annihilated, the rings of Saturn, the moons of our earth and of Jupiter, would crumble to pieces and fall inward upon the bodies they now attend, a mass of ruins. So, too, with the planets themselves. Small as are the asteroids, each has within itself a local centre of attraction enabling it to preserve its form and substance, despite the superior attraction of the larger bodies by which it is everywhere surrounded. So it is throughout our world. Look where we may we see local centres of attraction towards which men gravitate, some exercising less influence, and others more. London and Paris may be regarded as the rival suns of our system, each exercising a strong attractive force, and were it not for the existence of the counter attraction of local centres like Vienna and Berlin, Florence and Naples, Madrid and Lisbon, Brussels and Amsterdam, Copenhagen, Stockholm, and St. Petersburg, Europe would present to view one great centralized system, the population of which was always tending towards those two cities, there to make all their exchanges, and thence to receive their laws. So, too, in this country. It is seen by all how strong is even now the tendency towards New York, and that, too, in despite of the existence of local centres of attraction in the cities of Boston, Philadelphia, Baltimore, Washington, Pittsburg, Cincinnati, St. Louis, New Orleans, Augusta, Savannah, and Charleston, and in the numerous capitals of the States of which the Union is composed. Were we to obliterate these centres of attraction and place a centralized government like that of England, France, or Russia, in the city of New York, not only would it grow to the size of London, but soon would far exceed it, and the effect would be the same as would be produced in the astronomical world by a similar course of operation' (Carey, 1858, p. 42–43). Ernst Georg Ravenstein (1885) was the first person to apply the concepts of Newtonian physics to the study of migrating flows between British cities ('the laws of migration'). In the twentieth century, the idea of applying gravitational theory to spatial

the only possible approach to spatial interaction. Alan Wilson (1970), for example, demonstrated that spatial interaction models can be obtained from a problem of mathematical optimization, maximizing a particular entropy function.

Again dating back to the 1970s, a new important branch of spatial interaction models emerged, linked to discrete choice models (Hensher and Johnson, 1981; Pitfield, 1984; Leonardi, 1987). The advantage of this new approach lies in the fact that its models are not limited to constructing a simple analytic descriptive structure, but offer a justification of the patterns of interaction observed, referring to the processes of individual choice, thus relating the observed flows to the behaviour of the individual actors and assuming that the former are the result of the latter.

Models of spatial interaction, an extremely wide and varied category, have become the most important tools for the study of interrelations between the distribution of activities and the distribution of flows of movement in an urban or regional system, in particular following the first and fundamental model by Ira Lowry (1964);[110] they are still being developed according to new and varied lines of research and following different approaches (for a collection of new trends see, for example: Boyce, Nijkamp and Shefer, 1991; Fujita, Krugman and Venables, 2000; Reggiani, 2000). There are numerous applications of spatial interaction models, from transport planning to the location of housing and businesses, to the analysis of the demand for services and more still (for a more detailed description, see for example Campisi, 1991; Rabino, 1991; Nijkamp and Reggiani, 1992a; reviews of spatial interaction models and of the theories such are based upon can be found, for example, in Bertuglia et al., 1987; Fotheringham and O'Kelly, 1989; Bertuglia, Leonardi and Wilson, 1990; Bertuglia, Clarke and Wilson, 1994; Sen and Smith, 1995).

In the context of spatial interaction models, with particular reference to urban and regional sciences, the flow from a zone of origin i to a zone of

---

interaction, resumed by William Reilly (1931; see Chapter 1), was developed, in particular, following the first congress of the Regional Science Association in 1954, and has resulted in a vast series of studies and models (for example, Isard, 1960; Wilson, 1970, 1974; for a review see Sen and Smith, 1995).

[110] Lowry created his model for the Regional Planning Association of Pittsburgh, within the framework of a general attempt at planning for the reorganization of the urban areas of the United States, shaken by a disorganized industrial development. The originality of Lowry's idea lay in using an economic theory (the theory of the urban economic base) as one of the elements to model spatial flows, thus creating an economic-spatial type of urban theory, and in expressing it all in a mathematical form. It was the first urban simulation model to become operative.

destination j is commonly expressed in terms of a mobility rate P_{ij}; in the case that we are going to explore, it is calculated according to a discrete choice model, starting from a comparison between the costs and the benefits of a possible move from zone i to zone j that a single actor makes, to decide whether to make the move or not. In this way, the elements of socio-economic importance at the origin of the localization can be moved from a macro level, based on a gravitational type of law, or on the maximization of a spatial entropy, to a micro level, based on the maximization of an individual utility (following the introduction of these models, however, Giorgio Leonardi, 1985, demonstrated the asymptotic equivalence between the theory of random utility and the maximization of the entropy).

In this approach, the use of the multinomial logit model is very common. This is due both to its relative simplicity from a computational point of view and to the fact that it respects an important condition, often posed as a real axiom of the theory of choice models, called 'the independence of irrelevant alternatives', according to which the actor chooses between the alternatives (the zones) available by comparing them in pairs only. More specifically, it is assumed that the value of the ratio between the probabilities that two alternatives (two zones) have of being chosen depends exclusively on the utility associated to each of the two and that is as independent from the utilities associated to other zones as the total number of zones; as a consequence of which, alternative choices (zones) may be introduced or eliminated without having to recalculate the utilities of all of the other available alternatives.[111]

A multinomial logit choice model is fundamentally based on the following assumptions:

1. An actor acts rationally, choosing to move from zone i to zone j in the considered territory, only based on the calculation of a function u_{ij}, called the utility function, that he aims to maximize.
2. The utility function is assumed to be the sum of a deterministic part, regarding the attributes of the zone, and a stochastic part, regarding how each individual actor perceives individual utility. This means that the choice of destination to which the actor attributes maximum utility can only be expressed in probabilistic terms, according to a given probability distribution.

---

[111] The independence from irrelevant alternatives has often been judged as too restrictive an assumption, in particular when the actor has to compare two very similar alternatives of utility. This has led to the development of several variants of the choice model, such as the dogit model (Gaudry and Dagenais, 1979), the multinomial probit model (Daganzo, 1979) and the nested logit model (Ben-Akiva and Lerman, 1979), which we will not be discussing here. Interested readers should consult the specific literature.

3. All actors are identical and act independently.

According to the multinomial logit choice model, it is assumed that the probability P_{ij} of the move from zone i to zone j depends exponentially on a term $-c_{ij}$, that represents the deterministic part of the utility function (basically, the costs of the move) and that it is normalized with respect to all the potential destination choices, in such a way that P_{ij} falls between 0 and 1. By introducing weights W_k attributed to the k zones in an area, which represent the attractiveness of the zone k as it is perceived by the actors, the probability P_{ij} of moving from zone i to zone j (i.e. the attraction of zone j with respect to zone i) is expressed as:

$$P_{ij} = \frac{W_j e^{-\beta c_{ij}}}{\sum_k W_k e^{-\beta c_{ik}}} \qquad (22.2)$$

The parameter β that appears in (22.2) is a term associated to the sensitivity of the user to the cost of transport: the higher the value of β the lower the average distance travelled by the user and the higher the overall accessibility of the system (Campisi, 1991).

Nijkamp and Reggiani (1992a) demonstrated that, given certain conditions for the utility function, a multinomial logit choice model inserted into a spatial interaction model in discrete time can display chaotic behaviour similar to that which emerges in the logistic map.

Assuming, for the sake of simplicity, that $W_k = 1$ in all zones, and indicating the utility of the choice of zone j at time t with u_j the formula (22.2) can be transformed into the probability that zone P_j attracts population, independently from the zone i of origin, in the following form (Leonardi, 1987; Nijkamp and Reggiani, 1991; Rabino, 1991):

$$P_j = \frac{e^{u_j}}{\sum_k e^{u_k}} \qquad (22.3)$$

Now, because the magnitudes in play are functions of time, we can study the variation of P_j over time by calculating its derivative. Let us assume that the utility of the alternative j grows linearly over time; we assume, that is, that the derivative $\dot{u}_j$ of the utility function of a zone j is constant and we indicate its value with α_j. After several operations, differentiating (22.3), we obtain:

Application to spatial interaction models

$$\dot{P}_j = \alpha_j P_j(1 - P_j) - P_j \sum_{k \neq j} \alpha_k P_k \qquad (22.4)$$

In discrete time, (22.4) becomes, assuming $\alpha_j \cong u_{j,t+1} - u_{j,t}$:

$$P_{j,t+1} - P_{j,t} = \alpha_j P_{j,t}(1 - P_{j,t}) - P_{j,t} \sum_{k \neq j} \alpha_k P_{k,t} \qquad (22.5)$$

i.e.

$$P_{j,t+1} = (\alpha_j + 1)P_{j,t} - \alpha_j P_{j,t}^2 - P_{j,t} \sum_{k \neq j} \alpha_k P_{k,t} \qquad (22.5')$$

Equations (22.4) and (22.5), (22.5') describe the trend, in continuous time, and in discrete time respectively, of the speed of change of the probability of moving to zone j from any other zone k in the area in question.

It is immediately apparent that the second members of (22.4) and (22.5) are both the sum of two terms. The first term has the standard form of the logistic function in (22.4), or the logistic map in (22.5). The second, negative term expresses the interaction between the probability of moving to zone j and all the probabilities of moving to each of the other k zones, an interaction that is represented in the form of the sum of the products between P_j and each of the P_k, with α_k that acts as weight coefficient of P_k.

In analogy to the prey–predator models discussed in Chapters 11 and 12, P_k, which attracts individuals towards zone k and therefore tends to reduce the number of individuals that move to zone j, can be interpreted as representing the population of a predator species whose actions reduce the population of the prey species P_j by means of parameter α_k. The form of this interaction is the same type as that which appears in the basic Volterra–Lotka model (12.1) and in its variants used in urban planning (13.1), (13.2) and (13.4). The model, regardless of whether it is formulated in continuous time (22.4) or discrete time (22.5), can therefore be seen as if it were expressing the competition between zones like the sort of competition between species. The population of each zone is the result of two terms: growth according to a logistic law (continuous or discrete) and the competition between zone j and all the other zones to attract the actors.

Simulations of territorial dynamics in the field of urban and regional sciences, using a model constructed as in version (22.5') provide results that obviously depend on the values of α_j (one for each zone j). Nijkamp and Reggiani (1992a) present the simulation of the dynamics envisaged by the spatial interaction model (22.5') for an area split into three zones and for different values of the α_j parameters. The growth and/or diminution dynamics that are obtained show several aspects in common with the logistic type, which was predictable given the same form of the second member of (22.4) and (22.5). For small values of α_j, the model displays the trend of the logistic function tending towards states of asymptotic stability. The most interesting aspect to observe is that for large values of α_j, on the other hand, the simulations generate chaotic behaviour, similar to that displayed by the logistic map for high values of the growth parameter λ.

23 Chaos in systems: the main concepts

To conclude this part, we would like to briefly review and summarize some of the most important points discussed.

As we have seen, in specific circumstances, i.e. for certain values of parameters, the long-term prediction of the evolution of a nonlinear dynamic system can be achieved without error or ambiguity. In the case of the logistic map, this is true for values of the growth parameter λ that fall between 0 and 3, in which we have one fixed point attractor. This is the case we have called Case 1. Under these conditions, the most simple, the system inevitably evolves towards a previously known future, regardless of the initial conditions, even towards the extinction of a population, which occurs when $\lambda < 1$.

In other circumstances, the same nonlinear system, even if its formulation is very simple, becomes unpredictable. In Chapter 21, regarding the logistic map, we indicated these circumstances as Case 3, corresponding to $\lambda \geq \lambda_\infty \cong 3.56994$. For a dynamical system in this condition, we refer to deterministic chaos. Chaos, because the evolution is unpredictable and subject to great variability even following small changes to the initial conditions, although remaining contained in a limited region of the phase space; deterministic because there is always a law (a mathematical model) that dictates its evolution.

In the chaotic situation of Case 3, the system appears to evolve erratically, lacking any final state towards which to aim, even asymptotically, according to a dynamics in which the population never reassumes a value that it has already touched, and this occurs despite the fact that the evolution of the system is governed by the same law that in other circumstances envisages a stable final state.

As is evident from Figures 22.2–22.4, the chaotic zone includes 'traces' of order within it. There are streaks of the Feigenbaum tree in which the possible final states towards which the population tends during the course of its evolution, i.e. the number of fixed point attractors, are only a few, two, four, eight, but never one alone. Predictability returns, and a multiplicity of the possible states between which the system tends asymptotically to oscillate remain, but these final states *exist*, while in chaotic evolution they don't. The appearance of traces of order in the Feigenbaum tree is also reflected by the fact that the Lyapunov exponent is negative for the values of λ that give the areas of order, as shown in Figure 22.5.

Lastly, there is one other aspect that we would like to underline, which will be the subject of Part 3 of this book. Beyond the zone of the unique attractor point (complete predictability) and that of chaos (unpredictability), there is also an intermediate zone: that which in the logistic map is defined by the interval $\lambda_1 < \lambda < \lambda_\infty$ (where we would like to remind readers once again, $\lambda_1 = 3$ and $\lambda_\infty \cong 3.56994$) and that we have called Case 2. Here we do not have chaotic unpredictability, nor however do we have a simple asymptotic evolution towards a unique state that attracts the system, from any point in the phase space (i.e. from any set of values of the state variables) it originates. In this intermediate zone, there is a final (asymptotic) condition, that the system tends towards, which is made up of different states. The important aspect is that the number of the states and the numerical value of each of them (i.e. the 'identity' of each of them) depends on the specific value given to the growth parameter (Figures 22.2–22.4). Small variations of λ in the vicinity of each of the terms of an infinite sequence of critical values λ_n can generate considerable rearrangements of the final configuration, such as the doubling of the number of states between which the population oscillates or the appearance of sets of odd numbers of states.

The dynamics that the logistic map generates is such that the population system, evolving in an environment with limited resources, behaves as if, for each specific value of the growth parameter λ that falls between λ_1 and λ_∞, it *autonomously* finds, without any intervention from external forces, a configuration made up of *several* stable states between which it periodically oscillates; different states, that uniquely depend on the value of λ, as if these states represented the most convenient situation for the system, towards which the system in question inevitably tends. The population system autonomously 'decides', without the intervention of any external agent, if there are two, four, eight or more of these states, as a function of the growth parameter λ. In other words, the system *organises itself*, autonomously building its own configuration of stability.

These changes in the configurations of the (asymptotic) states of final equilibrium can be seen as real phase transitions that the system undergoes when the parameters vary a little around a threshold value λ_n. The expression 'phase transition' has been borrowed from physics within the scope of which it indicates particular processes that occur in correspondence with conditions of instability, the passing through of which leads to a reorganization of the elements of a system. It can regard, for example, the passages of state of a material that occur in correspondence with small variations in temperature in the vicinity of particular critical values: a crystalline solid melts when the temperature goes beyond a specific melting temperature; a liquid boils if the temperature exceeds a certain value, called the boiling point; a ferromagnetic material ceases to be such if its temperature exceeds a particular value, specific to that material, called Curie's temperature, etc. Similarly, the concept of phase transition can be transferred to the logistic map, and to the dynamics of systems in general, when sudden changes in the dynamics appear in correspondence to particular critical values of a parameter.

How should the above be interpreted in the case of the evolution of a population according to the model of the logistic map? We can see the process in this way: the system makes the population grow (cause) until it reaches values that are too high with respect to the available resources (effect); in this situation, the population tends to decrease (the effect mentioned before now becomes the cause of a new population value, that is the new effect); the smallest value that the population assumes now becomes the origin of new growth (the new effect has become a new cause) and so on. In this way, for a given λ comprised between λ_1 and λ_∞, the population ends up oscillating between several states. If, therefore, we interpret each state as the cause of the following one, we have a new nonlinear relationship between cause and effect, which may cause the second of two consecutive states to be greater or less than the first one, instead of the linear growth relationship, according to which one cause always corresponds to one single effect, as, for example, in exponential growth à la Malthus (Zanarini, 1996). We do not now have unpredictability, which would be absence of final configurations, but a final configuration that changes as a function of the parameter, redefining the final state of the system at each transition. This is not unpredictability, but 'surprise'. In the wealth of 'surprises' a new aspect that characterizes the dynamics of nonlinear systems emerges: complexity.

The question we can ask ourselves is the following: if chaos and complexity are already present in a simple deterministic system that basically appears to be fairly simple, such as the logistic map, what can we expect from systems that are much more detailed and complicated? With regard to systems governed by more complex laws, the presence of threshold values for parameters, passing through which phase transitions between simple evolution, complex evolution and chaotic evolution can be observed, is in general less evident *a priori*. The behaviour of a system made up of numerous nonlinear equations, containing various parameters, is generally unpredictable *a priori*, because it is extremely difficult, or even impossible, to identify the effect of the various parameters in a multifaceted system, whether such are considered individually, or are considered in their entirety. The only way to discover the dynamics that the model envisages is *a posteriori*, i.e. putting the model in question 'into action'.

We can therefore conclude that the eventual presence of chaos in the evolutive dynamics of mathematical models constructed in the form of dynamical systems is a further element that deprives the models of the value of methods for predicting the future in a strict sense. Not only, therefore, does the fact that we have to necessarily introduce simplifying hypotheses to such models, such as for example linearization, or choose which aspects should be considered and which to ignore because retained marginal (see Chapter 2), limit their significance and validity, but those same chaotic characteristics of the dynamics, at times generated by the models, make the prediction of the future a nonsensical idea.[112] No model can be assumed to be a 'real' and 'complete' representation of reality, able to 'predict' future developments, both because a model, intrinsically, is a partial representation of several aspects of reality, and because the dynamics that the model in question generate are at times unstable, chaotic and substantially inapplicable to the prediction of the future states of a system. In the majority of cases, models, as we have said on many occasions, are only tools of analysis, even though fundamental tools. And this is true not only for natural sciences but also, all the more so, for social sciences, where the identification of recurrences (or 'laws') is usually more difficult.

We can instead observe how the use of models actually enables us to highlight and to account for both the presence, if any, of determinism in the observed data, and therefore chaos and not randomness (see chapter 19), and any formation of complex structures in the evolutive dynamics of real systems, structures that are not *a priori* predictable, but that are in any event stable configurations of equilibrium. The model that manages to adequately

---

[112] The idea that the future does not submit to rational prediction may seem to be an obvious, if not trivial conclusion. However, it should be considered that it is in total antithesis to the beliefs of classical eighteenth and nineteenth century science, permeated with a rigid determinism, regarding the possibility of complete knowledge of the future, at least in theory. The romantic spirit of the late nineteenth century opposed said conviction, which we will better define in Chapter 26, and claimed man the freedom to escape the laws of physics, interpreted in an imperative and not descriptive sense, and as such felt to be unacceptably restrictive to man's free spirit. An expression of this contraposition, transferred to the level of a dispute between rationality (the mathematical model) and the freedom of man's actions (the unpredictable future), appears in the quote shown at the beginning of this part, taken from Dostoevsky, an author who was well aware of these issues, as his education had been predominantly scientific.

account for an apparent disorder in the data measured in the real system, to which it refers, therefore represents an effective deterministic law of the real system.[113] It is evident that *no* model can ever provide a *complete* and *exact* description (by 'exact' we mean 'with an arbitrarily high precision') for any real system, whether in natural or social science, and this is precisely because a mathematical model, in itself, has to make simplifications and inevitable adjustments (we are still taking about matching a mathematical formula to an observed phenomenon). Nevertheless, an effectively constructed mathematical model can be an extremely useful tool both for giving order to a set of apparently disordered data, by identifying a mechanism that in some way links one to the other, and for 'throwing light' on the possible situations that the mechanism identified can generate in the future evolution of the real system.

It is precisely the onset of complex structures in models made up of nonlinear dynamical systems, which are thus called complex, that will be the topic for discussion in Part 3.

---

[113] Dynamical systems can also be conceived in which the parameters (the coefficients of the differential equations) are not given with the precision that characterizes the models of classical mechanics. They are usually models that refer to systems that entail random uncertainties and fluctuations, generally linked to the intrinsic uncertainty of measurements or stochastic processes. Non-deterministic models of this type are called stochastic (Bellomo, 1982).

PART 3 Complexity

The next planet was inhabited by a tippler.
This was a very short visit, but it plunged the little prince into deep dejection. 'What are you doing there?' *he said to the tippler, whom he found settled down in silence before a collection of empty bottles and also a collection of full bottles.*

'I am drinking,' *replied the tippler, with a lugubrious air.*
'Why are you drinking?' *demanded the little prince.*
'So that I may forget,' *replied the tippler.*
'Forget what?' *inquired the little prince, who already was sorry for him.*
'Forget that I am ashamed,' *the tippler confessed, hanging his head.*
'Ashamed of what?' *insisted the little prince, who wanted to help him.*
'Ashamed of drinking!' *The tippler brought his speech to an end, and shut himself up in an impregnable silence.*

And the little prince went away, puzzled. 'The grown-ups are certainly very, very odd,' *he said to himself, as he continued on his journey.*

Antoine de Saint-Exupéry, *The Little Prince*

Complexity

24 Introduction

In Part 1, we presented various examples of very schematic models that provided fairly abstract and simplified descriptions of several types of phenomenology. These descriptions ranged from the simplest approximation, the linear case, to less drastic assumptions, that enabled us to perceive more sophisticated behaviour. In Part 2, we discussed several aspects of the dynamics of unstable systems, which, in certain circumstances, can give rise to chaotic behaviour. In particular, we examined one of the simplest nonlinear models, the logistic map, which evolves, depending on the growth parameter, towards a final stable state, or towards a series of final states that repeat periodically, or that demonstrate chaotic instability.

In Part 3, we will be discussing a particular aspect of the behaviour of nonlinear dynamical systems that is displayed in circumstances in which there is neither stable equilibrium, nor chaos. This is the case of systems that are so-called far from equilibrium (Nicolis and Prigogine, 1987), whose behaviour is even more surprising than that of chaotic systems; these are complex systems. Rather than providing a collection of examples and situations that display complexity, something that has already been illustrated in numerous general texts dedicated to this subject (for example: Nicolis and Prigogine, 1987; Waldrop, 1992; Casti, 1994; J. Cohen and Stewart, 1994; Gell-Mann, 1994; Coveney and Highfield, 1995; Bak, 1996; Gandolfi, 1999; Batten, 2000a; Bertuglia and Staricco, 2000; Dioguardi, 2000), we propose to focus on complexity from an abstract angle, and to develop several reflections on the meaning and the interpretation of complexity, as far as possible, from a general perspective.

On the basis of the above, in Chapter 25, we will start by having another look at reductionism and the concept of the linearity of real systems and of models, which is a fundamental part of the latter, that we have already encountered in the first two parts of the book.

25 Inadequacy of reductionism

Models as portrayals of reality

To describe, interpret and predict phenomena of the environment (or of the world, the universe or whatever we wish to call it) in which we are immersed, we generally use models, where the meaning if the term 'model', in a very general context, is a mental representation that, in a certain sense, replaces the environment (the world, the universe). In other words, since we have no means of going beyond the restricted window that our senses provide of reality, the latter is too complicated and substantially inaccessible to complete knowledge. All that we can do is to limit ourselves to constructing a representation of reality based on information gained from experience. In this vision, therefore, our knowledge is always relative to a perspective and is conditioned by a point of view, a product of the human mind, and not something inherent to the order of things. Science, therefore, does not study the physical world *per se*, but rather our way of depicting some regularities selected from our experience, that can be observed in certain conditions and from a particular perspective.

The models that we form of the world are not authentic copies of the latter, as we have already shown in Chapter 14; in any event, their correspondence with a 'real world', that can only be known by means of representations, can never be verified. The best we can do is to assess our model-representations of the world on a pragmatic level, judging whether they guide our understanding in a useful way towards the objective of an effective description of the observed phenomenology, i.e. towards a description that enables us to act in a way that we consider satisfactory (Bruner, 1994).

As we saw in the first two parts of the book, models are often formulated in the form of differential equations (in which case, as we know, they are called dynamical systems), linear ones being a particularly simple form of the same (see Chapter 4). A fundamental characteristic of a linear dynamical system is that if changes are made to the initial values of the variables (or input parameters), the changes in the values of the variables over the course of the dynamics that these originate from, follow exponential laws, and are therefore easily calculated and predictable (see Chapters 4, 11, 14 and 18). In general we could interpret a linear dynamical system as a *black box* in which small variations in the output variables are caused by small variations in the input variables, and similarly, large variations in the output variables are caused by large variations in the input variables. In a nonlinear dynamical system, on the other hand, small changes to the system's input *can* cause (and usually do cause) a large effect on its output, as occurs, for example, with chaotic systems.

Reductionism and linearity

Despite the fact that linear models are, in reality, an exception in the panorama of models that we can construct to describe reality, they have received particular attention, above all in the past, for a variety of reasons. One reason is that the mathematical education of scientists, particularly of those that study natural sciences, has been focused on linear mathematics for at least three centuries (basically from when the significant development of analysis, undoubtedly the most effective mathematical tool for natural sciences, began). On one hand, the reason for this was that linear mathematical techniques are simpler than those of nonlinear ones, as we have seen in Part 1, and on the other hand (perhaps above all), because linear mathematics is undoubtedly easier to grasp immediately than nonlinear mathematics. A second reason for this leaning towards linear mathematics is that there are several cases in which linear models provide effective descriptions of certain phenomena; the brilliance of the latter is such that it overshadows the other cases, the majority, in which linear models are not applicable and lack adequate descriptive techniques.

The techniques of linear mathematics are effective in all cases in which the dynamics of a system can be described simply as the sum of the dynamics of the subsystems (or single components) that comprise it. This method, reductionism, is strictly linked to the linearity of the dynamics; it is, so to speak, the other face, as we saw in Chapter 1. If reductionism is effective, in fact, this means that we can break down the systems, study the micro dynamics of the parts and obtain the macro dynamics of the system as the sum of the micro dynamics. We thus assume that the microscopic behaviour of a subsystem of small dimensions is directly proportional to the macroscopic behaviour of the system of large dimensions, by virtue of which the relation between dynamics at the microscopic level and that at the macroscopic level is equal to the relation between the two dimensions (or between the two scales). These hypotheses are identical to the assumption of a linear relation between cause at the micro level and effect at the macro level.[114]

Angelo Vulpiani (1994) proposed an interesting relationship between the propensity to linearize the descriptions of phenomena and a precise physical fact. We live on a planet that is a relatively cold body, therefore many of the natural systems that we observe are close to a condition of stable equilibrium, which often allows us to consider them 'almost' linear, ignoring any existing nonlinearities. It is precisely the fact that the majority of these systems are 'almost' linear and display, therefore, almost regular behaviour, that has enabled us to recognize the existence of laws of nature that do not change over time and that can be described in mathematical terms. We could possibly even go so far as saying that the same mathematics that has developed over the course of mankind's evolution as one of the ways in which the human mind works, has assumed its current form, precisely because the experience of each one of us is such that it is relatively simple to acknowledge the 'almost' linearity, when such is present in physical phenomena and to describe it effectively, while nonlinearity, despite the fact that its effects are evident, rarely, or rather almost never, provides us with a useful description for forecasting.

Nonlinearity generates unpredictable effects, while linearity does not, therefore a linear tool, even if substantially inadequate to describe natural and social phenomena, with the exception of a very limited number of cases, erroneously appears to be more useful and more correct than a nonlinear one, because the latter does not allow us to make predictions, whereas the former does. Basically we like, and we need, to make predictions. This has often ended up forcing the hand of intellectuals, pushing them to search for linear descriptions of phenomena, even though the latter were effective in only a small number of cases.[115]

---

[114] For example, linear hypothesis is the basis of Fourier's well-known theorem, according to which in certain conditions, the evolution of a magnitude can be expressed as the sum of (in principle) infinite sinusoids, each of which evolves independently from the others, as if they do not interact with each other. By virtue of this theorem, if a system can be transformed into a series of input components, it can be studied, focusing only on each component and the final overall behaviour is the sum of the partial behaviours of the components. This technique is known as spectral analysis and it is widely applied in various fields, such as, for example, the analysis of signals.

[115] It is a bit like the story of the drunk who, one night, is looking under a street lamp for his house keys that he has dropped on the ground. A passer-by asks him where he lost his keys and he replies 'over there', indicating a place a bit further away in the dark. The passer-by asks perplexed: 'Well why are you looking for them here, if you lost them over there?' And the drunk: 'Because there's light here and I can see, it's dark over there and I can't see anything!'

Linearity reigns where descriptions in these terms are easily formulated and work well. Beyond this, in the domain of nonlinearity, it appears, at least on first sight, as if the behaviour of the systems cannot be described by comprehensible rules. Thus, for example, the mechanisms that control the dynamics of the terrestrial atmosphere are well-known, but the dynamics that originate from it are very different to linear ones (see Chapter 16 on the Lorenz equations), which makes it almost impossible to make weather forecasts beyond a few days, unless on purely statistical grounds (such as it is hot in the summer and cold in the winter,...). This is due to the fact that, even though the laws that control atmospheric movements and thermodynamics are known, as they are nonlinear, we have situations of instability, in which small factors, such as small local fluctuations (or even the inevitable measurement approximations), can be enormously amplified over time. The system's evolution, therefore, although piloted by known mechanisms, appears unpredictable, as if it were subjected to the effect of a free inscrutable will acting arbitrarily.

A reflection on the role of mathematics in models

Mathematics, at least in one of its many facets, can be defined as the art of creating models, extremely abstract and simplified models, models, so to speak, in black and white, that describe the deepest essence, the skeleton (or, rather, what seems to us to be such) of a real situation. Of the many ways possible, this art has evolved by choosing a path that adapts to the interaction between man and the outside world or rather with the image that he 'distils' from the outside world. Mathematics applied to the description of phenomena is basically an effective set of tools and techniques that, to a certain extent, help keep the real situation under control. But it is also, and above all, the mental approach that is at the root of these tools and techniques. We construct a model that 'corresponds' to a choice of the aspects of the nature or the society that we are interested in and we express it using mathematical formalism (a formal system). Then, and this is one of the most interesting aspects, we construct a theory of *the particular* mathematics used in the model (for example, linear or nonlinear ordinary differential equations, partial differential equations, cellular automata, tensor calculus, fractal geometry, etc.), or we use one created independently of the specific situation. The use of mathematical theory therefore allows us to

Linearity is like this: it provides nice easy solutions that are often off-target with respect to the problems they are attempting to solve, and therefore substantially wrong.

operate on the model in question and to deduce its outcome. These results, referred to the real situation through a process of decoding, give concrete and above all effective information that can often not be obtained from observations alone.

In general, there is a dynamics typical for the model–reality relationship that evolves through a series of phases (Figure 25.1).

We codify the elements of the real system, which we believe to be the most important, in a formal system, making the necessary assumptions. Then we manipulate the formal system in various ways, seeking the consequences of the assumptions made; this operation is called 'implication'. The aim of the implications is to simulate the changes observed in the real system. We actually aim to establish relations between the changes observed, at the origin of which we assume there is a form of causality. If we find implications that correspond to the events observed, then the formal system is adequate and we use it, decoding it, to represent the events that in this way we can relate to one another. If the causality that links (or that we assume links) the phenomena can be successfully 'explained' by the set of encoding, decoding and implication processes, then we have built a model. Each time that we attempt to give a meaning to the world, we are attempting to construct a relation of this type.

This ability to describe reality provided by mathematics, that we apply when we construct and use a model, becomes increasingly surprising as we gradually consider more sophisticated examples. It often occurs that calculations made on a sheet of paper anticipate observations, in the sense that apparently incredible theoretical results are confirmed after experimental analysis. Given due consideration, the effectiveness of a tool as abstract as mathematics to describe reality appears surprising.[116] This is particularly

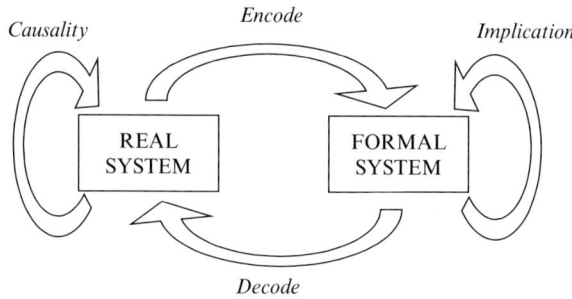

Figure 25.1 Relation between a real and a formal system (or model).

[116] The American, Hungarian-born, physicist Eugene Paul Wigner, Nobel prize for Physics in 1963, in a renowned conference held at New York University in May 1959, entitled *The Unreasonable Effectiveness of Mathematics in the Natural Science*, the text of which was

true in several of the natural sciences, but, at least in part, it is also true in social sciences; the latter also, as a matter of fact, increasingly make use of mathematical tools.

Social systems are usually characterized by very large fluctuations and by a high degree of nonlinearity. In general, it is very difficult to identify the presence of regularity in them. Social systems, therefore, do not lean towards treatment in mathematical terms similar to that which can be performed on a part of natural systems. The most significant intervention of

published the following year, questioned why an abstract discipline such as mathematics worked so well when applied to the description of physical phenomenology. At the conference, he stated: 'The enormous usefulness of mathematics in the natural sciences is something bordering on the mysterious and [...] there is no rational explanation for it. [...] It is true, of course, that physics chooses certain mathematical concepts for the formulation of the laws of nature, and surely only a fraction of all mathematical concepts is used in physics. It is true also that the concepts which were chosen were not selected arbitrarily from a listing of mathematical terms but were developed, in many if not most cases, independently by the physicist and recognized then as having been conceived before by the mathematician. It is not true, however, as is so often stated, that this had to happen because mathematics uses the simplest possible concepts and these were bound to occur in any formalism. As we saw before, the concepts of mathematics are not chosen for their conceptual simplicity – even sequences of pairs of numbers are far from being the simplest concepts – but for their amenability to clever manipulations and to striking, brilliant arguments. [...] It is important to point out that the mathematical formulation of the physicist's often crude experience leads in an uncanny number of cases to an amazingly accurate description of a large class of phenomena. This shows that the mathematical language has more to commend it than being the only language which we can speak; it shows that it is, in a very real sense, the correct language. [...] A much more difficult and confusing situation would arise if we could, some day, establish a theory of the phenomena of consciousness, or of biology, which would be as coherent and convincing as our present theories of the inanimate world. Mendel's laws of inheritance and the subsequent work on genes may well form the beginning of such a theory as far as biology is concerned. Furthermore, it is quite possible that an abstract argument can be found which shows that there is a conflict between such a theory and the accepted principles of physics. The argument could be of such abstract nature that it might not be possible to resolve the conflict, in favour of one or of the other theory, by an experiment. Such a situation would put a heavy strain on our faith in our theories and on our belief in the reality of the concepts which we form. It would give us a deep sense of frustration in our search for what I called 'the ultimate truth.' The reason that such a situation is conceivable is that, fundamentally, we do not know why our theories work so well. Hence, their accuracy may not prove their truth and consistency. Indeed, it is this writer's belief that something rather akin to the situation which was described above exists if the present laws of heredity and of physics are confronted. [...] The miracle of the appropriateness of the language of mathematics for the formulation of the laws of physics is a wonderful gift which we neither understand nor deserve. We should be grateful for it and hope that it will remain valid in future research and that it will extend, for better or for worse, to our pleasure, even though perhaps also to our bafflement, to wide branches of learning' (p. 7). The theme of the effectiveness of mathematics in describing nature has been discussed by numerous authors: for example, Changeux and Connes (1989), Barrow (1992ab), Israel (1996, 1999), Boncinelli and Bottazzini (2000), Ekeland (2000). We will return to this subject in Chapter 31.

mathematical techniques in social sciences, according to contemporary scientific vision, does not refer to the formulation of laws, as scholars have repeatedly tried to do since the birth of social sciences (see Chapters 1 and 9). The laws and technical aspects of these disciplines are in reality formulated and discussed often without the use of mathematical formalism; for example, the law of equilibrium between supply and demand in economics or, extending the scope somewhat, Mendel's laws in biology, or the phonetic laws in glottology are all laws characterized by a very precise content from a conceptual point of view, whose enunciations and applications use very little, if any, mathematical technique. Today, mathematics applied to social sciences is particularly useful above all when we are faced with certain aspects common to numerous scientific disciplines, such as, for example, transversal concepts regarding the evolution of systems, the onset of chaos and the phenomenology of complexity. Mathematics in this case acts as a tool to make connections, to construct analogies and to recognize and discuss common elements in different ambits.

A reflection on mathematics as a tool for modelling

Sometimes, if the model is too complicated, calculations are difficult or impossible; other times, particularly if the relations within the model are all linear, calculations are simple. Think for example of the birth and the development of geometry. Even though we live on an Earth that is not flat, if not in first approximation on small surfaces, Euclidean geometry, probably founded for reasons strictly linked to land surveying, developed and was successful. Flat Euclidean geometry works very well for practical purposes, so much so that for almost two thousand years it was considered to be the 'true' geometry, a unique and unquestionable description of one of the aspects of physical reality. Only in the past two hundred years or so has its interpretation been reviewed and it has become a particular case, we could say a linear case within a wider framework, in which the spaces are curved. At the same time, it has lost its meaning of a 'true' description of physical reality, and has assumed that of a simple hypothetical-deductive system based on postulates, and not on observations. Nevertheless, this has not diminished its importance in terms of its practical applicability. It was, and still is, a tool that can be effectively used in almost all ambits in which lengths are comparable with the scales typical of the senses that man is born with and in which the curvature of the space considered can be presumed to be negligible.

If we lived in a physical space in which the two, three or more dimensional curvatures were more evident, for example on an Earth with a much smaller radius, or in a very intense gravitational field, we might have developed a geometry different to Euclidean linear geometry right from the start. In the same way, if we lived on a very hot planet, where the fluctuations due to the movement of particles (molecules, ions, ...) were much higher, we might not have developed linear mathematics in such a marked way. On one hand, in fact, the nonlinearity of natural phenomena is too difficult to describe with the type of (linear) mathematical tool that we have developed, and on the other hand, in a context where the abstractions originating from the limited presence of regularities are too difficult, if not impossible, perhaps we would not even have created a tool like mathematics to make abstractions. Mathematics, as we have constructed it, would probably have been useless and perhaps would have died at birth. However, perhaps, on that very hot planet, with excessive fluctuations caused by high temperatures, life would not have even existed, as too large a fluctuation would not have enabled the molecular aggregates that are the basis of biochemistry to form, and therefore we would not exist (Vulpiani, 1994).

This last point inevitably leads us back to the beliefs of the anthropic principle in natural science (particularly in cosmology). According to one of the forms in which this principle can be enunciated, physical reality is precisely what we observe, from among the infinite universes possible, because it couldn't be any other way: if it wasn't so, we living beings wouldn't have formed and we wouldn't be here to observe it.[117] Applied

---

[117] The anthropic principle represents a new formulation of the relation between mankind and the universe as such is intended in a scientific vision. It was introduced in cosmology, between the 1950s and 1970s, thanks to the work of a diverse group of scientists from various schools, and right from its appearance, it has been the subject of heated debate in both scientific and philosophical circles. The anthropic principle takes its roots from the observation of the physicist Paul Dirac, in the 1930s, of a strange and surprising mathematical relation between certain physical magnitudes, whose sense was unclear: the square root of the estimated number of particles present in the universe is equal to the ratio between the electrical force and the gravitational force that act between two protons, a ratio that is expressed by means of the values of the basic constants of physics. The problem regarded in particular the fact that, while the ratio between electrical and gravitational forces is a universal constant that does not change over time, the number of particles, on the contrary, changes during the course of the evolution of the universe. Dirac concluded that the ratio between electrical and gravitational forces as well must not be constant in cosmological time and that some of the laws of physics should be reviewed. Later, in the 1950s, Brandon Carter and Robert Dicke proposed a new interpretation, according to which, in reality, the values of the basic constants are not constant at all in the scale of time of the evolution of the universe, and therefore the relation observed by Dirac, in reality, is not *always* true, but is only true *in this moment in time in the history of the universe*, with the current value of the numbers of particles in the universe and therefore of several atomic elements, amongst which carbon, which represent the constituents of living organisms. Dirac, like anyone else, therefore, couldn't help but observe the cited relation, precisely because it is associated to evolutive

to the principle of the meaning of mathematics, the anthropic principle would respond that mathematics is what it is, i.e. linear, and is not another, precisely because, if this wasn't so, we wouldn't be who we are and we wouldn't be here talking about mathematics.

Social systems are characterized by the almost total lack of linearity, which has constituted and constitutes a huge obstacle to the application of effective mathematical laws which, as we have said on many occasions, reflect an original linear approach, inapplicable to these systems even in first approximation. In brief, in the case of the social sciences, there is no correspondent to the problem of the pendulum and the two-body problem in physics; the simplest cases are already many-body problems. The effect of nonlinearity on social processes is related to many different phenomena. By way of example, consider the effects of scientific-technological innovation on economic development (Bertuglia, Fischer and Preto, 1995; Bertuglia, Lombardo and Nijkamp, 1997; Batten et al., 2000). By definition, the concept of innovation implies something new and when this appears for the first time, it appears unexpected, as if it were a random phenomenon; this also applies to the emergence of new ideas, opinions and fashions, and to the appearance of new products and production processes.

---

processes that lead to the existence of forms of life capable of verifying it. This led to the formulation of the anthropic principles in its weak form, according to which, if conditions in the universe, *at this moment in time and in this place* are ideal to host life, this is explained by the fact that, if this was not the case, there wouldn't be any form of (intelligent) life in it to question these coincidences; if those relations didn't exist and the basic constants did not have those particular values, then life would not exist and we would not exist, in this moment in time and in this place, to ask those questions (Barrow and Tipler, 1986). Note that the anthropic principle is not related to the metaphysical notion that sees nature constituted as it is, in order that man, a creature which, having a soul, bears in him the divine imprint and is thus superior to nature, can subjugate it. It rather uses the effect to explain the cause: the justification of that relation and of the value of the basic constants of physics is found precisely in the fact that we are here to question ourselves about their meaning. There are various 'strong' forms of the anthropic principle, which do not only consider our space-time collocation in the universe, as the weak form does, but sustain the point of view according to which the universe *in its entirety* is what we observe from among all of the infinite possible universes, because it is precisely this universe that has given origin to *conscious* life; i.e. the universe *must* possess those properties that allow conscious life to develop at some stage in its history (Barrow and Tipler, 1986). So man would not be a sort of secondary, random product of the evolution of the universe, but the outcome of a cosmic evolution that tends towards this end. The anthropic principle, particularly in its strong forms, expresses an anthropocentric viewpoint that is not accepted by everyone. For example Jack Cohen and Ian Stewart (1994) write: 'Some people go beyond this to argue the "strong anthropic principle" that Planck's constant has the value it does with the purpose of making intelligent beings possible. We'd rather not go that far: it sounds much like the remark that nature made noses so that you can put spectacles on them' (p. 49–50; see also Penrose, 1989, and Regge, 1999).

The mathematical modelling of social systems in general terms encounters great difficulties in taking into account the changes (technological innovations, new scientific ideas, new social characters, etc.) that are absent from linear mathematics and substantially untreatable in exact terms also in nonlinear mathematics. One of the reasons for these difficulties is linked to the fact that modelling innovation means accounting for it in the model *before* it appears, which deprives it of its novel and random content. Frequently, in stochastic models, programs called random number generators are used for this purpose (and others as well), which in some way simulate a random phenomenon described by a uniform probability distribution. However, another fundamental cause of difficulty lies precisely in the concept of randomness itself, or more specifically in the concept of the sequence of random numbers. While from an intuitive perspective it is a relatively clear concept, in theoretical terms it starts to present difficulties, but when it has to be translated into an algorithm, as an element of a stochastic model of a computer program, it becomes totally approximate and substantially unsatisfactory at a formal level (Dendrinos, 2000).

The search for regularities in social science phenomena

The history of the sciences (more correctly, we should say 'of science', in all its different forms) is very often depicted as a long struggle with the apparent disorder of phenomena in continuous change, in which, however, some regularities do transpire. On occasion, in particular in certain fields, such as those of the so-called 'hard' sciences, this has enabled us to formulate laws, genuine mathematical abstractions, based on the assumption, if not proven wrong by subsequent observations, that what has occurred in the past will also do so in the future.

On other occasions, however, the regularities that we thought we had caught a glimpse of were not sufficient. This has often happened in economics, in psychology, and in almost all of the social sciences. The latter, in fact, have to face several kinds of difficulty that are less evident in natural sciences. In social sciences, we rarely, if ever, observe situations that repeat themselves without changing; in practice there are no identical subjects, and furthermore, objective practical difficulties make it very hard, if not impossible, to conduct laboratory experiments. A new economic policy, for example, cannot be experimented in practice to observe its consequences, because it cannot be implemented eliminating the simultaneous actions made by other actors, thus isolating the effects of the policy under experimentation, without considering, moreover, the risks connected with experimentation itself. A new economic policy can be analysed from a theoretical perspective, can be examined using mathematical simulation tools, different scenarios of its consequences can be studied, but experimentation on a population is extremely hard to carry out.

When the evolution of a system is uncertain, the main question regards the time-scale on which we can extend any presumption of certainty. Natural and social sciences are both affected by uncertainty, but one of the most important differences between the two fields regards the time-scale of the phenomena treated in which it manifests itself.

Therefore, the mechanism of the Enlightenment wanted to interpret the reality of the world like the working of a huge clock, the movements of whose parts are linked together, and are known with absolute precision, in the past as in the future. It cannot be denied that this interpretation of the world has, on occasion, been very successful, such as in astronomy, for example, where Newton's deterministic schema, applied to the calculation of the backwards movement of the planets, has enabled the observation of the positions of the planets in the past, even as far back as the invention of writing has reported, to be calculated with the highest levels of precision.

Newton's law of universal gravitation is a typical expression of the deterministic approach, as were the laws of Kepler some time previously, even though the latter had a purely empirical meaning, were not formulated in a real theoretical schema and were still partially influenced by Aristotle (Koyré, 1961). The theoretical framework was created by Newton, with the law of universal gravitation, only 80 years after Kepler, in 1609, had made the laws from which it originated known. Inventing and using the tools of differential calculus, Newton started to give physics that hypothetical-deductive form that still today distinguishes it from the other sciences. Newton's law of universal gravitation, however, is only effective in first approximation, in which there is an abstract universe made up of two bodies subject only to a reciprocal interaction at a distance, to describe which, Newton invented *ad hoc* the concept of the force of gravity. This concept allowed him, in fact, to construct a differential equation as a model of the dynamics of the two bodies which portrayed the idea of a force of gravity as a cause of the movement (the fall) of one body towards another with constant acceleration.[118]

---

[118] We are obviously referring to the second principle of dynamics: $F = ma$, in which acceleration a is the second derivative of space with respect to time and F is the force of attraction that acts between the two bodies of mass m and M. Newton assumes (law of universal gravitation) that F is: (1) inversely proportional to the square of the distance r between the two bodies, and (2) proportional to the product of the masses, in the following form:

In this simple case, the model is extremely effective. For two bodies the dynamics are abundantly clear, easy to calculate for any time interval, starting from known initial conditions, regardless of whether we are dealing with an apple that falls from a tree or the Moon that rotates around the Earth. The two-body model, based on the gravitational law, makes some approximations, as it only takes the main terms in play into account, a system of two bodies and not more than two, such as the Sun–Mars, or the Earth–Moon systems, but not the Sun, Mars, the Earth, and the Moon all together. In a universe of two single bodies, the law of dynamics imagined as we naturally see it, i.e. linear, is effective for the purposes of calculating the evolution of the system: Newton's law is linear.

However, there are numerous bodies in the universe, and only in first approximation can we choose two and consider the others very small and very distant, for the purpose of creating the right conditions for conducting experiments to verify the law of the dynamics, that describes abstract cases and that has been conceived on the basis of approximated experimental evidence. But if the bodies, beyond the pair considered, are numerous and their presence cannot be ignored, then in practice, we no longer observe regularity in the dynamics, the linear model is no longer effective and evolution over time becomes very difficult to predict.

In a universe with many bodies, the linear law is no longer effective to describe the global dynamics. Even if we break down the set of the interactions into single terms (single pairs of bodies), according to the reductionist approach, what we obtain does not enable us to effectively describe the global dynamics of the system. In the long term, the prediction even becomes impossible. As we said in Chapter 16, according to numerical simulations (not according to the solution of the equations of the many-body model) carried out by Jacques Laskar (1989), the solar system itself, for periods of time in the order of hundreds of millions of years, appears unstable and its evolution substantially unpredictable.

The situation that is almost always encountered in social sciences is even more complicated than that described for a gravitational many-body system. In this context, the reductionist approach is almost never useful, i.e. isolating very small systems that are easy to describe, disregarding, at least in first approximation, the influence of external elements, as occurs, on the

$$F = G\frac{mM}{r^2}$$

where G is the constant of universal gravitation. Solving the differential equation as written above (the first differential equation in history), Newton demonstrated Kepler's laws, which, in this way were deduced by the assumed hypotheses (1) and (2).

other hand, for the movement of the two bodies in reciprocal gravitational interaction, when interactions of a lesser intensity with other bodies are ignored. But beyond that, even if this makes sense, in social sciences, isolating specific relatively simple situations would result in (still) not having an accurate and effective analytical mathematical tool for description and prediction, like Newton's schema for astronomy.

For example, let us consider that which is commonly called the 'law of the equilibrium between supply and demand'. It makes sense only when the operators on the supply side are numerous and similar to one another, and the same goes for the operators on the demand side. When we isolate a single salesman or a single buyer (the analogy of the two-body problem with respect to the many-body one), we can no longer talk about equilibrium between a variable supply and a variable demand, and the law loses its significance. The question we have to ask ourselves, however, is whether in a real market with numerous actors it makes sense to isolate a single seller–buyer pair.

To consider one single pair, as in the case of the regime of bilateral monopoly, considerably changes the entire economic dynamics. Can the behaviour of the two members of the seller–buyer pair, in other words, be considered without taking the presence of other potential sellers and/or other potential buyers into account? Is it correct to consider the interaction between the two as a basic interaction to be introduced into a more general system, as Newton does with gravity? The answer is obviously no: in a market formed by a single seller and a single buyer, the negotiation mechanisms are profoundly different from those in a market in which numerous actors interact with one another. In a regime of bilateral monopoly, each of the two actors knows that he can *directly* influence the other, and that he, in turn, has to suffer the effect of the decisions of the other party; everything depends on the strategy of the two parties, on their negotiating skills and, naturally, on their strength.

Just like a many-body system with respect to a two-body one, the system made up of numerous actors is intrinsically different from the simple sum of many isolated pairs. In social sciences we therefore find ourselves in a situation in which mathematics, which for the most part is linear and is therefore effective to describe the regularities of natural nonlinear systems only as far as it is acceptable to break these down into simpler, isolated systems, presents even more problems when it is used as a tool to construct models capable of describing the dynamics observed.

26 Some aspects of the classical vision of science

Determinism

Before actually going into the merits of complexity, we would like to define a framework, albeit partial, of the scientific thought in the eras, prior to modern times, in which the concept of complexity was developed.

Let us return to astronomy. In the long history of attempts that man has made to formulate predictions for the future, a crucial moment, as we know, was Newton's discovery that the motion of the planets could be described by a particular type of equation, differential equations, that he himself invented, which, therefore, became the main tool, if not the only one, that enabled scientists to look into the future. From astronomy and, in particular, from the successes achieved in celestial mechanics based on Newton's brilliant intuitions (1687), classical science, and more specifically mathematical physics that originated from celestial mechanics, started their extraordinary development.

Mathematical science was the first of the sciences to achieve considerable success in the study and formulation of rigid deterministic laws of nature expressed in mathematical language. Subjecting the universe to 'rigid' laws means proposing a deterministic model of the universe. The laws of physics are precisely this: sets of relations that characterize natural states – those observed – and distinguish them from those not observed (Ekeland, 1995). The evolution of a system, from this point of view, is the passage of the system through successive states that actually occur and therefore are natural, excluding states that do not occur.

The model that the laws depict, as we have observed on a number of occasions, does not necessarily have to be linear; it can be so sometimes, but in any event, only in first approximation. From the Enlightenment era up until at least the beginning of the twentieth century, science often, or rather almost always, has confused that fact that direct links between cause and effect could be identified and expressed by laws, i.e. determinism, with the linearity of the laws themselves, i.e. the assumption that the link between causes and effects is a relation of proportionality. Attributing the properties of linearity to determinism leads to the assumption that a deterministic, thus linear, model must necessarily contain all of the information needed for an *exact* and *complete* prediction of the future, as occurs in linear models.

The confusion of these two concepts, determinism and linearity, was precisely one of the root causes of the philosophical and epistemological error committed by Pierre-Simon Laplace in his concept of the world. According to Laplacian determinism, if the equations of motion are known, nothing new can occur that is not already contained in the initial conditions (Robert, 2001). The famous assertion that Laplace makes in his *Essai philosophique sur les probabilités* (1814), an extremely widespread book among all the cultured persons of the nineteenth century: 'We ought then to regard the present state of the universe as the effect of its anterior state and as the cause of the one which is to follow. Given for one instant an Intelligence which could comprehend all the forces by which nature is animated and the respective situation of the beings who compose it, an Intelligence sufficiently vast to submit these data to analysis, it would embrace in the same formula the movements of the greatest bodies of the universe and those of the lightest atom; for it, nothing would be uncertain and the future, as the past, would be present to its eyes. The human mind offers, in the perfection which it has been able to give to astronomy, a feeble idea of this Intelligence. Its discoveries in mechanics and geometry, added to that of universal gravity, have enabled it to comprehend in the same analytical expressions the past and future states of the system of the world. Applying the same method to some other objects of its knowledge, it has succeeded in referring to general laws observed phenomena and in foreseeing those which given circumstances ought to produce. All these efforts in the search for truth tend to lead it back continually to the vast intelligence which we have just mentioned, but from which it will always remain infinitely removed. This tendency, peculiar to the human race, is that which renders it superior to animals; and their progress in this respect distinguishes nations and ages and constitutes their true glory' (Laplace, 1814/1967, p. 243).[119]

---

[119] In the introduction to her Italian edition of the *Opere di Pierre Simon Laplace*, Orietta Pesenti Cambursano observes how Laplace accepts determinism 'not so much because it leads to the ability to grasp the data (. . .) in all of its truth, but more because it is the only guarantee that science has to predict (. . .). If science wants to be true science, it has to present the events it refers to in a sequence based on the cause–effect relation (. . .) otherwise it risks the erosion of any system of human knowledge' (Pesenti Cambursano, 1967, p. 12, our translation). Although Laplace is seen in the history of science as the standard bearer of determinism, we know that he also played a fundamental role in a variety of fields, such as the development of probability calculus. In his cited works, he also states, specifies, comments upon and extends

Laplace believed that the laws of nature were deterministic, nature doesn't choose and doesn't make mistakes; the main purpose of science is precisely to identify, through calculus and analysis, the 'necessary' sequence of the events that occur in nature (Boyer, 1968).

What Laplace didn't take into account was that the nonlinearities of system dynamics make their evolution unstable and that, by virtue of this, 'small' uncertainties that are intrinsic to *any* measurement of *any* magnitude and that inevitably create a sort of area of uncertainty in the determination of the values of the state variables at a certain time, in general amplify over time. In nonlinear systems, therefore, practically in the totality of real systems, because the area of uncertainty amplifies over time, future predictions are less accurate the further into time we seek to look.

In Laplace's deterministic vision, there is a further element of error: the superior Intelligence that Laplace speaks of is intended to mean the ability to know the data, i.e. to *measure* it, superior to that of man, who is forced to circumvent the fact that it is impossible to have full knowledge of the present by using probability calculus (see note 119). If, says Laplace, an Intelligence that is much better than us at collecting the initial data could have full knowledge of the present, as it possesses a powerful and effective tool for the prediction of the evolutive dynamics that we also possess, i.e. the deterministic laws that, presumed to be linear, do not display any problems of instability, then the Intelligence would be able to clearly see into the future and into the past, without uncertainties or approximations. In theory, still according to Laplace, all we need to do is to eliminate the area of uncertainty and the linear dynamics would manage to predict the future without approximations. In other words, we have an excellent calculation tool, but our knowledge of the data is too limited and approximate to make a complete and exact prediction of the future; the best we can do to reduce (but not to eliminate) the uncertainty is to use a probability distribution.

In addition to the fact that the dynamics of real systems are not linear, this conception does not take into account the fundamental fact that

---

the law of large numbers formulated by Jakob Bernoulli in 1713. According to Laplace, however, the calculation of probability was not part of the understanding of natural events, unless it regarded measurement errors, the result of the imperfection of our senses (in his *Essai philosophique*, he goes as far as writing that probability theory is nothing more that common sense expressed in numbers); the concept of randomness, therefore, in this context, is not in antithesis with the deterministic vision. Given Laplace's vast range of interests and the depth of his reflections, Pesenti Cambursano (1967) again warns against erroneously attributing a dogmatic approach to Laplace with regard to the deterministic theory that, in reality, was alien to him.

measurement approximations are *intrinsically* non-eliminable and that this does not only relate to the ability to take good measurements, but to other practical and conceptual matters that causes an area of uncertainty in any measurement to *always* be present. On one hand, for example, there is the inevitable observer–observed interaction, which due to the effect of the measurement itself, causes an alteration of the state of the observed system with respect to before the measurement.[120] On the other hand, there is the fact that a deterministic mathematical law requires data in a numerical form, but an 'infinitely' precise number means a number with infinite decimal places, i.e. a real number, that would require infinite time not only to measure, but also simply to be formulated. Necessarily, at a certain point, even an Intelligence much better than us at making measurements would obtain a result of the measurements and would have to truncate the sequence of the decimal places, thus approximating the real number with a rational number and introducing a sort of area of uncertainty in the data, which, therefore, would necessarily increase due to the effects of nonlinear dynamics (we will return to this point in Chapter 31).

Given the above, one would question the sense and the real validity that can be attributed to a set of 'exact' deterministic laws, that are claimed to be able to predict the future without uncertainties, as long as in possession of complete data without approximations. These laws, in reality, would have a purely theoretical significance because they could never be realized, practically, in the impossible conditions of complete knowledge of the data that they require to 'work best'. All that both we and the Intelligence can do is to attempt to provide deterministic laws based on 'approximately exact' data, that provide 'approximately exact' predictions. Laplace's Intelligence would be much better than us at reducing the area of uncertainty of the initial data, but it could never eliminate it; as the deterministic laws that it also possesses, are generally nonlinear, also the predictions that it can formulate are destined to see the area of uncertainty grow over time, just like our predictions.

In other words, applying mathematical laws, that, as mathematics, are abstract (however formally perfect in their conception they might be) to the description of real dynamics leads to two orders of difficulty which not even the Intelligence capable of a complete knowledge is able to remove: (1) the

---

[120] This concept was completely unknown in Laplace's time. It was defined precisely and accepted reluctantly only with quantum physics in the 1920s and 1930s (Heisenberg's uncertainty principle and the principle of superposition of states). However, the concept is not limited to quantum physics and can be extended and applied in different forms to many (if not all) natural and social sciences.

area of uncertainty is intrinsically uneliminable in a measurement made in finite time and (2) the area of uncertainty amplifies over time.

The principle of sufficient reason

In his renowned affirmation, Laplace sought to contradict the finalism in the formulation of laws, which does not assign a law in a direct manner, but identifies it 'indirectly', assigning an 'end' that, through such a law, can be realized in a physical world. Determinism, on the other hand, is based on Leibniz's principle of sufficient reason, namely on the search for causes: 'The other is the *principle of sufficient reason*, by virtue of which we consider that no fact could be found to be genuine or existent, and no assertion true, without there being a sufficient reason why it is thus and not otherwise – even though we usually cannot know what these reasons are'[121]. It assigns a cause to every event and sees a meaning of the event itself in that.

Leibniz's principle of sufficient reason was one of the basic assumptions of the philosophy of Western civilization from the Enlightenment era to very recent times. It expresses one of the fundamental requirements of scientific thought of the classical era: if the real phenomenon can be expressed by a set of concepts, the relations between the data must find an adequate correspondence in the 'logical' relations of said representative concepts. Because in daily experience the links between phenomena appear to be much more evident when there is proportionality between the causes (at least the apparent ones) and their effects, than the numerous cases in which said proportionality does not exist, then it ends up seeming 'logical' to assume as a general rule that small causes *always* determine small effects and that large causes are *always* at the origin of large effects. In this context, a small effect can only have been provoked by a small cause, and a large effect can only have been provoked by a large cause. Furthermore, in the system of concepts, the conditions of external determinism have to be

---

[121] 'Et *celui de la raison suffisante*, en vertu duquel nous considérons qu'aucun fait ne saurait se trouver vrai, ou existant, aucune énonciation véritable, sans qu'il y ait une raison suffisante pourquoi il en soit ainsi et non pas autrement. Quoique ces raisons le plus souvent ne puissent point nous être connues' Leibniz's original quote, in French, from *La Monadologie*, 1714, paragraph 32; English translation by George MacDonald Ross, 1999, available on the website http://www.philosophy.leeds.ac.uk/GMR/hmp/texts/modern/leibniz/leibindex.html, History Of Modern Philosophy, University of Leeds. Leibniz wrote *La Monadologie* in Vienna, for Prince Eugenio di Savoia; it was first published in Latin in *Acta eruditorum* in 1721 with the title *Principia philosophiæ seu theses in gratiam principis Eugenii conscriptæ*. The original in French was published in 1840 by Erdmann, in his edition of *Œuvres philosophiques*.

mirrored; therefore in any physical–mathematical theory that translates a set of phenomena into a system of differential equations, the arbitrary integration constants have to be unequivocally determined by the initial conditions, which are assumed to represent the causes that determine the phenomena (Enriques and de Santillana, 1936).

The principle of sufficient reason also sustains that natural processes usually correspond to certain maximum or minimum conditions (Leibniz even sees a finalistic aspect of nature in this), a concept that has been considered as an indication of which direction to take in the search for laws which, according to the vision of the classical era, 'guide' the development of phenomena, such as, for example, the renowned principle of least action.[122] This approach was very effective in leading to many discoveries that uphold a large part of physical mathematics, both classical and more recent.

Starting from the principle of least action, linked first to the figure of Pierre de Fermat, and then later to Pierre-Louis Moreau de Maupertuis and William Rowan Hamilton, the concept of geodetics was created, as a line of minimum length travelled by light in a three-dimensional space, and we obtain one of the central elements of Albert Einstein's theory of general relativity, where the concept of geodetics is extended to a curved four-dimensional space-time, becoming one of the fundamental elements in the description of nature provided by contemporary physics (Einstein, 1922; Landau and Lifshitz, 1970).

It should be noted (Atlan, 1994) that, in a certain sense, determinism counters the idea that phenomena are fully subjected to the domain of randomness, and it is precisely because of this that sometimes it has to be included in the foundation of a scientific-like cognitive process that refers to reason. The simple fact of undertaking a 'scientific' search for causes, or for mere reasons or even only for natural laws, in fact, implies that we assume the existence of deterministic causes at the origin of recurrences, even if such are not yet known. If a deterministic approach didn't exist, then there would no longer be any reason behind scientific research. As Henri Atlan (1994) sustains, from this point of view, the notion of randomness, in reality, immediately appeared to be complementary to absolute determinism (see p. 254, note 119). Randomness regards above all our knowledge: it

---

[122] The principle of least action, one of the cornerstones of science's classical deterministic vision, was reviewed and discussed in-depth by the mathematician Ivar Ekeland (2000) in the light of contemporary philosophical–scientific developments. Ekeland, although an optimist with regard to the potential offered by science to improve human conditions, is sceptical of a unified vision of the world that the principle of minimum action appeared to propose to scientists of the classical era.

is an epistemological notion that we apply, using probability calculus, to circumvent the difficulties of recognizing deterministic recurrences and laws, thus attempting to reduce our *ignorance* of causes explicitly proportional to the phenomenon; or, we add, to get around the *lack* of causes proportional to the observed phenomena, as occurs with chaotic situations in which we usually refer to the butterfly effect (Rasetti, 1986).

The classical vision in social sciences

The same rigidly deterministic approach in mechanics, as already discussed in Chapter 1, is also reflected in socio-economic sciences, founded in the mid-eighteenth century, when it was attempted to apply the methods of physical mathematics that had demonstrated themselves to be so effective in the description of certain classes of physical phenomena, to politics. In Enlightenment culture, in fact, politics is seen as a sort of art that the 'enlightened' sovereign has to exercise for the good of his subjects, to perform which he must use the new tools made available by science. This is how political arithmetics came about (Crépel, 1999), namely the science, according to Denis Diderot in the *Encyclopédie*, the purpose of whose operations is research useful to the art of governing the peoples (Dioguardi, 1995).

In particular, this complete faith both in the existence of laws in economics similar to the physical relations that control the material universe (which, incidentally, justifies the introduction of the concept of economic science after the mid-eighteenth century) and in the fact that they are 'correct', insofar as they are the work of God, characterizes the school of thought of the physiocrats in economics. Once the laws had been discovered, the appropriate human behaviour simply entailed following them. What the state has to do is to allow each of us to follow our 'natural' tendencies instilled by God[123] (Lekachman, 1959).

---

[123] Physiocracy rapidly became fashionable throughout Europe in the second half of the eighteenth century. To illustrate the contents of the above text, below are several excerpts from the book entitled *The Physiocrats* written by Henry Higgs in 1897, which includes a conversation between Catherine of Russia and Paul-Pierre Mercier de la Rivière, one of the main disciples of Quesnay, the founder and the most illustrious figure of the school of physiocracy.

'The next eminent Physiocrat to require mention is Mercier de la Rivière (1720–1794), a magistrate who filled for some time the post of Governor of Martinique, and wrote an important treatise, already referred to, *L'Ordre naturel et essentiel des sociétés politiques*, 1767, which Adam Smith has described as "the most distinct and best connected account of the doctrine" of the sect.' (p. 38) [...] 'The question addressed to the Physiocrats was, "If

Adam Smith's work was characterized by a similar approach. The latter had borrowed, in part, his ideas from the physiocrats, with whom he had had direct and personal contact, in particular with Quesnay (Lekachman, 1959). Smith and the other classical economists wanted to interpret the market as being supported by rigid natural laws that act in a deterministic way, put into motion by the action of something similar to real forces, in a similar way to that of the mechanistic schema initiated by Newton around a century earlier and which was being fully developed in the physics circles of the era.[124] The first economists, in Smith's era, in fact, focused on the concept of individual egoism as a driving force in history (Dobb, 1973). This led to the general concept of an economic system put into motion by

your system says 'Hands off!' to the state, and begs it to 'let things alone,' what do you consider the functions of the state to be?". Mercier de la Rivière attempts to create a philosophy of the state. Newton and others had discovered great laws governing the harmonious order of the physical world. There were surely similar laws governing the moral order of the social world, and the motto of the book is a sentence from Malebranche's *Traité de Morale:* "*L'Ordre est la Loi inviolable des esprits; et rien n'est réglé, s'il n'y est conforme.*"'
(p. 39) [...] 'Mention has already been made of the advances of Catherine of Russia to Mercier de la Rivière, but these seem to have been little more than a womanly whim for the fashion of the moment, and to have had little practical result. When the philosopher arrived at her Court at Moscow she had an interview with him, which Thiebault reports as follows:
– Sir – said the Czarina – could you tell me the best way to govern a State well?
– There is only one, Madame- answered the pupil of Quesnay- it is to be just, *i.e.* maintain order, and enforce the laws.
– But on what basis should the laws of an empire repose?
– On one alone, Madame, the nature of things and of men.
– Exactly, but when one wishes to give laws to a people, what rules indicate most surely the laws which suit it best?
– To give or make laws, Madame, is a task which God has left to no one. Ah! what is man, to think himself capable of dictating laws to beings whom he knows not, or knows so imperfectly? And by what right would he impose laws upon beings whom God has not placed in his hands?
– To what, then, do you reduce the science of government?
– To study well, to recognise and manifest, the laws which God has so evidently engraven in the very organisation of man, when He gave him existence. To seek to go beyond this would be a great misfortune and a destructive undertaking.
– Monsieur, I am very pleased to have heard you. I wish you good-day.
She sent him home richly rewarded, and wrote to Voltaire: 'He supposed that we walked on all fours, and very politely took the trouble to come to set us up on our hind legs.' (p. 48–49)

[124] It should be noted that there are several substantial differences between Smith's vision and that of the physiocrats regarding the meaning attributed to natural laws. Smith considered them expressed in terms of real market forces, essentially the competition, that act like the forces of physics and tend to establish determined 'natural values' of equilibrium. The physiocrats, on the other hand, conceived the existence of a 'natural order' of divine origin, thus transcendent, that expressed itself in terms of regular and repetitive models of measurable flows or exchanges.

forces (or impulses), whose movement is controlled by specific economic laws, the discovery and definition of which became the fundamental objective of classical economics – the idea of the potentially creative force of individual egoism: in reality this is the essence of Adam Smith's metaphysical construction of the invisible hand (Dobb, 1973).

Characteristics of systems described by classical science

The theories that draw their origins from the classical school, both in natural and in social sciences, are characterized by several fundamental aspects that in modern language can be indicated by the terms equilibrium, linear causality and negative feedback.

Equilibrium. In Part 1 (Chapter 5) we looked at several examples of physical systems that evolve tending towards a stable state of periodic motion or asymptotic equilibrium. We also examined an example of a dynamical system described by a relatively simple mathematical law such as the logistic map, identifying an initial area of asymptotically stable equilibrium in its evolution. This was the fundamental vision of classical and neoclassical economics: markets and economies tend towards states of equilibrium. It was assumed that markets functioned like mechanisms aimed at the formation of states of equilibrium (market clearing), in such a way that, when environmental factors change the level of demand, for example, this automatically triggers changes in prices which restore the equilibrium between supply and demand, i.e. the previous one or a new equilibrium, unless some type of monopolistic power interferes, impeding said readjustment. In this schema of equilibrium, formulated by Adam Smith, we can identify an analogy with Newton's theory, in which the systems automatically correct deviations from the trajectories of equilibrium determined by basic laws. We thus have a vision of society as being made up of systems whose natural state is equilibrium, in which, therefore, any deviation is damped by negative feedback mechanisms (which we will look at later), such as Smith's 'invisible hand'. When the changes are too great, in order to recover its initial equilibrium, the system tends towards a new state of equilibrium.

An analogy can be seen between Smith's concept of the invisible hand that restores equilibrium to the markets and Charles Darwin's theory of evolution (1859). Darwin saw the evolution of life as a process of natural selection; competition for resources and for survival act like an invisible hand that determines a dynamic equilibrium. This classical vision of equilibrium does not change even with the neodarwinists, who relate said selection to random changes at the genetic level. According to the neodarwinists, the species that survives natural selection are still seen as systems that chance has changed in such a way that they are more adapted to environmental changes than the species that do not survive.

Linear causality. In the classical vision, the assumption that systems tend towards equilibrium was supported by another assumption: that there is a direct link between the cause that acts on the system, originating from an environment external to the system, and the change in the system's structure. This led to the idea that, given the same causes, at least in principle, systems always react in the same way. From here, it is then a short step to arrive at the idea that we can identify real laws in the observed recurrences that *guide* (not that *describe*) the evolution of systems, and that therefore we can effectively act on some element external to the system with the intention of provoking the appropriate causes with the certainty of achieving the expected result. We assume, that is, that we can pilot the system's evolution by means of the causes of known consequences, that are always the same type and, above all, that are proportional to the intensity of the cause (Bohm, 1957).

Just as in a linear physical system, a change in the force applied to a body provokes calculable consequences, usually proportionately related to the cause that produced them (see, for example, Chapter 5), in the same way, in classical economics, if the levels of demand in a market change, this is seen as the direct cause of a specific variation of prices.

The assumption according to which we believe we are able to pilot a system's evolution, once we know the laws on which it is founded (the laws seen as cause, not as description of recurrences in phenomena), is the same assumption that is found at the basis of a widespread vision of the nature of management, the idea that managers can successfully control the evolution of the organizations that they oversee, by taking 'sensible' decisions based solely, or mainly, on their knowledge of future changes in the environment. This management concept, similar to the classical vision, marked by linearity, is known as strategic management (Robbins and Coulter, 1998). Alongside it, there is another concept known as ecological management, which sustains that a decision determines only a particular initial array of competencies within an organization, but then inertia and ignorance tend to hold the array still in its initial state; when environmental changes occur, only organizations, which by chance have the appropriate array of competencies, survive. This Darwinian vision, however, did not have much success against the strategic vision, according to which it was believed that the most powerful coalition within an organization chooses a future state for the organization and then works in such a way as to ensure that said choice is realized. In any event, the underlying idea is still that the most important links between long term causes and effects can be identified.

The idea, basically a prejudice, of direct proportionality between cause and effect, has often led to the assumption of the existence of causes linearly related to the effects observed in practice, if not even directly proportional to the latter. There is no doubt that this way of seeing things has often been successful; we can even go as far as saying that, at least partially, it is at the origin of Enlightenment science and therefore of *tout court* science as well. The search for 'physical' causes where it was thought that only superior, and thus, inscrutable free will acted, was one of the fundamental elements that characterized the origins of science, initially as an activity that aimed to control nature, in a cultural environment dominated by ideas that were ascribable to magic, mysticism or various forms of metaphysics. In this way, seeing into the future ceased to be the task of the oracles and became the field of investigation of system dynamics, as it is called in modern language.

Nevertheless, there are lots of situations where the principle of a linear relationship between cause and effect, ascribable to Leibniz's principle of sufficient reason, cannot be applied to. These are situations in which macroscopic effects originate from simple fluctuations, perhaps very small, of a magnitude that is close to a critical value, so much so that said effects appear to have no effective causes. Habit or prejudice pushes us to search for linear causes and, when we don't find them, to continue to look for them in the wrong direction (as in the story of the drunk illustrated in note 113 on p. 242).[125]

Negative feedback. Both Newton's and Darwin's visions are based on the existence of control mechanisms that act like negative feedback cycles, i.e. mechanisms that, when the system moves away from equilibrium, either due to the effect of an external force or due to normal fluctuations inherent to its dynamics, dampen the effects of these, bringing the system back towards its initial equilibrium. The latter regards a system's stability that we discussed in Part 2: in particular, negative feedback means asymptotic stability (Chapter 18). Systems in which negative feedback mechanisms are

---

[125] Often, in certain environments external to contemporary scientific culture, the difficult or lack of identification of causes proportional to observed effects leads to the resumption of concepts that are more pertinent to metaphysics than physics (the latter considered only in the etymological sense); this lack of identification, on occasion, is even assumed to be a demonstration of the presence of superior and inscrutable wills (Cohen and Stewart, 1994).

either exclusively or prevalently present are easier to treat in mathematical terms, because their linear approximation is simpler.

And this is not all. These types of system are also more easily identifiable in the environment; in a certain sense, by guaranteeing the system's stability, albeit asymptotically, the negative feedback defines it, because the system's characteristics remain (asymptotically) unchanged, even if it is subjected to environmental stimuli that, acting on it, push it towards a change in its configuration. Identifying a stable system is simpler than identifying a system which changes its configuration continually and unpredictably; this explains classical science's particular interest in systems of this type.

The idea of negative feedback was extensively developed after the second world war, within the field of cybernetics (Ashby, 1957). Cybernetics, in reality, can be considered a science that anticipated complexity in the investigation of dynamical systems, precisely because it was the first to make use of concepts such as isolated or closed systems that regulate themselves by means of internal feedback cycles. As we will see in the following paragraphs, complexity has overtaken cybernetics because it makes use of new concepts such as, in particular, self-organization and emergence (see Chapter 27); in other words, because it considers systems that evolve towards new states that do not have negative feedback cycles.

In a system controlled by a form of linear causality and maintained in equilibrium by control mechanisms such as negative feedback cycles, there is no room for surprise, i.e. for an unexpected change in the internal structures of the system, or for a paradox, i.e. the emergence of structures that are different to what we were expecting if we returned towards equilibrium; in a system of this nature, opposing forces are settled in a harmonious equilibrium.

The way in which the world was seen, according to the vision of classical science, that we discussed before, is based on the idea of a net distinction between a real situation that objectively exists and a subjective observer who constructs a representation (model) of that objective real situation, observing it without perturbing it. The successful scientist, according to this vision, is he who constructs a body of knowledge, i.e. a set of propositions, on how nature and society function. In the same way, we could say that the successful manager is he who is able to construct accurate representations of his own organization and of its environment, and who takes the best decisions on the basis of predictions that he makes regarding how that environment will change.

One of the most important outcomes of what was discussed in Part 2 regarding the instability of dynamical systems, and, in general, the new vision introduced by the concepts of chaos and complexity, is precisely the substantial review, if not the complete abandonment, of the principle of sufficient reason, one of the fundamental principles of classical science. In Chapter 27 we will be illustrating some of the aspects of the framework of the sciences that is obtained from the new ideas that overcome the rigid definition of classical science; we will look into the theme of complexity in its own right, presenting one of the most typical elements that characterize it with respect to the vision that was upheld in the past: the self-organization of systems.

27 From determinism to complexity: self-organization, a new understanding of system dynamics

Introduction

According to Newtonian theory, as we know, space is assumed to be an absolute entity and time a magnitude independent of the former and reversible. All of the processes described within the Newtonian framework are reversible. In other words, there is no way of distinguishing between the motion of a body subjected to the action of a force and that of the same body subjected to a force opposite to the previous one, but with time that goes backwards; both appear to be natural. As we know, this schema is completely based on the concept of space provided by Euclidean geometry; it was accepted, used and developed for over a century, after Newton created it, and became one of the distinguishing elements of Kantian philosophy.

In the nineteenth century, the theory of thermodynamics was developed by Rudolf Clausius, Lord Kelvin and, later, Ludwig Boltzmann, in which, unlike the Newtonian schema, time plays a fundamental role because the natural processes it considers are irreversible. In the vision adopted in this new theory, over the course of time, the systems lose the energy that they possess in a 'noble' form, i.e. mechanical energy, that may change from kinetic to potential and vice versa. The noble energy degrades into heat, namely a form of energy that is no longer able to fully return to its mechanical energy form. This is the second principle of thermodynamics. This is what makes the evolution of systems irreversible; the transformation of mechanical energy into heat occurs in ways other than those of the reverse transformation; in addition to this, explains Boltzmann, simultaneous to the degradation of the energy from mechanical to heat, the systems themselves become increasingly disordered. In this new framework, the process in the opposite direction, i.e. what we would obtain going backwards in time, seems unnatural as it involves the complete transformation of heat into mechanical energy and the simultaneous increase in order of the system, without generating other effects on other systems connected to the former.

A ball that bounces according to the Newtonian scheme bounces elastically and always reaches the same maximum height; the process is the same even seen backwards in time. In the thermodynamic schema, on the other hand, the maximum height of the bounces decreases over time due to the gradual transformation of the total mechanical energy of the ball into heat (more precisely, in an increase of the internal energy of the ball system), during the subsequent inelastic bounces. This process, seen backwards in time, would appear, on the other hand, like the motion of a ball that, bounce after bounce, cools and gets increasingly higher, drawing mechanical energy *exclusively* from its 'heat content' (i.e. from the internal energy), which progressively diminishes. A process such as this, alien to anyone's experience, would appear to be completely unnatural.[124] Recalling what we discussed in Part 1, the first case, the ball that bounces elastically, is similar to that of the pendulum that oscillates without friction, described by equation (5.2), while the second, the ball that bounces to progressively lower heights, is similar to that of the damped oscillation of the pendulum, described by equation (5.4).

In the first 20 years of the twentieth century, Einstein developed the ideas of the theory of relativity, once again changing the role of time and demonstrating that both space and time (not just space) are relative to the observer who describes them, that they are inseparably linked between them in a single entity called space-time, the geometry of which is not Euclidean, but is curved due to the presence of masses, and it is precisely this that causes the motion of a mass towards another, something that in the Newtonian schema, was called gravitation.

A few years later, in its new quantum paradigm, physics demonstrated that matter and radiation are both entities that enjoy both the characteristics of corpuscular particles, and those of waves, and that they display the properties of either particles or waves depending on how the observer

---

[124] Precisely the fact that no process has *ever been observed* in which work is performed drawing energy from a single source of heat, becomes, in thermodynamics theory, a general affirmation of *impossibility* and is one of the various equivalent formulations that can be given to the second principle of thermodynamics.

interacts with them. In this new framework, there is a sort of compensation between the maximum precisions with which the position and speed of one of these elementary particle–waves can be measured, insofar as it is the act of observing that perturbs the state.

This is how several absolutely revolutionary ideas entered scientific thought. First of all, the idea that there is a fundamental and intrinsic indetermination in the measurement of the state of a particle–wave, eliminates the concept of material point dear to Newtonian mathematical physics. As a consequence of this, the idea that it is possible to measure both the position and the speed of a body simultaneously and with arbitrary precision is refuted. Secondly, it imposes the idea that an observer cannot be detached from the real situation he is describing. The act of measuring in itself perturbs the state of what is being measured, which leads to the abandonment of the idea that we can obtain a totally objective description of a reality detached from the observer and in itself objective. Lastly, a further and fundamental revolution is given by the new role that chance and probability now assume, no longer the expression of our ignorance of the details of the phenomena, but intrinsic and unavoidable elements that characterize what we perceive as the system's evolution.

These new revolutionary concepts of contemporary science are absolutely fundamental to the cosmic scale, particularly for the theory of relativity, or to the atomic and subatomic scales, particularly for quantum physics, but appear less relevant to daily experience, characterized by the scales of human senses. This is the reason why they attracted the attention of scientists only in the last century.

In a totally separate context we can look at how quantum physics (as well as the theory of relativity) imposed a reconsideration of the concept of causality, which, however, does not regard the linear aspect of causality; the foundation of quantum physics, in fact, is always linear-deterministic, as the motion is described by a second order linear equation, the Schrödinger equation, which is simply a distant descendant of Newton's equation. The methods of the theory of relativity and those of quantum physics, at present, appear to be specific to their respective sectors and are difficult to extend to other disciplines, despite the fact that several interesting attempts have been made to apply these methods to social sciences (Dendrinos, 1991a).

The new conceptions of complexity

In a later period, when the above-cited theories had already been extensively developed, modern scientific thought gave rise to a further new and revolutionary set of concepts, usually indicated by the expression (the science of) complexity. Complexity, which has been developed in the past few decades, unlike the theory of relativity and quantum physics, appears to be relevant to an enormous range of phenomena on all scales, including those that are more familiar to us.

The fundamental characteristic of complexity is the fact that, in the study of the evolution of dynamical systems that complexity deals with, the nature of the system in question is usually irrelevant. In the vision proposed by complexity, we can identify forms and evolutive characteristics common to all, or almost all, systems that are made up of numerous elements, between which there are reciprocal, nonlinear interactions and positive feedback mechanisms. These systems, precisely for this reason, are generally called complex systems.

It is precisely the transversality of the aspects that are displayed in complex systems, and that characterize the nature of such, that make the phenomenology of complexity relevant to all scales, and not just the cosmic or subatomic ones, as occurred, at least initially, with the theory of relativity and with quantum physics, even those that are more directly linked to our senses. This is true for a wide range of scientific fields, even if distant from one another. For example, complexity attempts to account for collective phenomena such as the turbulence in fluids, atmospheric dynamics and the evolution of weather, the evolution of life, the formation of organized structures in all types of society, or the phase changes that occur, both in matter and in society, where by 'phase changes' we mean, in social sciences, economic, social, political or cultural revolutions or, in general, the appearance of new forms of social aggregation.

The question of how systems organize themselves in response to an action of the external environment is as old as science itself, but the study of how systems can spontaneously self-organize is relatively recent. The forms, both in the strict sense and in a broader sense of characteristics that we identify in the world around us, are only a subset of all of those theoretically possible. Why then do we not observe a larger variety of forms? Traditional science has attempted to account for the forms observed, using a reductionist approach, i.e. seeking laws applicable to the single components of the system, such as gravitation, chemical links, etc. The science of complexity, by contrast, takes mainly systemic properties into account and shows how a spontaneous process of self-organization can take place when a dynamical system finds itself in a state that is distant from equilibrium, without any external force acting on the system.

Classical Darwinian theory conceived the evolution of life as the competitive selection between the various living forms, which are in competition with one another for the limited resources that exist in the environment; neodarwinian theory proposes that the fundamental cause of biological differences is random mutations at the genetic level. In the new perspective that the science of complexity proposes, in contrast, life is seen as evolving by means of continuous tension between competition and cooperation. Competition alone is not sufficient to successfully guide complex systems: greater success is enjoyed, in this vision, when there are forms of cooperation between the different elements (individuals) (Axelrod, 1984, 1997; Holland, 1996, 1999; Allen, 1997b; Staricco and Vaio, 2000).

Complexity contradicts the elements that characterize the classical vision illustrated in Chapter 26, that formed the basis for Newtonian theory on physics, Darwinian theory on biology and classical economic theory. It shows that systems are creative *only* when, in the phase space in which we represent their dynamics (see Chapter 8), we find a particular zone, far from equilibrium, which, using a famous expression that has nowadays become commonplace (Langton, 1990), is known as the 'the edge of chaos'[127] (Lewin, 1992; Waldrop, 1992; Kauffman, 1995). Furthermore, complexity also shows that causality is not linear, but rather that it acts in such a way that the links between causes and effects in the long term dissolve and, in principle, cannot be identified. Lastly it shows that positive feedback cycles are essential to the evolution of systems and that, therefore, they need to be considered in the control of the same. Positive feedback cycles, by amplifying the effects of the perturbations ('the new elements'), act in such a way that the paradoxes cannot be resolved in a harmonious and static equilibrium, regardless of whether such is asymptotic or otherwise, but are maintained, creating the tension that guides the system's evolution, which leads to the appearance of new forms that are totally unrelated to the stable equilibrium in which the unperturbed system finds itself.

Ilya Progogine, 1977 Nobel Prize winner for chemistry, and other scientists from his school (Prigogine and Stengers, 1984; Nicolis and Prigogine, 1987) showed how some chemical systems display particular characteristics when they are far from equilibrium, a condition in which they are called dissipative systems. Dissipative systems import energy and/or information from the environment, which, dissipated in the system in question, cause its readjustment into a state characterized by new structures that do not exist

---

[127] On occasion, other authors use different expressions such as, for example, 'edge of disintegration'. We will be using the expression cited in the text which we consider to be the most appropriate, as well as the most widespread, following the success of a number of books on the theme of complexity that make wide use of it, such as Kauffman (1995).

in the state of equilibrium. In a dissipative system, the symmetry and the uniformity of the state of equilibrium are lost in favour of a new structure that is created at the expense of the energy absorbed and/or of the information that has reached the outside of the system and that are dissipated in its interior. The dissipative activity acts just like part of the process of creation of a new structure in the system, which is thus called dissipative. Unlike dissipative structures, on the other hand, stable equilibrium in a system is reached by the same by freeing energy, not absorbing it: stable equilibrium is so precisely because it occurs in a condition of minimum energy.

Dissipative structures emerge though a process of self-organization that takes place spontaneously in correspondence to certain critical values of the parameters that control the system. In this process the elements of the system organize themselves spontaneously to produce a new schema, according to a real *bottom-up* process, that is to say a process in which the general framework emerges as a consequence of the formation of details.

Self-organization

The mechanism at the origin of self-organization can be explained as follows. Random changes, disequilibrium or something else, drive the system to explore new positions in the phase space close to that occupied, without being subjected to an external force. The changes act by therefore pushing the system on a different trajectory towards an attractor that represents the new self-organized state. In other words, what happens is that various types of fluctuation allow the system to exit the basin of attraction of an attractor and enter the basin of attraction of another attractor. As time passes, the system can evolve towards a stable organization, i.e. towards an attractor and remain there, or it can jump between the basins of attraction of various different attractors. The study of the properties of self-organization of a system is therefore equivalent to the examination of the attractors that its dynamics presents.

The concept of self-organization is fundamentally different to that of selection. For a system in evolution, the attractors are the only available stable states. By self-organizing, the system passes from one attractor to another, *spontaneously* crossing the border between one basin of attraction and another, and therefore only due to the effect of fluctuations in a condition of disequilibrium in which the system finds itself, without the intervention of any external force. Self-organization only considers the *spontaneous* movement of the system from one attractor to another attractor, movement that is not caused by any thrust coming from outside the system. The dynamic schema that characterizes self-organization, therefore, persists regardless of changes in external forces.

Selection, on the other hand, is a choice between different stable states, therefore of equilibrium, that are in competition with one another; this choice takes place with reference to criteria that are external to the system. Selection is a movement in the phase space in a particular direction that maximizes a fitness function of the system, defined in a certain way, which expresses the degree of the system's fit with the external environment. This movement takes place due to the effect of the action of the environment on the system, like a force that pushes the system towards a new equilibrium.

With self-organization, a new order of the system 'emerges', an order of the non-equilibrium, a non-static order, different to the condition of asymptotic stability, in which the entire evolutive dynamics tends to dissolve into a stable equilibrium. The essence of self-organization is the appearance of a new system structure without any external pressure, apart from a simple flow of some form of energy (and/or information). In other words, the constraints that lead to the organization of a new structure are internal to the system in question, result from interactions between the components and are usually independent of the specific physical nature of the components. This, in particular, is what makes the study of complexity of fundamental importance to science: the phenomenology that is displayed is independent not only of the particular system, but also of the nature of the system. The fact that a system exists as such, therefore, with an internal network of interactions, is the origin of certain of its characteristics that we indicate with the term 'complexity'.

As we said in Chapter 1, the general term 'system' is intended to mean a set of parts in reciprocal interaction, that act in such as way as to render said set identifiable with respect to the external environment from which it is separated by recognizable borders. We can identify various types of systems, from those in which the interactions are fixed, in such a way as the components are rigidly connected to one another, as occurs, for example, in an engine, to those in which the interacting parts are *almost* totally free, such as the molecules of a gas, each of which interacts in a very weak manner with the other molecules. In this context, the most interesting systems are those that display both fixed and changing interactions, such as, for example, a cell or a human society characterized by efficient means of communication and transport.

The system displays certain properties that are called 'emergent' because they are not intrinsically identifiable in any of its parts taken individually, but only appear when observing the system in its entirety. With self-organization, new structures sensitive to the values of the control parameters 'emerge': a new order in which the schema produced by the elements of the system cannot be explained in terms of the individual action of said elements, i.e. it cannot be traced back to the behaviour of the single elements of the system. Something new appears, typical of systems in a condition of non-equilibrium: synergy between the elements.

The analytic vision is unable to account for this, as it interprets the system, in its entirety, simply as the sum of the single parts it is composed of; on the contrary, we need to consider each particular component of the system as if it were integrated in a network of relations that link it to all of the others.

The analytic vision is no doubt effective for many purposes, but it is often insufficient because it does not take into account certain characteristics of the system that do not belong to the single components but that pertain to the whole of the components and their interactions. The analytic vision needs to be supported, not replaced by a synthetic one, in which the system is no longer the simple sum of the interacting parts, but acquires new characteristics due to the fact that it is a whole.

This is what makes us distinguish between a wood and a set of trees, a house from a pile of bricks, an engine from a set of metal parts, an organized society from a group of individuals that do not communicate with one another, a living organism from a broth of protein, etc. This holistic perspective is needed to provide a more general and complete level of description that is inaccessible if based on the simple examination of the components one by one, in which the system acquires new emergent properties. How could we think of the ecosystem of a wood based on the trees alone, the concept of a house based on the bricks, the functioning of an engine based on the separate parts, social organization founded on communication based on a set of isolated individuals that do not communicate with one another, or life based on just proteins?

The new schema that emerges from self-organization is an unstable structure that can easily be dissolved, if the critical values of the system's control parameters are changed. A structure of this nature is called dissipative. While the structure of a system in equilibrium cannot be easily destroyed, or rather it can be maintained without any effort, a dissipative structure, on the contrary, requires an effort to maintain it, but relatively little effort to change it.

When we introduce nonlinearities into a system in the form of positive feedback cycles, the result that we obtain is that of keeping the system far from equilibrium, in a state of instability; in other words, it is put into a condition of amplifying the small changes and it is prevented from adapting itself to the surrounding environment.

It is precisely the amplification of these small changes that causes the change in the system's structure in its entirety. The stability, acting in the sense of maintaining the system as it is, damps, suffocates the changes and renders them local; action in conditions far from equilibrium, on the other hand, destabilises the system and makes it receptive to change.

The idea that systems can self-organize, derived from the holistic vision that must be applied to the same, represents a new revolutionary perspective in the study of the evolution of systems themselves. Nevertheless, even though it represents one of the main cores, the said idea does not exhaust the theme of complexity, which presents other distinguishing features, which we will discuss in the next chapter.

28 What is complexity?

Adaptive complex systems

Over the past 30 years, there has been considerable progress in studies and research in a number of separate areas that are, however, united by the expression 'complexity science'. Said research work has led to the development of useful techniques and has generated significant and explicative metaphors. However, to date, the work conducted in this field has not been founded on a coherent 'complexity theory' even though it is a commonly used expression (even we will be using it in the following sections), nor has it been formalized into a system of axioms and theorems that can be generally applied across the various disciplines. Our further investigation of the subject of complexity will not be based on an abstract view of the latter; we will be looking more at how it manifests itself.[128]

---

[128] There is a very long list of books and periodicals that have been and are published on the various branches of complexity science; interest readers will find numerous references in the bibliography of this book. Equally, the list of research institutes that have directed part or even all of their activities to this area would be extremely long. We would like to provide the reader with the names of a few research centres, chosen on the basis of the fact that they provide specialist information on the subject of complexity in the form of working papers or journals monitored by an editorial committee, that are accessible on the Internet without the payment of a subscription fee.

Santa Fe Institute (SFI): is a private non-profit organization, based in Santa Fe, New Mexico, USA, founded in 1984, expressly and exclusively dedicated to research on complex systems; it can be considered the most important research centre on complexity in the world, and incorporates a very large number of researchers active in a diverse number of fields (mainly mathematics, physics, biology and economics), belonging to various institutes and universities, most of which are American; it regularly publishes books, journals and working papers on the subject of complexity; the institute's bulletin is available on http//www.santafe.edu

Monash University, Victoria, Australia publishes an on-line refereed journal, formerly published by *Charles Sturt University* Bathurst, Australia, dedicated to the various branches of complexity science, called *Complexity International*, available on http//www.complexity.org.au.

There is a category of systems that display self-organization and dynamic behaviour that is qualitatively different to both that of simple stability and chaos. At present, said systems cannot be characterized in mathematical terms in a general sense; we will restrict ourselves to saying that their foundation cannot be a linear dynamic, as the latter is too restrictive, as we have seen in the first two parts of the book. Systems of this kind are called adaptive complex systems, and are characterized by the fact that their nonlinear dynamics can give rise to self-organizing and self-reproducing structures, which are capable of interactions that imitate the behaviour of living beings.

From a complexity perspective, adaptive complex systems can be defined as follows: 'An adaptive complex system is an open system, made up of numerous elements that interact with one another in a nonlinear way and that constitute a single, organized and dynamic entity, able to evolve and adapt to the environment' (Gandolfi, 1999, p. 19, our translation). The behaviour of adaptive complex systems displays one of the main aspects of complexity: the response of a system to changes in the environment, i.e. its adaptation to such and to the stimuli that are generated by such. Human beings find adaptive complex systems particularly interesting because each of us, just as all other forms of life, can be considered to be an adaptive complex system, both in terms of the body, from a biological perspective, and also, in particular, with regard to our conscious thought processes (Albin, 1998).

As regards the behaviour of adaptive complex systems, several general characteristics can be identified that we have briefly described below. An adaptive complex system:

1. Is made up of a large number of elements (agents) that interact reciprocally, which cannot be modelled into a linear schema; the elements are connected together in such a way that the action of each element can provoke more than one response in each of the other elements.
2. Interacts with other adaptive complex systems, which, as a whole, constitute the environment in which it is immersed and to whose stimuli it reacts.
3. Acquires information on the systems that make up its environment and on the consequences of its own interaction with said systems; in other

The New England Complex System Institute (NECSI), based in Boston, Massachusetts, USA; (*www.necsi.org*). it is a consortium of scholars from industry and university, whose research is directed, in particular, to the study of complexity in management activities; it publishes various journals, including the on-line referred world wide web-based journal *Interjournal*, available on http://www.interjournal.org.

words, it is sensitive to the information that it receives from the environment (environmental feedback).
4. Identifies regularities in the flow of information that it acquires, and based on such, develops a model (or a schema) that attempts to 'explain' the regularities identified.
5. It acts in relation to the systems that make up its environment, on the basis of the model that it has developed, observes the responses its actions provoke, as well as the consequences of said responses and uses this information to correct and improve the model in question; in other words, it uses the environmental feedback to learn and to adapt.

Basic aspects of complexity

We can identify several different types of complexity, but the fundamental underlying mechanism is always the instability of the system, the main effects of which, in certain particular circumstances, are the appearance of bifurcations (in the sense attributed to the term by René Thom, 1975) in the system's orbits, during its evolution. The system can therefore evolve towards numerous different states, i.e. towards different attractors, each of which is characterized by its own basin of attraction. In general, during the evolution of a system, complexity takes the form of a particular intermediate state between stable equilibrium and chaos, a situation in which the system displays a different behaviour, both from stable equilibrium tendencies, therefore unchangeable in the future, and from chaotic tendencies.

The evolution of a dynamical system generally depends on a set of control parameters. For low values of the same, we have a tendency towards stable equilibrium, whereas for high values we have the generation of chaos, in which, as discussed in Chapters 18 and 19, the dynamic trajectories are characterized by positive Lyapunov exponents. In an intermediate area of the control parameter values, the system displays the tendency to create new configurations, which are different both to those established by asymptotic equilibrium and those established by chaos, and to assume new functional properties that didn't exist for lower values of the parameters. The system displays emergent properties: there in that intermediate area between order and chaos, the area at the edge of chaos, the area of the sudden appearance (*emergence*) of unexpected properties, we say that a dynamical system displays a very particular type of behaviour that we call complexity (Figure 28.1).

When the system finds itself at a point in the phase space in which its properties change unexpectedly, this is commonly known as a critical point,

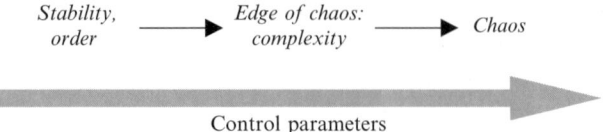

Figure 28.1 Stability, complexity, chaos.

in correspondence to which a phase change can occur that can push the system into a chaotic state or block it in a state of stable equilibrium.

We discussed and examined the dynamics of a mechanism of this nature in Chapters 21 and 22, in which the logistic map was presented as a sort of paradigm of a dynamics which, depending on the value of the growth parameter, changes from a tendency towards stable equilibrium to chaos, via an intermediate stage of complexity, characterized by the presence of a sequence of bifurcations.

A system's complexity manifests itself as the collapse of a state, following internal instability or an action external to the system, and with the adjustment of the system to a new state; the dynamics of a complex system, therefore, appear to be characterized by trajectories that are sensitive to perturbations of intrinsic or external origins (in real systems, in practice, inevitable). Obviously, this brings up the problem of the predictability of the system's evolution. The vision that was adopted for a considerable period of time, from the era of Leibniz's principle of sufficient reason to a few decades ago, is that the origin of the phenomenology that we attribute to complexity is not inherent to the dynamical system, but rather constitutes an undesired consequence of the presence of poorly identified causes or variables that act in a deterministic way. This is in line with the analytic–reductionist approach that had characterized almost all sciences.

As various scholars have pointed out (for example, Davies, 1989), in the past, reductionism represented a real conceptual obstacle that, by ignoring the system's ability to spontaneously create new forms of internal organization, precluded the development of a complexity theory. In the current, post-reductionist era of physics, we no longer look at the single molecules to explain large-scale self-organization, or even the action of a particular molecule on its neighbour. Our attention is now focused on global matters, such as the boundary conditions, the constraints, the distances from equilibrium and so on. It is undoubtedly still true that the properties of the molecules, considered individually, and of their interactions, are ingredients required to obtain a full explanation. Alone, however they are simply not sufficient.

What brings together the various fields of science that adopt the vision proposed by complexity, regards, in particular, several methodological aspects. Firstly, new formal techniques are used, such as formal logic, cellular automata, topological models, etc. These are techniques that generally attempt to 'capture' some of the system's most abstract properties, in areas in which the application of standard techniques is insufficient to produce effective descriptions. Secondly, we are dealing with systems in which the behaviour we are interested in emerges in contingent situations, that is to say, unpredictable and subject to change following even small variations in the control parameters. Thirdly, modelling techniques are used that are not able to formulate quantitative predictions on the system's future evolution, but that can describe several statistical properties and can attempt to predict the evolution in qualitative terms.

In complex systems we can identify several basic common characteristics:

1. The feedback networks are active and give rise to effects when they act close to a critical configuration of the system, in which the latter is close to a phase change, in a situation that is called the 'edge of chaos'.
2. The creative process that takes place at the edge of chaos is intrinsically destructive, because it causes the disappearance of states of equilibrium, and it is fundamentally a paradox, because the new states of the system (configurations) that are created are unexpected and unpredictable.
3. Neither the creative process, that takes place at the edge of chaos, nor its results, can be predicted or planned because the long-term outcome cannot be known; the links between the actions and their outcomes in the long term dissolve and none of the outcomes is under control any longer. The fact that a system finds itself at the edge of chaos doesn't mean that it is at the mercy of randomness or anarchy: the new configurations that emerge at the edge of chaos are the expression of an implicit order that originates from inside the system itself. At the edge of chaos, cooperation does not take place in accordance with a pre-established pattern; there is no one element (agent) that is stronger than the others that imposes its will. We are rather dealing with a bottom-up type process, in which the elements of the system (the agents) face the situation spontaneously, and organize themselves. This self-organization gives rise to new emergent strategies and the interaction itself creates schemas that no agent, on its own, can predict.

'Emergent property' therefore means that a new property of the system appears that was impossible to predict as the overall result of the individual interactions, considered one by one, between elements. In other words, the appearance of emergent properties means that it is impossible to reduce the general schema of the system's evolution to the sum of the behaviours of the individual agents (see, for example, Holland, 1996, 1999).

An observation on complexity in social systems

The concepts illustrated, originally formulated for natural sciences, can easily be applied to social systems as well, i.e. to various kinds of organized societies. This is possible if we consider that each society is made up of a network of individual actors, which interact both with one another, in accordance with the rules of the internal organization of the society, and with the actors of other societies that constitute the environment.

The new vision proposed by complexity becomes particularly important in practical terms if applied to the description of the dynamics of societies and of human communities, ranging from the larger ones, such as populations and urban systems (starting from the first pioneering works by Jay Forrester, 1969, 1971), to those of smaller dimensions, such as the various forms of local organizations (Byrne, 1998; Herrmann, 1998; Christensen, 1999).

In particular, the area of study that regards the management of companies, business management, has received a lot of attention from scholars, both before and above all after the subject of complexity became more prominent (Levy, 1994; Alvesson, 1995; Weick, 1995; Overman, 1996; Stacey, 1996; Thiel, 1998). Traditional management activities have generally conceived linear interactions between the business enterprise and an objective external world, identifying the manager as the person that has to foresee, assess and decide. In the light of new concepts, however, an organization can be seen as a complex system which, in order to survive and develop, is continuously in search of new ways to interpret the events of the external world and which, as a consequence of the feedback it receives from the environment regarding its actions, self-organizes itself, displaying emergent properties, so as to render its interaction with the environment in which it finds itself as favourable as possible. As Gianfranco Dioguardi writes (2000): 'We are (...) in the presence of component elements, of interactions and feedbacks, of phenomena of self-organisation, in a situation of eternal evolution that is constantly displayed in the history of businesses and that emerges from an exchange of both knowledge and of concrete economic facts between the internal and external environments, due to its turbulent nature and therefore difficult to analyse and plan' (p. 82, our translation).

One of the main preoccupations of management science is therefore the understanding of how people interpret reality and how said interpretation can define a context for action. From this perspective, complexity theory presents itself as a reference framework that helps us follow the evolution of organizations and the complex interactions between actors. Conceiving organizations (business, social or other) as adaptive complex systems enables the manager to improve his decision-making and supports the search for new solutions. In nonlinear systems, long-term forecasting is intrinsically impossible; as we have said on many occasions, even significant changes can intervene unexpectedly. This means that flexibility and adaptability are characteristics that are essential to the survival of the organization, just as they are to biological life.

In a complex world, management strategy must therefore be a set of processes aimed at understanding the behaviour of the external environment and controlling the actors that act within the organizations, identifying the dynamics of the organization-system and attempting to provide resources and incentives for future moves (Levy, 1994). Complexity science and management science, in this way, have an area in common, namely the study of phenomena of emergence, an aspect that is not adequately perceived in traditional management and that, here as in other fields, originates from the consideration that the system is more than merely the sum of its parts.

Some attempts at defining a complex system

Chaos theory, just like the thermodynamics of systems far from equilibrium, has been extensively studied by physicists, chemists, and mathematicians. Even complexity theory, which has been around for only 30 years, has been extended to various disciplines such as physics, biology, sociology, economics, etc. Complexity theory sustains that in our daily lives we are surrounded by systems that possess nonlinear feedback mechanisms that can display chaotic behaviour or adaptive behaviour. Systems are assumed to evolve in an open phase space, without a general pre-established schema, but by means of a spontaneous system of organization.

To understand the dynamics of systems, we have to overcome the reductionist, Cartesian-based vision adopted by the classical science of the seventeenth century, as we have repeated on several occasions in this book; complexity emerges only if the systems are described from a holistic perspective. Complexity is generated by the failure of the attempt of the Newtonian paradigm to be a general schema through which to understand the world.

If we simplify things somewhat and, for the sake of clarity, ignore the need for rigour, we can say that a complex system, stimulated in some way, does not appear to be inert, as if it were impervious to the abandonment of the condition of stable equilibrium, nor does it display the tendency to react in a random way, without any coordination between its parts, i.e. as if it were lacking a mechanism that controlled it in a deterministic way. A complex system, on the other hand, *appears* to react to a stimulus in a coordinated way, as if it were animated by a will that enables it to find new answers, i.e. new configurations, when faced with new stimuli; said new configurations, however, are unpredictable to us, which masks the determinism that the system's dynamics is subjected to. We have placed particular emphasis on the verb *to appear*; complexity is not just an intrinsic property of a system, it also, and particularly, depends on who is observing it. We will be returning to this fundamental point on several occasions in this section.

It is clear that this new approach to describing a system's dynamics turns several fundamental assumptions on which conventional concepts are based, upside down. Firstly, complexity theory assumes that evolutive systems are not those that tend towards equilibrium, but rather those that find themselves far from equilibrium, close to disintegration, at the edge of chaos. It is also assumed that in this state, the causality is such that the links between cause and effect are lost in the complexity of the system's evolution, and, therefore, the future cannot be predicted.

This is not only due to the possibility of chaotic evolution, that in itself is an antithesis to the rigid predictions of determinism, but also to the fact that between the evolution towards a state of equilibrium and towards a state of chaos, there is an intermediate situation, in which systems do not evolve towards chaos, but towards new forms of internal self-organization. It is precisely the amplification of the perturbations caused by positive feedback that end up being vital to the system's evolution, while on the other hand, negative feedback traps systems in conditions of static equilibrium, with no chance of evolving. Lastly, it is assumed that complex systems evolve in reciprocal interaction with the environment, thus contributing to constituting the universe in which the same act.

We have still not defined complexity, but we have limited ourselves to saying what the characteristics of dynamical complex systems are, and how complexity manifests itself in them. Continuing on these lines, we can also say what complexity *is not*. Complexity is not complication. A complex system is different and is more than a complicated system. A complicated system is predictable if we have thorough knowledge of all of its components and of all of the relations that unite said components. On the other hand, a complex system is not predictable, if even our level of knowledge is the same as that of a complicated system, i.e. if we have thorough knowledge of all of its components and of all of the relations that link them, because it is a system that is able to independently adapt itself, i.e. to self-organize. Complexity is not a stable equilibrium and is not chaos; it is a third condition in which the system is creative, as if it were displaying intelligent behaviour in adapting to environmental stimuli. A complex system appears to possess the ability to evolve independently, to adapt itself and to improve. Complexity, according to Stuart Kauffman (1994), is a situation of transition, poised between the rigid order of determinism and the anarchy of chaos.

Complexity nests in a shady area of our comprehension of systems. In general, it appears that we associate complexity to anything that we find difficult to understand (Flood and Carson, 1986). The complexity that we identify in systems must not be confused with our ignorance of the systems in question; the two concepts, complexity and ignorance, are substantially different. On one hand, the very complexity of a system may be a cause of ignorance; for example when the resources available for the comprehension of the dynamics displayed by a system of some nature are too limited against the variety of phenomenologies that the system displays, and do not enable us to create an effective model. On the other hand, ignorance, the cause behind errors or imperfections in the understanding of the phenomenologies or in their interpretation, can lead to the construction of models in which the typical mechanisms of complexity are included in order to attempt to describe phenomena that a better knowledge would enable us to describe in other terms. The ignorance of the mechanisms at the origin of a system's dynamics can, for example, lead us into not seeing simple cause–effect relationships or, even, on the contrary, in seeing them when there is no correlation between the components to suggest their existence. Ignorance, therefore, can be the cause of an erroneous interpretation regarding the complexity of a phenomenon that can be explained without having to refer to nonlinear feedback, self-organization or emergence.

It should be said that a system's complexity depends on the models that we construct of the systems in question. This means that the complexity of a system varies considerably depending on the model adopted and that it is always possible to increase a system's complexity by taking more and different aspects into account, and going into further detail. According to Bruce Edmonds (1999), complexity is that property of system models that makes it difficult to formulate a general behaviour for the system, even when we possess almost complete information on the basic (atomic) components of the system in question and on their relationships. In this statement, complexity is related to the increased difficulty of formalizing 'everything' with respect to formalizing the basic components, from the perspective of a representative model. More than an abstract definition, Edmonds proposed an operative characteristic of the concept of complexity, which is susceptible to different interpretations depending on the context, in particular as regards the exact meaning of the expression 'makes it difficult to formulate a general behaviour'. This perspective, in reality, is only applicable to cases in which we possess complete information on the individual basic components of the 'whole' system, but it helps to clearly separate the concepts of complexity and ignorance; ignorance is when the behaviour of a system appears to be unpredictable due to our imperfect knowledge of its basic mechanisms; on the other hand, complexity is when the unpredictability of the system is due to our imperfect understanding of how the individual system components behave in relation to one another within a whole.

And that's not all: complexity relative to a model's structure, according to Edmonds, also depends on the purpose for which the model has been conceived, as the details that are taken into consideration and the assumptions made depend on this. The level of detail at which the model is conceived depends on the use that the model has been constructed for, but the same level of detail, and therefore the system's complexity, also depend on *how* the model is represented, i.e. what language (meaning formal system, i.e. a set of symbols and rules that link said symbols to one another) is adopted to construct it. Remember that we intend the term model to mean, in this context, any representation that we make of the surrounding reality, not only and not necessarily a dynamical system. The model that we construct in the formal system that we choose to use to 'capture' the properties of the system, is only a partial representation of the system. We could say, therefore, that complexity appears to be similar to that property of a real system that manifests itself in the fact that no formal system is adequate to describe (to 'capture') all the properties of the real system and *all* of the aspects of its dynamics.

The dynamics of a complex system elude a complete representation that efficiently describes *all* of its aspects; some formal systems highlight certain elements, other formal systems capture others. Being a complex system transcends, in other words, our skills of formal description that we have developed for systems that do not display properties such as self-organization, emergence, etc. Complexity means, from this point of view, that we can find different irreducible (incommensurable, so to say) ways to interact with the systems, in the sense that we can create different effective models of a complex system, but the different formal systems needed to describe each particular aspect of it are not compatible with one another.

The complexity of a system and the observer

We possess numerous forms of language: verbal language, the language of art, mathematical formalism, etc. Each of these is capable, better than the others, of representing the models that we make in a particular environment. But that's not all; in addition to the above, language itself is an element that stimulates the variety and the organization of models. Possessing an effective language, in fact, means also possessing a precious tool for investigation, to 'dig' into the idea that we are constructing of reality, i.e. basically, to enhance and to provide detail to the model.

In this regard, we note the distinction that John Casti (1986) made between *design complexity*, namely the complexity of a system as it is perceived by the observer, and *control complexity*, namely the complexity of the observer as perceived by the observed system. In other words, as we stated at the beginning of this chapter, according to Casti, a system's complexity is a concept that is related to the complexity of the person observing it: the observer himself is a complex system, and this influences the level of complexity that he perceives in the system he observes.

This vision supersedes the concept of complexity as an intrinsic characteristic of a particular system. Complexity now becomes, in addition to the above, a characteristic of the *interaction* between two systems, both of which are complex. According to Flood and Carson (1986), complexity is associated both to us 'people', and to the systems as they are perceived by us 'people'. This point, which can be applied to an observer–observed pair, becomes particularly important in terms of social systems, which are much more difficult to describe than systems in natural science, so much so that to date there is no mathematical formulation of social laws that can claim to be rigorous and reliable at an equal level to, for example, numerous formulations concerning physics, despite the countless attempts to treat social systems using methodologies at least similar to those used in physics (Cohen, 1993).

Complexity, therefore, also depends on *who* is observing the system and on *how* this person observes it, which complicates matters because it makes it difficult to unequivocally define complexity as a magnitude that can be measured objectively.

Complexity is a concept that we would like to be comparative; we would like to be able to state that a system is more complex or less complex than another system, on the basis of quantitative measurements of complexity that enable us to establish a value according to a suitably defined scale. But this is not possible, at least for the moment, precisely because complexity is a concept that is extremely difficult to define precisely, although, intuitively, it is not difficult to imagine. Dozens of different definitions and different ways to measure complexity have been proposed (Blum 1989; Lloyd, 1989; Zurek, 1989; Gell-Mann, 1994; Horgan, 1995; see also Edmonds, 1999; Bertuglia and Staricco, 2000; Dioguardi, 2000), which refer to concepts such as entropy, information, causality, etc. More often than not, however, these definitions are not sufficiently generalized, but are only applicable to a limited context, or are not very rigorous, or even vague or approximate, particularly regarding the possibility of quantifying complexity defined as a measurable magnitude. Moreover, some of these definitions characterize something that we intuitively see as simple as complex or, on the contrary, that deny the character of complexity to phenomenologies that we see as 'clearly' complex.

The problem of defining complexity, as we have seen, is linked to which aspect (or aspects) of a system the observer considers (predominantly or exclusively) and, if he can, measures; it depends on the observer and on the level of complexity that characterizes the latter, and also, as a consequence, on the language used to construct the (complex) model of the system. This does not in any way mean that complexity is a subjective concept: linked to the observer means linked to the interaction between the observer and the observed, not simply dependent on the opinion of the observer. In this regard, there is the risk that, as there is no rigorous definition of complexity, and as there is neither a method to measure it nor a unit of measurement for it, complexity does not generate a complete and coherent scientific theory, but ends up being a sort of philosophical dimension and nothing more. It is a risk that we must be more than aware of; complexity is not *only* an epistemological matter, in the same way that neither the theory of universal gravitation nor the law of natural selection are.

The complexity of a system and the relations between its parts

There is one key objective element that is common to all complexity theories. If we refer to the Latin etymology (and further back to the Greek) of the term, *complexus* means intertwined: to have complexity, there have to be several component parts connected in such a way that it is difficult to isolate them; connected in such a way, that is, that an interpretation of the phenomenology that they present in an analytic perspective is not sufficient for a correct description, but a holistic vision needs to be adopted. Intuitively, therefore, a system appears to be more complex, the more parts we manage to distinguish and the more numerous the connections between these parts, that we manage to identify. However, both the number of the parts and the number of connections between these depend on the observer and on the way in which the observer looks at and describes the system he is observing.

'More parts' means that the model that represents the phenomenology is more extended, which therefore requires, if the model is written in the form of a program for a computer, more time for its study and a longer processing time. As the components of a complex system cannot be separated from one another without destroying the character of the system itself, the analytical method, i.e. breaking the model down into independent elements, cannot be used to simplify the model. This implies that complex entities are difficult to model, that the models are difficult to use for forecasting and control and that, to put it simply, the so-called complex problems cited above are difficult to treat. This demonstrates why the word 'complex' is often associated with the word 'difficult'.

Complexity can be characterized by means of the two dimensions of *differentiation* and *connection*. Differentiation means variety, heterogeneity, it means that the different parts of the system behave in different ways. Connection, on the other hand, concerns the constraints that link the component parts to one another, it concerns the fact that different parts are not independent from one another, but also the fact that the knowledge of one part can enable us to determine the characteristics of other parts. In the latter sense, connection is also a concept that is similar to redundancy. On one hand, differentiation leads to disorder, to the growth of the entropy, as occurs in a perfect gas, in which the position of a molecule is independent of that of the others. On the other hand, connection leads to order, to the diminution of the entropy, as occurs in a crystal in which the position of a molecule (or of an atom) is totally determined by the position of the molecules (or atoms) nearby. Complexity arises when both of these aspects are present: neither total disorder, that renders the system describable only statistically, nor perfect order, that renders the system totally describable and its dynamics predictable in purely deterministic terms.

Complexity, as we have said, is situated in an intermediate position between order and disorder, or more precisely, when the system finds itself at the 'edge of chaos'. To use a significative metaphor (Gandolfi, 1999), order, complexity and chaos are like three physical states in which the system 'water' can be presented: (1) in the solid state of ice: this is total order, the crystal, the translational symmetry (or periodicity); (2) in a liquid state: the molecules are linked together, but each of them can move with respect to the others, with the result that the water system appears changeable and is able to adapt itself to the form of the container (the environment); symmetry is broken, this is complexity, the water system is in a condition of precarious equilibrium that exists only for restricted intervals of the temperature and pressure values; (3) in the aeriform state of steam: this is unpredictability, disorder, despite the fact that the laws that describe the behaviour of each molecule are known; this is chaos.

It is interesting to observe how, on one hand, total order is characterized by the existence of symmetry, i.e. by invariance with respect to a group of transformations, but on the other hand, how complete disorder is also characterized by symmetry. In the first case, there is symmetry in the position of the single parts of the system: the larger the group of transformations with respect to which the system is invariant, the smaller the part of it needed to reconstruct the whole. In the second case, there is a statistical type of symmetry. The origin of the symmetry is not the positions of the individual elements, but the equal probability for each element of the system of finding itself in any particular position. A perfect gas is statistically symmetrical: any position can be occupied by any molecule with the same probability of any other molecule in any other position. Similarly, we note that the empty space[129] is at the height of its symmetry and homogeneity, as it is invariant with respect to any transformation and any of its parts can be used to reconstruct the others.

In one case, the symmetry is due to the static nature (or immobility), to the stable equilibrium, in which the positions are maintained constant (possibly asymptotically). In the other case, the symmetry is statistical: the component parts of the system lack a fixed position and there is only symmetry in the spatial or spatial–temporal averages.[130]

---

[129] In this context we are obviously referring to a Euclidean–Newtonian abstract three-dimensional linear space, ignoring both any consideration that refers to curved spaces due to the presence of masses, a concept pertinent to the theory of general relativity, and any conception of vacuum, from quantum physics and the theory of fields, contexts in which the vacuum assumes a meaning that is very different to the one we are referring to here (Parisi, 1998).

[130] The case in which we have total statistical symmetry is that in which the temporal averages are equal to the spatial ones, the case of a so-called ergodic system. Think, for example, of an ideal school class (a space of n students), in which the n students are all averagely good in all subjects, without distinctions. The arithmetic average of the marks of each student, in a year and in all subjects (temporal average) is equal to that of all of the other students and is equal to the arithmetic average of the marks of all n students in all subjects in a single day (spatial average).

Complexity is neither one case nor the other. Complexity is characterized by the breakdown of the symmetry, both that of the perfect order, and that of the total disorder, due to the fact that no part of the system is able to provide sufficient information to predict, even statistically, the properties of the other parts. This is precisely where the difficulty of modelling complex systems lies. In reality, however, the very idea of complexity, seen as an intermediate condition between order and chaos, also depends on the level of representation. What appears complex in one representation can appear ordered or disordered in another representation with a different scale. For example, a system of cracks in a surface of dry mud may appear complex, but seen from afar (*zoom out*), the surface appears to be flat and homogeneous, i.e. ordered, i.e. translation invariant, while, seen from close up (*zoom in*), considering each particle of mud individually, it appears to be a totally disordered system, newly translation invariant. A further dimension that characterizes space and time also needs to be considered: the scale (Havel, 1995). A system is more complex the more it appears to be invariant to changes in scale: a crystal, whose structure is repeated at different scales, and that is therefore 'symmetrical' to changes in scale, is more complex than a molecule, in the same way in which a fractal is more complex, for example, than a triangle.

Complexity increases both as the number and/or variety of the elements that constitute the system increase, and when the number of relations that link them, i.e. the degree of integration between the elements, increases. The variables in play are, therefore, the differentiation, expressed by the number and the variety of the elements and the connection. These same variables, however, should be seen in different dimensions: the spatial one (the geometric structure), the temporal one (the dynamics) and that of the scale. Therefore, to prove that there is an increase in the complexity of a system, it has to be shown that, on one hand, the number and/or variety (i.e. the system's degree of differentiation) increases with respect to one of these dimensions, and on the other hand, the connection in it (i.e. the system's degree of integration) increases with respect to one of these dimensions.

In reality, neither the differentiation nor the connection are defined objectively, as both depend on how an observer relates to the observed system. As we have already mentioned, the degree of variety and the constraints that an observer recognizes in an observed system depend on how he observes and on how much he distinguishes. In a certain sense, we can say that they depend on his 'resolving power' with respect to that system. What the observer does is identify the differentiations that in some way he judges to be more important in the system's phenomenology and create generalized categories, in which he classifies the phenomena that he sees as being similar to one another, ignoring the differences that exist between the members of those categories (Heylighen, 1990; for other observations Dendrinos, 1997).

Depending on what distinctions he makes, the observer can perceive a greater or lesser degree of differentiation and connection, and therefore see the complexity increase or decrease. We could also say that this can depend not only on the observer's perception but also, implicitly, on his *personal* choice (or definition) of the unit of measurement of complexity and the measurement technique that he adopts.

Therefore there are subjective elements of evaluation alongside objective ones in the choice that the observer makes of which aspects of the system to model. The reliability of the model, therefore, also depends on the degree of interdependence between the system's characteristics included in the model and those which, consciously or otherwise, are ignored. As we are not able to produce a complete model of the system (a 'complete' model does not exist), the introduction of the different dimensions, cited above, enables us to at least better perceive the system's intrinsic complexity.

The huge variety of fields to which the theory of complexity can be applied, thanks to the generality of the concept, appears to be one of the most important factors that unites the different branches of science. The very study of nonlinear models to describe complex phenomenologies, in recent years, has found a broad spectrum of application, both in natural sciences and in social sciences.

In Chapters 29 and 30, we will be presenting several examples of how the new concepts of complexity can be applied, in different fields, to interpret the evolutionistic phenomenologies of systems.

29 Complexity and evolution

Introduction

In this chapter we will be presenting several applications of the ideas proposed by the new approach of complexity to the evolution of systems, focusing particularly on biological systems and on the appearance of life following a spontaneous process of self-organization.

Is there is a general tendency for complexity to increase as a system evolves, just like, in a closed system, there is a tendency of the entropy to increase? This question assumes particular importance, for example, in biology, regarding the evolution of living species, but not only this. The evolution of a complex system is spontaneous, whether it is a natural system, like a galaxy, a planetary system, a biological organism, a crystal, a liquid or a real gas in which the interactions between molecules, etc. are not negligible, or whether it is a social system, such as a human population, an economic aggregate or a language (intended in the strict sense, as a complex of phonemes, words, rules of syntax, etc.). Systems are usually assumed to evolve over time towards an increase in complexity, but little or nothing is known of the mechanism that could cause the evolution in that direction, rather than towards a simplification or a reduction in complexity. To complicate matters, remember that the problem of a correct and effective definition of complexity is ever present, without which it is difficult to identify the growth of complexity in systems over time.

The three ways in which complexity grows according to Brian Arthur

The general issue of the evolution of complexity is discussed, for example, by Brian Arthur (1994), who presents three different ways in which complexity grows.

The first way is encountered in certain systems, such as a part of biological or economic systems, in which complexity grows due to the increase in the number and the variety of the species, that takes place due to the effect of interaction with other systems. These are systems made up of entities, species or organisms that coexist with one another, together constituting a set of populations that reciprocally interact. Systems of this nature are called coevolutive systems. Typical examples are ecological and economic systems. In certain specific circumstances, in fact, the appearance of new species or of new goods and/or services can give rise to new niches (ecological or market), which are available to new species or new goods and/or services, that are located in said niches, and that previously would not have found a place in the system (biological, economic, etc.) and therefore would have been eliminated from the scene. In their turn, the newly arrived species and the goods and/or services that arrive in the system, contribute to the formation of new niches for other new elements that locate themselves there, in a general increase in the variety of elements in the system that, in most cases, is classified as top-down. We therefore have a continuous increase in the number and the variety of the component parts starting from the first that occupy an empty space.

In this regard, the case of the so-called Cambrian explosion is well known, the rapid and unexpected increase in the variety of biological species that happened around 550 million years ago, at the beginning of the Palaeozoic age, when a biological creativity that had never been observed before gave rise to all of the most important phyla (with the exception of the vertebrates), and gradually to the genera, families and species. As the ecological space became occupied by the first genera, reducing the dimensions of the available niches, the low-level varieties increased. The genera differentiated into families and species that, in this way, better adapted to the niches that were formed and that the species themselves contributed to forming (Kauffman, 1995).

The second way in which complexity grows according to Arthur can be seen in systems that evolve autonomously towards an increase in complexity, through an increase in the sophistication of their structure. The system, in this case, displays an increasing number of subsystems or subfunctions or even of subcomponents, to overcome constraints to the effectiveness of its behaviour or to improve its functioning or even to face new difficulties. Think, for example, of the evolution of human language, in general, but also of the individual languages, starting from the first rudimentary onomatopoeic sounds up to symbolic and structured language, and of the development of this over time due to the constant acquisition of new elements for different reasons, at times in contrast to each other. On one hand, in fact, there is a need to communicate new concepts, with increasingly precise subtleties; on the other hand, there is a general tendency to economize the energy employed in communication, the tendency, namely, to reduce the mental and physical effort employed in communication to a minimum, even taking into account the need for a certain level of information redundancy (interested readers can consult popular works such as Martinet, 1960, and Barber, 1964, and at a more technical level, works by Lyons, 1968, and Horrocks, 1987).

In systems in which the complexity increases in the two ways cited above, it can sometimes occur that, even though in a general growth trend, there are phases of degrowth, on occasion even rapid, of the complexity. Think of the development of the verbal and nominal inflections that have characterized all of the Indo-European languages. The presence of these elements, even though considerable in the past, has diminished in some groups of modern languages. This is true, in particular, of Latin and Germanic languages; think of modern English, which is almost entirely lacking in grammatical genders, in which verbal inflections are reduced to five forms, and in which adjective agreements have disappeared, even though such were present in prior periods of the history of the English language, in particular in *Old English* in the early Middle Ages in Britain. But this is also true of Neo-Latin languages, such as Italian, whose morphological–syntactical structure is less complex than that of Latin. Nevertheless we observe that this morphosyntactic impoverishment, caused by the tendency to economize, that we mentioned previously, does not influence the effectiveness of the language for communication purposes, because other aspects of linguistic structure replace them, i.e. other grammatical subsystems, that increase their complexity; for example, the increased use of prepositions to compensate for the loss of nominal inflections, or more rigid and more numerous positional rules to compensate for the loss of the agreements (Lyons, 1968).

As we have observed on a number of occasions, complexity science studies characteristics that are common to the evolution of systems, regardless of the specific nature of the latter. This is particularly evident, if we consider certain subsystems. On one hand, for example, in the rebound at the beginning of the Mesozoic era, after the extinction of the Permian, the genre of reptiles displayed an increased complexity. This same complexity, however, diminished with the disappearance of the dinosaurs of the Cretaceous period, at the end of the Mesozoic era, around 65 million years ago, to make room for the increase in the variety and the complexity of the species of mammals that came to occupy the ecological niches that had previously been the appanage of large reptiles (Kauffman, 1995). In the same way, the complexity of the subsystem of the morphosyntatical rules of

Latin diminished in favour of the increased presence of positional rules in Neo-Latin languages, with respect to those of Latin.

The third way in which complexity grows, according to Arthur, regards systems that evolve by means of a mechanism that is different to those of the previous one, called *capturing software*. A system 'discovers' one of its elements or one of its subsystems and finds a way to use it for some very basic purpose, i.e. it 'learns'. These elements or subsystems, however, are characterized by a set of operating rules that control the way in which they can be combined or used; we are dealing with a real form of interactive grammar that permits various combinations of simple elements or subsystems, and that the system, as it evolves, learns to use. In the end, when the system has fully mastered the grammar through which it can use the subsystems, it programes the elements, or the subsystems, as if it were software, hence the name given to the process, in order to use them for more complex purposes (i.e. more varied, more detailed, more numerous, ...) than the elementary ones with which the process started.

A typical example of this process can be recognized in technology seen as a complex system. Electronics, which represents a subsystem of it, originated from the study of several natural phenomena connected with electricity. Once the grammar of electromagnetism, i.e. the laws that describe its functioning, had been thoroughly learnt, the system of modern technology developed the technology of electronics i.e., it learnt how to program the functioning of the elementary components, the electrons, for purposes that were much more varied and sophisticated than those that were envisaged when the study of electricity had just started.

The same situation arises in numerous other sectors of scientific and technological research; for example, the development of quantum physics, which originated from a few experimental facts that had caught the attention of physicists towards the end of the nineteenth century, subsequently gave rise to a vast and extremely detailed grammar. We observe how this mechanism could also be applied to complexity science itself. At present, complexity appears to be at the first stage, that of the observation and the study of a few elements. Once we have thoroughly learnt the grammar of complexity, that is, once a real theory of complexity has been fully developed, then we will probably be able to 'program' its functioning for various and numerous purposes. This would mean using complexity to its best advantage, i.e. managing complexity.[131]

---

[131] In other words, one of the methods of complexity, or better, one of the schemas in which it develops and evolves, taken as a concept in itself, could give rise to a study, the subject of which is the complexity of complexity, i.e. basically metacomplexity.

Once again, linguistics provides us with an interesting example as regards this process (Barber, 1964). It is generally believed that the first men, some hundreds of thousands of years ago, discovered that a few rudimentary sounds could be used to communicate different concepts and pieces of information in various human communities, which started to develop a grammar of rules for their use. Over time, this system of phonemes and grammatical rules became a complex interactive system, i.e. a real language. The human society system had 'captured the software' and developed a program, language, that was extremely effective for the highly complex purposes of communicating concepts that could not be expressed onomatopoeically.

One particularly interesting aspect of the relationship between complexity and evolution is that regarding the evolution, in its strict sense, of living species and the appearance of life on Earth (Kauffman, 1994, 1995). Evolution in living species is an extremely powerful force which, given sufficient time, is able to create an extraordinary ecological complexity starting from a few simple elements. Evolution is the process that led to the creation, if not of all, certainly of the majority of known complex systems, studied in social sciences (human and animal sciences, cultures, economies, languages, etc.), and of many of the complex systems that are the domain of the natural sciences, in particular biology (cellular systems, organisms, ecological environments, etc.). The understanding of the mechanisms of evolution is therefore a fundamental step towards understanding the way in which complex systems function.

Nevertheless, we encounter a number of obstacles when we attempt to investigate further into evolution. One of the greatest obstacles is the fact that we only have one example of the result of evolution: what is in front of our eyes everyday, i.e. that of life on Earth. But this is not the only obstacle. A second difficulty, for example, can be attributed to the fact that evolution cannot be followed for very long periods of time. Despite these limitations, however, biology has succeeded in formulating the basic principles of evolution by analysing the static results of evolution without conducting experiments. The concepts developed by Darwin (1859) regarding evolution by means of natural selection, that he intuitively sensed without being able to observe them directly, still represent the core foundation of evolutionist theories.

The Tierra evolutionistic model

As we have already discussed in Chapter 2, when direct experimentation is not possible, we try to simulate the real situation by using a model formulated on the basis of appropriate hypotheses and assumptions. Thomas Ray (1992, 1994) has developed a model of evolution called Tierra, based on the mechanism essential to the origins of the evolution of species: replication with random errors. The Tierra model takes its form from a particular computer program able to produce other copies of itself in the RAM memory of the computer, copies that, in turn, replicate themselves. At each replication, however, there can be a random introduction of errors, in the form of bits that have a different value to the original one. The replicated programs play the role of new creatures; they occupy the available RAM memory, which, in turn, plays the role of the environment, whose resources, i.e. the capacity of the RAM, are limited. The errors introduced, that each program, by replicating itself, passes on to the new programs (the new creatures) it creates, are the equivalent of genetic mutations. Selection is made by the limited nature of spatial resources, i.e. the capacity of the memory, and energetic resources, represented by the time that the CPU requires to replicate the program.

The need for spatial and energetic resources differs in the various creatures, because the random genetic mutations inserted by the parent program are different; this necessity varies from one program to the next according to the bit configuration of each. When the memory is full, a 'killer' program appears that eliminates the old creatures to leave room for the new ones that replace them. Starting from the rudimentary father creature (program), we therefore observe the development of an entire ecology of genotypes (the subsequent descendants) that emerge spontaneously. The most efficient genotypes at replicating are those that, following the random mutations, leave a higher number of descendants and thus become those most present in the population that occupies the RAM. This whole processes represents a metaphor of evolution and of competition for resources.

Evolution improves the adaptation of creatures to the environment. Initially, when the environment is a uniform space, available in abundance, the creatures do not interfere with one other and the most efficient replication is performed simply by optimizing the replication algorithm. Subsequently, when the amount of available memory (the environment) has been filled with creatures, a new phenomenon occurs. The programs (species) present interact, thus becoming important components of the environment. If the environment is limited in some way, selection favours smaller creatures, because their duplication requires a lower calculation time than that needed to duplicate larger creatures.

What we observe is that the maximum efficiency of the replication mechanism is achieved when each individual takes into account what exists around him, through a real learning process. This mechanism enables the creatures to be kept small; they do not need to hold all of the information needed, as long as the latter is available in the environment. Evolution therefore drives the creatures to activate exploitation strategies and defence strategies against exploitation, i.e. it gives rise to the development of real information parasites, which multiply more quickly because they use the information contained in the creatures they exploit (the hosts), and thus keep their dimensions small.

The joint parasite–host dynamics occurs in Volterra–Lotka cycles (see Chapter 12). Subsequently, however, the evolutive competition gives rise to creatures that are immune to exploitation by the parasites. These new creatures increasingly develop through the sharing of information, i.e. setting up real forms of very strict social cooperation that, in the end, leads them to be the winning creatures that occupy all of the available environment, thus causing the extinction of the parasites.

This model therefore shows how, despite the fact that it does not form part of the hypotheses, the best efficiency of a species from an evolutive point of view is obtained when different creatures cooperate by sharing information. Basically, what happens is that not only the most efficient evolution ends up depending on a species already present in the environment, but also that the most effective survival strategy is neither competition for resources, nor parasitism, but cooperation.[132]

The appearance of life according to Kauffman

Even before the evolution of the species, the appearance of life on Earth can be interpreted from the perspective of the evolution of a complex chemical system in a state of non-equilibrium. The received view about the origin of life maintains that DNA and RNA molecules are the best candidates for the first living molecules due to the self-templating character of the polynucleotide double helix. Much research has been carried out to demonstrate the prebiotic synthesis of critical biomolecules such as amino acids and

---

[132] The theme of the transition from competition to cooperation as a more efficient way to survive in a world dominated by the egotism of its actors (whether such are superpowers, businesses, individuals, or other), which exclusively concentrate on their own utility without the interference of any higher authority, is discussed by Robert Axelrod (1984, 1997), in particular with reference to the prisoner's dilemma, a well-known problem in the theory of games widely studied in mathematics and used as a model in economics and sociology, proposed in the 1950s, right in the middle of the cold war, within the framework of research conducted with a view to possible applications to international political strategies (see also Holland, 1996).

nucleotides, and to achieve template replication of arbitrary RNA sequences. This work at the moment, however, has not yet been successful, but it might well be in the future.

Stuart Kauffman (1994, 1995) proposed the idea that life appeared following a phenomenon of self-organization that occurred in a situation of non-equilibrium. Before the origin of life, he suggests, there was a form of prebiotic chemical evolution that gave rise to the formation of the variety of existing organic compounds, in a general state of non-equilibrium due to the flows (exchanges) of matter and energy to which these systems of molecules were subjected.

From Darwin's era, we have been used to considering living organisms as sort of strange devices that have organized themselves more or less randomly, through a sequence of numerous mutations, in themselves random, the vast majority of which unfortunate. Natural selection is the mechanism that is able to bring order to the disordered set of attempts produced by the mutations, eliminating the less efficient organisms in terms of competition for resources (i.e. the less adapted) created by unfortunate mutations, from the scene. Again according to Kauffman, however, Darwin intuitively sensed only a part of the mechanism of life: selection due to competition for resources as the sole cause that produces order. In reality, life, seen as 'self-reproducing chemical systems capable of evolution' (Kauffman, 1994, p. 90), is by no means a random phenomenon, but could be considered as an *emergent collective property* of sufficiently complex chemical systems in a state of non-equilibrium, due to the presence of sets of catalytic polymers. The latter, as Kauffman proposes, would have inevitably contained autocatalytic sets, systems, that is, in which each molecule sees its own formation catalysed by some other molecule in the set. The appearance of autocatalytic systems is seen, in this context, as a kind of phase transition that occurs in a system that is self-organizing. As Kauffman writes, 'The new view of the origin of life, is based on the generalization of small autocatalytic sets. I believe good theory exists to support the possibility that sufficiently complex sets of catalytic polymers will almost inevitably contain *collectively autocatalytic sets*. In such a collectively autocatalytic set of monomers and polymers, no molecule need catalyze its own formation; rather, each molecule has its formation catalyzed by some molecule in the set such that the set is collectively autocatalytic. The set constructs itself from exogenously supplied monomers or other building blocks.' (Kauffman, 1994, p. 88).

A simple image will help better understand the idea. Let us consider, for example, 10,000 buttons on the floor and begin to connect them at random with a thread. Every now and then let us pick up a button and see how many buttons are connected to it. This connected set is called a component (or a cluster) in a random graph made up of N nodes (or points) connected at random with E edges (or lines). It can be shown that when $E/N > 0.5$ most of the nodes are connected in one giant component i.e. when E/N passes 0.5 most of the clusters become cross-connected into one giant cluster. When this giant web forms, the majority of buttons are directly or indirectly connected to each other. In other words, when E/N passes 0.5, a nonlinear process takes place resembling a phase transition; the size of the maximum cluster increases abruptly, just like when separate water molecules freeze to form a block of ice. In a similar way, in reaction graphs, when a sufficiently large percentage of reactions is catalysed, i.e. we have the formation of a sort of network of interconnected nodes, a giant component crystallizes. As the maximum length polymer increases, the number of types of polymer increases exponentially, but the number of types of reactions between them increases even faster. Thus for any fixed probability that an arbitrary polymer can act as a catalyst for an arbitrary reaction, there are so many possible reactions per polymer that at least one reaction per polymer is catalysed by some polymer. The autocatalytic process has taken place, autocatalytic sets crystallize. The phase transition, the 'crystallization', has occurred. (For technical details on the biochemistry of the autocatalytic processes that have led, amongst other things, to the formation of DNA, see Kauffman, 1986, 1994, and the references indicated therein.)

Life might emerge, therefore, as an inevitable consequence of polymer chemistry. At a critical diversity of monomers and polymers the system will contain collectively autocatalytic sets: a phase transition is surpassed. Collectively reproducing metabolisms, able to grow, survive environmental changes and able to evolve in the absence of a genome are expected. According to this perspective, life, therefore, is not a random phenomenon, but appears as the necessary outcome of a situation of instability in a self-organizing chemical system. 'If so, the routes to life in the Universe may be broad boulevards, not back alleys of thermodynamic improbability' (Kauffman, 1994, p. 89). This is the meaning of the title of Kauffman's very successful book, *At Home in the Universe* (1995). The universe is our home and we are its inhabitants by right, because we appeared not due to the highly improbable fortuitous event of a combination of molecules, but due to a necessary order given by the union of energy and matter in a system that found itself in a condition of non-equilibrium, at the edge of chaos.

Naturally, Kauffman's proposal is only a proposal, an interesting theory to be discussed. Nothing is certain yet. This theory is destined to remain such until such time as a complete theory of self-organization and complexity is formulated.[133]

---

[133] A similar type of process, i.e. phase transitions, also occurs in socioeconomic processes, and is common in human behaviour in general. When interactions between the components of the system are sufficiently dense and when these interactions add up in such a way as to generate large scale correlation, then a different entity emerges on a higher level of organization than its constituents, obeying certain laws of its own. This is what happens, for instance, when a group of people meet to discuss an issue of common interest. As the intensity of their interaction increases, clusters of 'like-minded' people begin to emerge spontaneously, linking up with other 'like-minded' clusters creating even larger clusters (Batten, 2000ab). Phase transition has also been applied to the description of urban development, interpreted as a phase transition in a graph of minor centres, which, when the graph becomes sufficiently connected (i.e. it crystallizes), give rise to a new system from the emergent properties: the city.

30 Complexity in economic processes

In this section, we will continue to look at how the approach of complexity has been applied, presenting several examples of models of economic processes constructed according to this new vision of systems. Economics, in fact, is actually one of the fields of research in which the concepts of instability, chaos, and complexity that we have discussed in this book, can best be applied, and it is one of the areas that has most attracted the attention of scholars who have adopted the new concepts proposed by complexity science (see, for example, Barkley Rosser, 1991; Brock, Hsieh and Le Baron, 1991; Benhabib, 1992; Nijkamp and Reggiani, 1993, 1998; Krugman, 1996; Thiel, 1998; Batten, 2000a).

Complex economic systems

One of the basic assumptions in economics, in certain situations almost an axiom, is, to put it briefly, that it is not so much current facts that determine expectations for the future, as expectations for the future that determine behaviour and therefore current facts. As a consequence of this, on one hand, the present acts on the future, through actions decided and undertaken, but on the other hand, the future acts on the present, because the objective of the actions that are undertaken are based on future gain. Future prospects and hopes, in fact, are what is behind present actions, aimed at maximizing future gain. Therefore we are dealing with a typical feedback cycle in the evolution of an economic system, in which the output, the future, returns certain information to the input, the present, through expectations. In addition to the presence of feedback cycles, another characteristic specific to economic systems is that in them, and more generally in social systems, the actors are often very numerous, irrespective of situations of monopoly or oligopoly.

In economics we therefore encounter systems with lots of interacting elements, in which there are also feedback cycles. It is natural, therefore, to expect that the evolution of economic systems over time can display the phenomenologies typical of chaos and complexity, such as for example the presence of a chaotic attractor towards which the unstable orbits direct themselves and along which they travel in a more or less cyclical way.

As David Ruelle (1991) proposes, we could construct a parallel between the economy of a society at different stages of technological development, on one hand, and a dissipative physical system subjected to the action of external forces, such as a viscous liquid heated from underneath, on the other. The analogy, obviously, is purely qualitative; we maintain our distance from a physicalism that has been superseded by the times. At low levels of technological development, we can think of the economy in a stationary state, like a liquid weakly heated from underneath. If the flow of heat is not sufficient to trigger convective motion, the heat spreads into the still liquid and disperses into the air by means of conduction. This situation corresponds to classical economics at the end of the eighteenth century: Adam Smith's invisible hand that regulates prices, by matching supply and demand in a static economy. At more advanced levels of technological development, we can expect to see oscillations that are more or less periodical such as economic cycles that correspond, according to the above analogy, to nonturbulent connective motion that a liquid subjected to a source of heat displays. This is the economics of the period just before the mid-twentieth century, that, just to be clear, of Keynes and Schumpeter. At even more advanced levels of technological development, such as those typical of a modern advanced society, we have a turbulent economy characterized by irregular variations and by a strong dependence on initial conditions. This case corresponds to that of a liquid heated from underneath by a very intense flux of heat, such that the liquid is kept boiling.

Just as in the motion of a liquid, we can identify, even in turbulent conditions, relatively stable forms such as vortexes, that appear, live for a period of time, transform into other vortexes or die out; in the same way, in economic processes and more generally in social processes, we observe phenomena of self-organization, i.e. the formation of particular structures, whose extension is limited and whose functioning is complicated. The economic cycle theories of writers such as Marx, Keynes, Schumpeter and Samuelson (Allen, 1959) do not envisage situations analogous to turbulence and, therefore, need to be reviewed in the light of the modern interpretation of the complexity of economic systems. Many economic mechanisms can produce oscillations in the values of the macroeconomic magnitudes;[134] they appear both in capitalist economies without any form of regulation of

---

[134] For the sake of brevity, hereinafter we will call these oscillations 'cycles' in line with current use, even if the periodicity that the term presumes is anything but rigorous.

free market competition, and in socialist ones with rigid centralized planning, and can result from nonlinear interactions between different economic and political variables, as was already pointed out in the studies of Marx, Keynes and Schumpeter.

Economic analysis, as we are aware, shows that the historic evolution of economic systems presents cycles that are more or less periodic, but also shows that said cycles, in reality, are not identical in their repetition, unless approximately; each cycle develops in a different form from the previous one. In particular, the cycles, just like the various types of fluctuations and variations that are displayed in the growth of capitalist economies, such as for example Kondratiev's waves, superpose a general growth trend. This is one-way historical evolution, that cannot be ignored; according to Schumpeter (1934, 1939), the cycle is actually the form in which growth presents itself. The engine of change in economics is the search for private gain and, due to this, the economy does not only generate that sort of perpetual cyclical motion, but evolves at the same time, on occasion changing its structure, according to a process that can be considered analogous to that, in biology, of a single species that changes its own structure. It is a process, therefore, that can be considered similar to a real morphogenesis, or to the selection of new species through competition for resources (Goodwin, 1991). Without this growth process, that gives the evolution of economic systems their particular nature, the real, periodic, stable cycle could be described, at least in first approximation, in terms of a simple Volterra–Lotka model, such as that illustrated in Chapter 13.

The persistence of economic growth without shocks, even for a long period of time, obviously doesn't guarantee that this type of growth will continue indefinitely; there is always the possibility that an unexpected morphogenetic change may occur. Alongside growth, a core element in economic evolution, we have structural change, which is related to the system's structural instability or to its dynamic instability or to both. In traditional economics, however, there is no theory that efficiently interprets the irregular movements observed in the dynamics of economic data, using any endogenous mechanism. One of the most interesting aspects that we would like to highlight is that these irregularities, however, can also occur for entirely endogenous reasons; namely due to mechanisms, internal to the system in question, independent of external causes, mechanisms that can remain dormant for a long period of time and then emerge unexpectedly in particular circumstances, causing violent and sudden changes in the structure of the system itself, the so-called catastrophes. Phenomena of this nature were a mystery up until the recent development of the theory of nonlinear systems.

Synergetics

A few decades ago, when new ideas on chaos and complexity had started to diffuse in scientific circles, the theoretical physicist Hermann Haken proposed, through a series of works, a general theory on the dynamic behaviour of complex systems in certain situations: synergetics (Haken, 1977, 1983, 1985, 1990). More specifically, synergetics concerns the cooperative interaction between many subsystems that can be the source of the macroscopic behaviour of the system. The synergetic approach is a direct result of the holistic vision of systems in which collective phenomena are displayed that we have called 'emergent'. It is natural to consider the economy as a typical field of application for the ideas of synergetics (Zhang, 1991). Economic synergetics looks at the spontaneous formation of new macroscopic structures associated to discontinuous processes and shows how the parts of an economic system can organize themselves into new structures in certain particular times of instability. The ideas of synergetics can be efficiently applied to problems of economic analysis that regard structural changes (bifurcations, catastrophes, the onset of chaotic dynamics), just as they can be applied to the study of the role of small random fluctuations (the so-called stochastic noise) in economic structures, in relations between fast and slow economic processes, and in relations between economies at macro and micro level.

In our era, the complexity of economic reality, and of socio-economic reality in general, is growing continuously. There are numerous factors behind this phenomenon: from the increase in the speed of communications, to the growth of capital and knowledge at the disposal of evolved societies. This leads to an increasing difficulty in finding a description for the evolutive processes in economics. Synergetics shows how the theory of nonlinear systems provides, in reality, a possible common tool for the analysis of the complexity of economic evolution. Synergetics generally concerns nonlinear and unstable systems, even regardless of the specific aspects of economic systems, and focuses on certain particular nonlinear phenomena that manifest themselves in said systems, such as structural changes, bifurcations, and chaos.

Even though societies can be very different from one another, they can always be described using several common variables. The synergetic approach can be adopted to examine the possible behaviour and the possible structures of socio-economic systems, considered in their temporal dynamics, referring to different combinations of said variables (Zhang, 1997). They can be grouped into several general sets, such as, for example: (1) monetary variables, such as prices, salaries, interest rates, etc. (2) material variables, such as infrastructures, capital, natural resources, etc. (3) variables linked to knowledge, such as level of education, skills, technology, etc. (4) demographic variables, such as birth and death rates, the population structure in terms of age, sex or other; (5) sociocultural variables, such as legal provisions, tax systems, habits and tradition. Now, all of these variables depend on time according to different speeds of change, according to the value of the other variables, or of some of them. In general, we can describe the dynamic interactions between the socio-economic variables considered, that we will indicate with $x_1, \ldots, x_n$, in the form of a system of first order differential equations like (8.1) in Chapter 8, thus forming a dynamical system:

$$\frac{d}{dt}x_1 = F_1(x_1, \ldots, x_n)$$

$$\ldots$$

$$\frac{d}{dt}x_n = F_n(x_1, \ldots, x_n)$$

The main objective of synergetics in economics is, therefore, to investigate how these variables act on each other and how this influences, through the n nonlinear functions F_i, in which $i = 1, \ldots, n$, the evolution of socio-economic systems.

The instability of economic systems was already considered in the studies of Marx, Keynes, and Schumpeter, although these economists stated different causes as the origin of such. As Wei-Bin Zhang (1991) observes, the synergetic vision of economic development, in certain situations, is not that far from that of Schumpeter. For example, the shock of innovation, to which Schumpeter attributes the cause of the qualitative changes in economic systems, plays a similar role to that of the flows of energy in the description of the economic evolution provided by synergetics. In both approaches, an economy without innovation remains in its state of equilibrium, while innovation can lead to chaotic behaviour. The synergetic approach and that of Schumpeter in particular, however, even if similar in terms of perspective, differentiate with regard to several aspects; this leads to different levels of understanding of the phenomenology studied. In synergetic economics, the role played by mathematical techniques is fundamental: mathematics is the main tool to express and to treat what we mean by instability, cyclical development, chaos, etc. with precision, while there is nothing comparable to this in the works of Marx, Keynes, or Schumpeter.

Even though synergetics derives from physics, it is by no means a physical theory that attempts to reduce complex phenomena to laws of matter. Right from the start, Haken emphasized how the physical systems he considered displayed behaviour characterized by aspects similar to those of the phenomena of collective behaviour that are observed in a wide range of natural and social systems. Many elements in synergetic theory were actually developed starting from the observation and the study of numerous specific cases in a large variety of sectors: from physics to sociology, to psychology, to artificial intelligence, to spatial economy and regional and urban science. The guiding idea, in each case, is the search for the properties, at the macro level, of a complex system, starting from the components at the micro level.

One of the original aspects of synergetics is therefore the fact that it can be applied to a large number of fields. As Léna Sanders (1992) observes, we are no longer simply transferring methods developed in physics to social sciences; synergetics conceives a direct similarity between natural and social systems, regardless of their comparability at the microscopic level. As we have already observed in Chapter 27, almost all systems, in fact, at the macroscopic level are structured in a similar way, therefore the mathematical concepts developed to describe their dynamics can be applied to both natural and social systems (Weidlich, 1991).

The synergetics approach has been successfully applied to the analysis of the dynamics of urban and regional systems, in particular to cities, considered as complex systems that self-organize following cooperation between the components[135] (Pumain, 1997; Staricco and Vaio, 2000). Monographic studies on the themes of synergetics and self-organization in urban and regional systems can be found, for example in Dendrinos and Sonis

---

[135] In this regard, Juval Portugali (2000) observes how urban organization following cooperation is, in reality, simply a particular case of a more generalized situation, in which ordered and self-organized cities can still be chaotic, due to the permanence of chaotic phenomena. There are two aspects of this: local chaos, at the micro level, and deterministic chaos of the complex city system, at the macro level. The local chaos originates from the irregular movement of the numerous components of the complex city system, such as the movement of vehicles on a thoroughfare in conditions of light traffic or the movement of individuals in an uncrowded square. In both cases, we are dealing with irregular movements that are not coordinated with other movements: the starting point for self-organization. Deterministic chaos, on the other hand, emerges as a consequence of self-organization (chaos from order), for example, when the various system components start to experience a push towards an attractor and, as a consequence, display coordination of the individual dynamics. Often, in cases like this, the systems are dominated, so to speak, by several parameters of order, whose behaviour is obviously chaotic. For example, for a certain period of time one parameter dominates over the others, then, suddenly, the dominant parameter becomes another. In this case the city passes rapidly from a structurally stable state to another structurally stable state. For example, a residential city (suddenly) becomes a commercial city. It can also occur, in this regard, that within a dynamically stable city, there are local chaotic areas, clearly defined from a socio-spatial viewpoint, which act against the maintenance of global stability.

(1990), Nijkamp and Reggiani (1998) and in Portugali (2000); reviews of works on complexity in urban and regional sciences can be found in Bertuglia and Vaio (1997a) and in Bertuglia, Bianchi and Mela (1998). In particular, the evolution of the dimensions of cities as a function of their location within an urban hierarchy, their localization in the territory and their socio-economic profile have been studied (Weidlich and Haag, 1983, 1988; Haag et al., 1992).

Synergetics applied to the urban field aims to identify the effects, at the macro level, of the micro-geographic actions of the urban actors, such as family nuclei, businessmen and administrators, and of their interaction, starting from the basic roles of the actors. This has been implemented, for instance, in the model of the system of French cities based on synergetics presented in Sanders (1992) and in the model of spatial self-organization of Allen (1997b), different versions of which have been applied to population dynamics and economic activities in the United States, in the provinces of Belgium and in Senegal (Allen, 1997a).

The synergetics approach has also been applied to the study of the processes of the diffusion of innovation in a given area (Sonis, 1988) and, in particular, to the study of the effects of socio-cultural and economic intervention on the diffusion of innovation (Sonis, 1991, 1995). The assumption at the basis of these studies is that the diffusion of innovation can be seen as the spatial-temporal realization of a dynamic choice between 'old' and 'new' alternatives, that arise due to the effect of the interaction between the actors in the area in question. This means that we pass from the micro level of the *homo oeconomicus*, who acts alone, driven by a search for personal interests, to the meso level of the *homo socialis*, who acts in search of his own interests, but exchanges information with his fellow men and acts on the basis of the information he receives (Sonis, 1990a). It is assumed, in other words, that there is a synergetic type of interaction between the diffusion of the competitive innovations introduced to the various functional niches, that is the result of numerous decisions at the micro level, and the socio-economic structure of the area considered, which evolves, at the meso level, following decisions at the micro level, resulting from the opinions of the individual actors that the same socio-economic structure contributes to forming.

Two examples of complex models in economics

Economic science can gain from developments in mathematics applied to the study of system dynamics, particularly from the theories of catastrophes, chaos, and complexity (of the many authors who have written on this subject, see, for example: Anderson, Arrow and Pines, 1988; Chiarella, 1990; Dendrinos and Sonis, 1990; Barkley Rosser, 1991; Hodgson and Screpanti, 1991; Zhang, 1991; Benhabib, 1992; Nijkamp and Reggiani, 1992a, 1993; Krugman, 1996, Arthur, Durlauf and Lane, 1997; Fang and Sanglier, 1997; Hewings et al., 1999). In this section we will be briefly presenting several recent models, as examples of the application of the new concepts that derive from complexity theory to economic systems.

The first model we are going to cite is that of Richard Goodwin (1991).[136] Goodwin proposes a cyclical dynamic model of the evolution of an economic system with two variables v, the rate of employment of the population of an employable age, and u, the share of the gross domestic product represented by wages. The model attempts to take a new approach to the understanding of cyclical economic development based on the ideas of Marx, Keynes, and Schumpeter. In particular, the author wishes to develop a unitary theory of short and long growth waves, but in conditions of continuous change with endogenous origins, and to be able to account for, using the concepts of chaos theory, part of the general irregularities of the economic temporal series, obtaining, in the place of the classical closed cycles, a strange attractor (represented in Figure 30.1), whose form recalls that of the Rössler attractor (Figures. 16.3 and 16.4).

Tönu Puu (1990), on the other hand, uses the model of the economic cycle formulated by Paul Samuelson (1939, 1947) and John Hicks (1950), which we mentioned in Chapter 9 (see p. 63, note 33) and shows that a simple accelerator can generate situations of chaos. This originates from the fact that Puu uses a third order relation for the function I_t that links investment to income Y_t, therefore a nonlinear function, in place of Samuelson's, which assumed a production structure in which capital is proportional to the income generated according to the following relation:

$$I_t = v(Y_{t-1} - Y_{t-2}).$$

Puu assumes:

$$I_\tau = v((Y_{t-1} - Y_{t-2}) - (Y_{t-1} - Y_{t-2})^3) \qquad (30.1)$$

---

[136] To avoid misunderstandings, we would like to point out that here we are not dealing with Goodwin's famous model of the economic cycle dated 1951 (in addition to Goodwin, 1951, see, for example, Allen, 1959). The model that we are referring to in the text is, however, basically a derivation of the latter, although with significant differences. Goodwin's article is also of particular interest, due to the considerations made by the author that originate from his personal acquaintance with Keynes and Schumpeter.

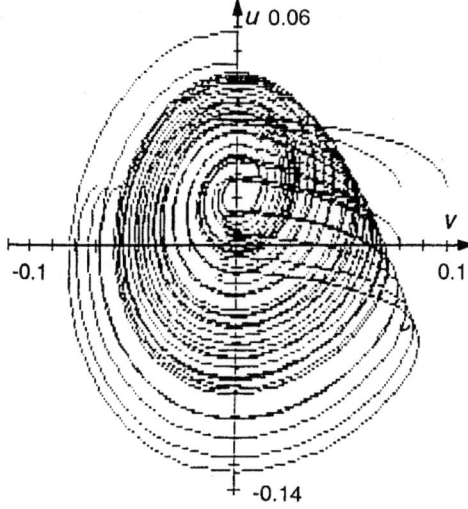

Figure 30-1 Strange attractor from Goodwin's model.
Source: Goodwin (1991, p. 143).

Equation (30.1), when $x_k = (Y_{t-1} - Y_{t-2})$, becomes equal to that of the logistic map (20.4) with the sole difference of the exponent: 3 instead of 2. Puu thus obtains that the dynamics of the economic cycle, under this assumption, can display a chaotic attractor that has many elements in common with the Feigenbaum tree in the logistic map presented in Chapter 22 (Figures 22.2–22.4), including therein those elements that we have indicated as aspects of order from chaos. This should come as no surprise; as we said in Chapter 22, the characteristics of the dynamics of the logistic map are common to the dynamics of all unimodal quadratic maps with a continuous first derivative. The map (30.1) is continuous and presents one single maximum value in the interval [0,1]; it can be shown that a third order maximum, and not a second order one, does not result in any substantial changes to aspects regarding the onset of chaos in the dynamics with respect to the quadratic case of the logistic map.

A model of the complex phenomenology of the financial markets

Often, incomplete information and irrational decision-making in economics are other factors that are considered a cause of instability in economic systems. This is particularly evident in the dynamics of the financial markets, often subjected to unexpected events that are not linked to causes that can be objectively identified in economic data, but that originate from events of a more psychological than economic nature.

This is even more evident in a specific financial market, the market where certain particular financial instruments called derivatives, such as futures and options, are traded.

Derivatives are typically highly speculative financial instruments that originate from other financial instruments, such as for example, shares traded on the stock market, or other goods of a different nature traded on other markets, and could not exist without them. Derivatives enable the investor to make a gain on price variations of the underlying instrument, let's say a share, risking a lower amount of capital than that needed to buy the share itself. In practice, simplifying somewhat, the idea is that the investor, on the basis of expectations relative to the future price of the underlying share, bets on the future rise (if he invests in a so-called *call option*) or on the future fall (a so-called *put option*) of the share price. The purchase of an option gives the investor the right, but not the obligation, to buy (call option) or to sell (put option) the underlying share at any time before the option expires (American option) or exactly on the option's expiry date (European option), at a price fixed when the option is purchased (the *strike price*).

To make this even clearer, let us consider the case of the European option. If on the expiry of the option, the market value of the share is higher than the sum of the price paid for the purchase of the option (the premium) and the strike price, then the investor that holds a call option gains, because, by paying the strike price, he becomes the holder of a share with a superior value (at least at that moment in time) to the total price he paid. If, on the other hand, the market value of the share is less than the sum of the premium and strike price, then the investor can choose whether to buy it anyway, and suffer a loss, or forsake the purchase, let the option lapse and thus lose all the capital invested in the option. Vice versa, the investor that possesses a put option when the option expires will collect the difference between the strike price, i.e. the pre-established sale price and the current value of the share, and will make a gain if said difference is higher than the premium paid for the option (in the case of futures, the mechanism is similar; the only difference with respect to options is that by buying a future, the investor does not have a right, but is committed to purchase, in one case, or to sell, in the other, the underlying share).[137]

---

[137] In reality the topic of derivative financial instruments is much wider and more detailed than we have presented here. For example, as we have said, there are also derivatives that

What we are interested in pointing out is that options and futures do not only *suffer* the expectations that the market has on the underlying shares, but because they exist as derivatives, they themselves, through a real feedback mechanism, *act* on the underlying shares, whose quotations are subject to the expectations of the market (this is commonly known as 'discounting') and, therefore, *also* feel the impact of the quotations of their derivatives (Dendrinos, 2000). In fact what happens is that the investor that foresees an increase in the share price buys a call option, whereas the investor that foresees a fall in the share price buys a put option.[138] But by doing thus, both contribute to forming the expectations of other investors in the market, and therefore contribute, through their decisions, to establishing the price trend of the underlying shares.

The relation between the price of the derivative instrument and that of the underlying share is highly nonlinear and describes a positive feedback cycle; the changing and often irrational psychology of the investor also plays its part, as investors usually act in conditions of incomplete information. The result of this is that the percentage variations during the life of a derivative are usually larger than those of the underlying share. On one hand, this gives the investor the opportunity to make higher gains with respect to those that he could obtain by buying the underlying share, but on the other hand, it presents the risk of losing the entire amount of capital invested when the derivative expires; a derivative instrument, in fact, does not represent the ownership of a share in a company, but merely a right (or a duty in the case of futures) to purchase (or sell), and at its expiry, leaves nothing in the hands of the investor that does not exercise it.

Price trends in financial markets have been analysed ever since the stock markets have existed, basically since the mid-sixteenth century (Bruschini Vincenzi, 1998). In the past few decades, new mathematical techniques have been applied in an attempt to identify recurrences and deterministic mechanisms in the trends of prices on the financial markets, borrowing and

---

refer to goods other than shares; furthermore, there are different mechanisms to define and trade so-called American, European and Asian options, and there is an important formula, called the formula of Black and Scholes (1973), which is widely used for the calculation of the theoretical reference value (the so-called *fair value*) of options. We will not be introducing any technical aspects that are not relevant to our discussion. Interested readers can consult current literature on this topic, for example Wilmott, Howison and Dewynne (1995) and Neftci (2000).

[138] In financial jargon it is normally said that, in the first case, the investor has a long position, or 'goes long', whereas in the second case has a short position, or 'goes short'.

transforming even approaches and concepts from the world of physics. This has led to the birth of a new discipline, called econophysics, whose objective is to apply the mathematical concepts, models, and techniques, that have enjoyed considerable success in physics, to the financial markets (Mantegna and Stanley, 2000; Mantegna, 2001).

In this regard, the collection of studies by Benoît Mandelbrot (1997a; see also Mandelbrot, 1997b, 1999ab), is particularly interesting, in which fractal geometry is used, developed mostly by Mandelbrot himself, and in particular the invariance of scale that it is based on, to obtain a model that describes the price trend of securities traded on the financial markets, as we already mentioned in Chapter 9 (on this subject, see also Peters, 1994, 1996; Franses and Van Dijk, 2000). In any event, we should point out that the opinion that there is an unknown, but identifiable deterministic law behind the dynamics of the financial markets, and that therefore the markets are not intrinsically stochastic, but deterministic, is anything but a commonly held belief. This is due, above all, to the lack, at the moment, of effective models and really significant results.[139]

The idea that the financial markets have a fundamentally reflexive and self-referential nature, as the expectations that the actors form are constructed on the basis of the anticipations that they make with regard to the expectations of other actors, generally without a real logical-deductive procedure based on economic data, forms the basis of a stock market model studied by a group of researchers (Arthur et al., 1997), from the Santa Fe Institute (see pg. 275, note 128). The authors propose a theory of price formation that considers a heterogeneous set of actors, which continuously adapt their expectations to the market that they themselves, all together, construct. The model refers in particular to the reality of the US stock market, in which the practice of direct speculative investment in financial markets is more widespread than in European ones. Said investments are made by means of frequent purchases and sales of financial instruments, in general via the Internet, so-called *trading on line*, by private investors called *traders*. Traders act in a way that is fundamentally different to that of institutional investors. In fact they usually act from a short-term perspective, often in an inductive way and, in any event, are more often driven by factors, often with emotional origins, that are generally not ascribable to economic and financial data (the so-called 'fundamentals') of listed companies. Usually a trader moves relatively small sums of money, very frequently, often several times in one day, in contrast to institutional

---

[139] Somebody ironically said that the only way to make money by applying the chaos theory to financial markets is to write books on the subject.

operators. It can occur, however, that traders, all together, move a considerable sum of money and that this, in certain situations, reaches critical values that trigger important movements in the market that are not based on purely economic data, such as the so-called speculative bubbles or a sudden fall in prices. Movements of this nature are difficult to predict, because they are linked to the spreading and the self-reinforcement of a feeling or a hope, according to a type of positive feedback mechanism, more than a rational decision taken following information based on new objective data.

Arthur et al.'s model (1997) considers traders as actors that act by continuously making hypotheses, exploring the market in the light of certain patterns of expectations, and that buy or sell securities on the basis of the conjectures that they form relative to the performance of the securities themselves. Traders act in an inductive way; the perfect rationality, which is at the basis of deductive reasoning, ceases to be the market engine. Traders produce individual patterns of expectations and constantly try them, adapt them, completely change them or abandon them to go on to new patterns. In this way, individual opinions and expectations act endogenously to the financial system and are in continuous competition with the opinions and expectations of others, within an environment formed of opinions and expectations that are related to one another and that evolve over time. The market, in this way, is moved by changing opinions, which adapt themselves endogenously, i.e. without the intervention of forces external to the actors in the market, to the environment (the 'landscape') of relations, the environment which, on the other hand, these same expectations have contributed to creating.

The implications and the effects of a system such as that described above have been explored by means of computer simulation, using an artificial environment called the 'artificial stock market of Santa Fe' created several years ago as a simulation model of the real stock market (Palmer et al., 1994). Arthur et al. (1997) have reached the conclusion that the financial market can operate in two different regimes. In conditions in which there is a low rate of exploration of alternative forecasts, the market stabilizes in a simple regime that corresponds to a sort of stable equilibrium alternative based on rational decision making, described in the standard theory as an *efficient market* (see p. 64, note 34). In a market of this nature, in fact, the mechanism – highly efficient – of circulating information does so in such a way that the information is immediately taken into consideration by each operator as soon as it is available (or, as it is commonly said, this information is 'discounted by the market'). The hypothesis of a totally efficient market means, however, in reality, that the information is immediately reflected in current prices and that the 'best' estimate of the intrinsic value of a share (in this hypothesis we presume it exists) is its current price, which is formed as the immediate result of the diffusions of the information (Batten, 2000a).

In the more realistic conditions of high rates of exploration of the alternatives, and of information that circulates in an imperfect way, on the other hand, the market self-organizes into a complex regime, in which a much more varied price behaviour emerges that displays the statistical type characteristics that are encountered in the real market. In this second case, alongside trader activities conducted according to the rules of technical analysis, in which the use of past time series (past performance) is an essential instrument for future predictions, we have the speculative bubbles and the crashes generated by the self-reinforcing and the rapid circulations of opinions and feelings. Phenomena of this type emerge when individual expectations mutually reinforce one another, ending up by distorting the role of market indicators.[140] The indicators, in this situation, no longer perform their primary role of reflecting the current situation, but also become signals that coordinate opinions and feelings, and end up contributing to the self-reinforcing and to the circulations of sets of opinions and feelings. Following the simulations, the authors sustain that the real markets, with irrational and inductive traders, operate according to the regime that has just been described, in which the markets themselves are characterized by a heterogeneity of opinions and by deviations to rational decision-making based on the values of fundamental data.

---

[140] The term 'market indicators' is given to the various quantitative magnitudes, measured in the financial markets, that are used by analysts to formulate predictions on future price trends. Technical indicators and fundamental indicators are usually distinguished between. The former includes, for example, average prices over time, calculated according to various definitions, and the relative indices of volatility; the latter includes for example the volumes traded by a certain share or eventual information on insider trading activities (in markets where such is permitted).

31 Some thoughts on the meaning of 'doing mathematics'

The problem of formalizing complexity

In this Chapter and in Chapter 32 and 33, to conclude the book, we would like to reflect for a moment on the fact that, as we have said on many an occasion, at the moment there is no formalization of complexity that enables it to overcome its current, rather confused state, which for now is known as the 'complexity approach', and to achieve the objective of first becoming a method and then a bona fide scientific theory. The complexity approach that has recently appeared in modern scientific circles is generally still limited to an empirical phase in which the concepts are not abundantly clear and the methods and techniques are noticeably lacking. This can lead to the abuse of the term 'complexity' which is sometimes used in various contexts, in senses that are very different from one another, or at times even erroneously, to describe situations in which the system does not even display complex characteristics.

Formalizing complexity would enable a set of empirical observations, which is what complexity is now, to be transformed either into a real hypothetical–deductive theory, similar to the role of physics when it is treated with mathematical methods, or into an empirical science, like biology, geology, economics, and sociology, in which the hypothetical–deductive method is rarely applied (there are no theorems in these fields). Here, however, we don't intend to discuss epistemological matters that are not the subject of this book. In this section, more than complexity in the strict sense, we will be looking at the fact that, at present, the study of complexity from a mathematical perspective is substantially vague, in terms of definition, ineffective in terms of methods and lacking in terms of results.

Therefore, at least for the moment, there is no unified theory of complexity able to express the structures and the processes that are common to the different phenomena, observable in different contexts, that can be grouped under the general heading of complexity, even though such belong to different areas of science. As we have repeatedly said in this final part of the book, there are several evident shortcomings in modern mathematics, which make the application of a complexity theory of little effect. Basically, this can be put down to the fact that mathematics is generally linear, where chaotic and complex phenomenology are regulated by positive feedback mechanisms, and can be modelled in the form of nonlinear differential equations which, usually, cannot be solved in formal or general terms but only through the use of approximated numerical methods.

As we mentioned in Chapter 27, one of the fundamental notions of the complexity approach is that according to which the idea that we have of a system's complexity is strictly linked to the interaction between the complex system and the observer, who is also complex (*design complexity* vs. *control complexity*). The problem of the interaction between the observed and the observer already exists in quantum theory; in fact it is one of the basic theoretical elements. However, quantum theory is substantially linear. The evolution of quantum systems is described, limiting ourselves to considering a non-relativistic approximation, by a linear equation with second order partial derivatives, Schrödinger's equation; therefore it is a substantially deterministic theory. It is true that in the description of quantum systems, probabilistic concepts are extensively used; however, the unpredictability that derives from said systems is not chaotic, but is due to the two following reasons. On one hand, it is postulated that it is theoretically (and not only in practice) impossible to know the values of *all* of a system's state variables *simultaneously* and with arbitrary precision, but it is possible to fully determine a subset of them only (Heisenberg's uncertainty principle); on the other hand, the probability associated to measuring the state of a quantum system can be directly related to the principle of state superposition, i.e. to the fact that it is postulated that a system, before being observed (i.e. before the observer interacts with it) finds itself in several different states simultaneously, one of which, according to a probability distribution, is 'forced' to appear like *the* state of the system from the act of measurement itself.[141] In the context of quantum physics, therefore, the concept of indetermination distinguishes itself in a different way both with respect to chaos and with respect to that intermediate, but clearly defined, area between determinism and chaos that we call complexity.

---

[141] These concepts that are out of the ordinary sense have given rise to the formulation of renowned paradoxes, such as that known as 'Schrödinger's cat', or the Einstein, Podolsky, and Rosen's paradox (1935). An original discussion of these paradoxes, and more generally of the current situation of quantum physics theory, can be found in Accardi (1997).

There are studies that are leaning towards defining rules appropriate to the description of complex systems; however, they are still not sufficient to formulate a general complexity theory, nor even to establish methods for the theoretical and practical determination of a measurement of complexity unequivocally applicable to systems.

We could say that the situation, in some ways, is similar to that which the natural sciences, in particular astronomy, experienced during the seventeenth century. Algebra had taken great steps forward, mainly thanks to the work of several scholars from the Bolognese school (general solution of third and fourth degree equations, first introduction of complex numbers, etc.) and of several French scholars (such as François Viète with the formalization of algebraic symbols) and British ones (such as John Napier with the introduction of logarithms). What was missing, however, was a tool that enabled the analysis of the 'continuum', that is to say even just the definition of concepts such as instantaneous speed, a straight line tangent to a curve at a point, the surface area of a non-rectilinear section, etc. Therefore the search for a new horizon in mathematical techniques and conceptions began, and was led, initially, by Pierre de Fermat and René Descartes and later by Newton and Leibniz. The work of these scholars led to the creation of infinitesimal calculus, which turned out to be a particularly effective tool to describe motion in several relatively simple cases, for example, in the case of the movement of two bodies in reciprocal gravitational interaction.

With regard to this objective, the algebra of the previous century was insufficient, as it was basically static in terms of vision and technique, and incapable of giving a meaning to the still unclear concept, but that was gaining ground, of 'instantaneous motion' (Boyer, 1968; Kline, 1990).

The fundamental aspect that distinguished the emerging branch of mathematics with respect to existing algebra was represented by the concept of infinitesimal and, strictly linked to this, the concept of the continuity of a curve. In this regard, the long diatribe that opposed Newton and Leibniz and the harsh criticism of the very idea of infinitesimal by the bishop and philosopher George Berkeley are very well known. The latter's criticism was in reality more than justified by the approximate and contradictory definition that Newton gave of the 'infinitesimal quantities', even though Berkeley recognized the usefulness of the technique of the so-called Newtonian fluxions.[142]

---

[142] 'And what are these Fluxions? The Velocities of evanescent Increments? And what are these same evanescent Increments? They are neither finite Quantities nor Quantities infinitely small, nor yet nothing. May we not call them the Ghosts of departed Quantities?' (quotation from *The Analyst* written by Berkeley in 1734).

The concept of infinitesimal, which had already been partially discovered, even though in a different form and context, by Archimedes, and later by Galileo Galilei and mathematicians from his school, such as Evangelista Torricelli and Bonaventura Cavalieri, is a typical example of how a happy intuition finds it difficult to take the form of a concrete theory and to be accepted by the scientific community (on this point see Giusti, 1999, and the discussion that the author presents on what a mathematical object is and how such emerges).

The development of calculus techniques that used concepts linked to infinitesimals was extremely important during the entire course of the eighteenth and nineteenth centuries, even though the theoretical foundations at the basis of the techniques developed had not been specified. Basically, infinitesimals were used without it being completely clear as to what their nature was. The study of the continuum did not make any significant progress until the work of Bernhard Bolzano, in the first half of nineteenth century, and of Karl Weierstrass some years later. The most significant contribution, however, was made from the early 1870s onwards by the work of Georg Cantor and Richard Dedekind, who, in a series of publications, created a totally new theoretical framework, in which the concepts of infinity[143] and continuum were included and given meaning (Dedekind, 1901, 1915; see also Hallett, 1984, and above, p. 197, note 71).

---

[143] More correctly, we should distinguish between *actual* infinite and *potential* infinite. Actual infinite is associated to the idea of a segment being made up of 'infinitely numerous' points. This is the problem tackled by Cantor and Dedekind, the arguments of which return to and go beyond known paradoxes, some of which come from ancient Greece, such as that of Zeno of Elea (Southern Italy), also known as 'the arrow' and 'Achilles and the tortoise', of which Aristotle writes in his book *Physics*, that can be summarized as follows. In order to traverse a line segment it is necessary to reach its midpoint. To do this one must reach the 1/4 point, but to do this one must reach the 1/8 point and so on *ad infinitum*, hence motion can never begin, because space can be infinitely divided. The argument here is not only that the well-known infinite sum $1/2 + 1/4 + 1/8 + \ldots = 1$ never actually reaches 1, more perplexing is the attempt to sum $1/2 + 1/4 + 1/8 + \ldots$ backwards. Before traversing a unit distance we must get to the middle, but before getting to the middle we must get 1/4 of the way, but before we get 1/4 of the way we must reach 1/8 of the way and so on, therefore we can never get started since we are trying to build up this infinite sum from the 'wrong' end. The paradoxical conclusion is that 'there is no motion'. Potential infinite, on the other hand, is associated to the idea of the growth of a function without a finite limit, a problem already tackled by Augustin Cauchy, together with that of the clarification of the concept of a finite limit of a function, in several fundamental works in the 1820s, in particular in the monumental work *Cours d'analyse de l'École Polytechnique* in 1821 (Cauchy, 1882–1970), in which he gave infinitesimal calculus its formal definition that, with the fundamental contribution that Weierstrass brought some decades later, it maintains even today.

We are now faced with the following problem. We are not able to describe chaotic phenomenology or even that type of organized chaos that is complexity by means of adequate general laws; consequently, we are not able to formulate effective long-term predictions on the evolution of complex systems. The mathematics that is available to us does not enable us to do this in an adequate manner, as the techniques of such mathematics were essentially developed to describe linear phenomena, in which there are no positive feedback mechanisms that unevenly amplify any initial uncertainty or perturbation. In which direction then should we search for an adequate solution? In the direction of the development of new mathematical techniques in the general context of the current foundations of mathematics or in the direction of a review of both the foundations themselves and of the meaning of mathematical methodologies that they form the basis of? In other words, is it a purely technical matter, i.e. limited to the development of new theoretical–practical methods to solve problems, or is it a conceptual issue, i.e. pertinent to problems regarding the very notion that we have of mathematics, and only at a later stage, and as a consequence of the latter, regarding technical aspects?

Probably, the problem of formulating a complexity theory on mathematical foundations should be tackled through a review of the foundations at the basis of currently used mathematical methodologies, particularly the concepts of number and of continuity. The very problem of understanding the continuum of a curve, just as that of the continuity of a set of real numbers, have still not actually been resolved. In particular, there are still open issues that date back to the scientific world of more than a century ago, regarding, for example, Cantor's continuum hypothesis and Zermelo's (or Zermelo–Fraenkel's) later postulate on infinite sets, known as the 'axiom of choice',[144] questions that are at the basis of what we mean by number, set (particularly infinite set), and mathematical object.

---

[144] In 1874 Cantor proved that there is more than one order of infinity, even going as far as stating that a real hierarchy of infinite orders of infinity can be constructed (see p. 197, note 71). The lowest order is the infinity called countable or numerable, which characterizes the set of natural numbers, as well as the set of rational numbers. Infinities of a higher order, for example the infinity that characterizes the set of real numbers, are uncountable or not numerable. In 1877, after three years of research without any conclusive results, Cantor hypothesized that there are no intermediate orders of infinity between the numerable one, of natural and rational numbers, and that of real numbers. In this way the latter is the order immediately superior to the numerable one. This hypothesis was called the 'continuum hypothesis', because, as it is possible to establish a one-to-one correspondence between the set of real numbers and the points on a straight line, the real numbers, so to speak, 'render the idea' of the continuity of a straight line (Cantor, 1915; Weyl, 1918; Hallett, 1984). In response to the antinomies such as that of Russell on infinite sets (see p. 336, note 156),

Mathematics as a useful tool to highlight and express recurrences

We have already discussed the matter of why we tend to apply a predominantly linear version of mathematics to our descriptions of the world in Chapter 25. Now we would like to return briefly to this point and reflect on the reasons why we use mathematical methods in natural and social sciences. The reason could be related to the habit, or possibly to the necessity, of looking for recurrences in the phenomenology that we perceive, formulating laws that summarize said recurrences. As the mathematician John David Barrow writes, 'the process of evolution has deemed it appropriate to employ a fair number of our resources in recognizing schemes: if we were not capable of recognizing regularities we wouldn't be able to store information about the environment and adapt ourselves to it, but if we were to store the most minute details of the structure of the world that surrounds us, we would waste energy accumulating useless information and also in this case run risks to our survival' (1994, p. 22–23, our translation).

which emerged at the end of the nineteenth century following the first attempts to construct a theory of arithmetics on strict logical–formal bases, Ernst Zermelo (1908) started to review the theory of Cantor sets, rejecting the intuitive approach adopted by these to the concept of a set and formulated, instead, a series of postulates at the basis of a new formal conception of sets. In particular the eighth of these postulates, the axiom of choice, was and is a particular subject for discussion, insofar as not all mathematicians agree in accepting it. It is stated as follows: if F is *any* family of non-empty pairwise disjoint sets $X_1, X_2, X_3, \ldots, X_n, \ldots$, then a set G *exists* (this is the point of contention) containing one element of each of the $X_1, X_2, X_3, \ldots X_n, \ldots$, for all the sets X_n of F. In rather more approximate terms, this means that, by choosing a first element from one set, a second element from a second set, a third element from a third set, and so on, the elements chosen represent, in their turn, a new set. If F consists of a *finite* number of sets X_n, then the postulate does not present any problems. For example, if X_1 is the set of even numbers and X_2 that of odd numbers (in this case F consists of two sets only), it would be sufficient to choose any one even number (for example 4) and any one odd number (for example 7) to construct G, which thus results as being composed by the numbers 4 and 7. The problem arises if the set F is made up of an *infinite* number of sets X_n; in a situation of this kind, the postulate ceases to be evident. At the basis of the very concept of number there is the theory of sets, in the ambit of which the axiom of choice plays a role similar to that in geometry of Euclid's fifth postulate on parallel lines. As it is a postulate, it can be accepted or not. By accepting it, or accepting its negation, we construct a so-called standard theory; if, on the contrary, we do not recognize either the postulate or its negation, we construct a so-called non-standard theory. Despite the fact that more than a century of research has passed, Cantor's continuum theory and Zermelo's axiom of choice still provoke debate. Two well-known theorems, the first by Kurt Gödel and the second by Paul Cohen (who was awarded the Fields medal in 1966 thanks to this), have demonstrated the complete independence of the standard theory of sets from Cantor's continuum hypothesis (Devlin, 1980; Hallett, 1984).

Mathematics is not just about numbers. As well as numbers, modern mathematics also looks at the relations between them. The passage from pure numerology to this new vision has derived from the realization that the most profound meaning is not in the numbers but in the relations between them. Mathematical investigation is precisely the exploration and the study of the different possible relations; some of them find a concrete and immediate application in the environment in which they are immersed, others just 'live' in the minds of those that conceive them. Again, according to Barrow: 'If we see things in this way, we understand why there has to be something "similar" to mathematics in the Universe we live in. We, like any other sentient human being, are none other than examples of organized complexity. We are stable complex schemes in the fabric of the Universe. As long as life exists, in any form, randomness and total irrationality need to be kept at a distance. Where there is life, there are schemes and whether there are schemes there is mathematics. If there is a seed of rationality and order in the potential chaos of the cosmos, then mathematics exists as well. A non-mathematical universe containing living observers could not exist' (1994, p. 24–25, our translation).

The search for recurrences that can be expressed in the form of mathematical laws, however, has not always been successful. Economists, for example, produce knowledge, which, however, does not lead to universal laws, valid at all times and in all places. They rather lead to relations that are valid momentarily, locally and partially, without producing a general formula in a rigorous language such as mathematics, as we have been doing in physics for the last four centuries. Economic knowledge, in fact, links certain specific behavioural hypotheses with certain particular conclusions, valid in a schematized environment, by introducing some sort of mechanism that can account for the connection. But the empirical validity of these types of mechanism depends both on the particular nature of the real behaviour of the individuals and on the institutional context within which these individuals act which prevents these mechanisms from being transformed into laws (Orléan, 1999). By systematically analysing these mechanisms, economists have constructed that sector of applied mathematics called social mathematics. The language of mathematics is useful because it is not ambiguous and it is precise: it enables hypotheses and results to be controlled in the ambit of a theory expressed in the form of a mathematical model. We must not forget, however, that these are forms of thought reconstructed in a 'mathematical laboratory', far from the changing and many faceted reality of society.

When we make social statements and claim to give them the universal validity that distinguishes what we call 'regularity', an expression commonly used in the context of the social sciences, or, borrowing the term from physics *tout court* 'law', we often observe counter-trends that set themselves against the former and that may also conceal them. So, if this is true, they are not authentic laws (even though, for the sake of convenience, we will continue to refer to them in this way). Karl Marx, for example, formulated in the *Capital* (1867) the law of the tendential fall in the profit rate, according to which, in a developing capitalist economy, the increase in competition 'necessarily' leads to a continuous increase in the amount of capital invested (more precisely, a continuous increase in the relation between means of production and the workforce employed) with the outcome, in the long term, of a progressive reduction in the profit rate, i.e. the relationship between the profit produced and the capital invested.[145] Marx, however, highlighted how there are a great number of counter-trends that oppose the law of the tendential fall in the profit rate, counter-trends that, at present and with hindsight, appear even more interesting to our understanding of the evolution of capitalism than the very law of the tendential fall in the profit rate. So, then, we cannot say that the tendential fall in the profit rate, even if we accept its existence in some form, qualifies as an authentic law, if it can potentially be contradicted.

Despite the fact that it is convenient to use laws formulated in a mathematical form (i.e. models) in economics and in social sciences in general, various conceptual obstacles hinder the transformation of the mechanisms proposed (or assumed as true) into laws, making the use of the laws, if taken to the letter, fundamentally unacceptable. André Orléan (1999) identified four basic obstacles. The first is the one that we have just mentioned with regard to the law of the tendential fall in the profit rate: the ambivalence of the mechanisms, i.e. the simultaneous presence of tendencies and effects in opposition to those envisaged by the law. A second obstacle is represented by the role of history, which profoundly changes the institutional structures and therefore makes it difficult to identify recurrences, due to the need to historicize the basic mechanisms. The third obstacle, which is more complicated than the other two, draws its origins from the fact that men learn, and therefore change their beliefs and their behaviour as a function of the knowledge they acquire. Simply the fact of stating an economic or social law can provoke new behaviour that invalidates the freshly formulated law. As Bernard Guerrien (1999) points out, there is a vicious circle between theories and facts in social sciences. The statements themselves are social

---

[145] The idea that the growth of an economic system leads to a reduction in profits was proposed in similar terms, by David Ricardo, in his *Essay on the Influence of a Low Price of Corn on the Profits of Stock* (1815) several decades before Marx. According to Ricardo, the growth of the economic system leads to the progressive utilization of increasingly less fertile land in order to handle the growing demand for agricultural products; in the absence of technological innovation, this leads to an increase in the unit cost of an agricultural product and therefore a fall in profits (what became known as the law of diminishing returns).

products; they are used by the social actors that formulate them and that can change their behaviour as a consequence of the freshly formulated statement, rendering it worthless and leading to the formulation of a new statement, which, in turn, leads to a change in the behaviour of the social actors and to the consequent formulation of a new statement instead of the old, and so on.[146] The fourth and final obstacle to the formulation of mathematical laws in social sciences regards the fact that human actions, such as for example buying and selling, are never strictly economic actions; economic interest and rationality are not, in reality, the only reasons behind an individual's actions.

In conclusion, in social systems, typical examples of complex systems, the recurrences (laws) that we attempt to highlight by stating them in mathematical form, are inadequate to be applied in the same way as they are in natural sciences, at least in the so-called 'hard' sciences. Nevertheless, we should not forget that often, in the natural sciences themselves, the laws formulated with mathematical tools are inadequate to describe a phenomenology that is effective in terms of forecasting, even if for reasons that generally, but not always, are different to those that are true for social sciences. On the other hand, as the history of science over the past four centuries has demonstrated, mathematics is just as often an extremely effective descriptive tool.

The time has come to ask ourselves the core questions that we have often touched upon during the course of the book: why is mathematics an effective tool to describe the world? And, in any event, to what extent is this true? Can the point at which mathematics ceases to be an effective tool be extended by means of an appropriate review of its foundations and methods? And finally, more significantly: what is mathematics, what is it about and what does 'doing mathematics' mean?

A reflection on the efficacy of mathematics as a tool to describe the world

Without going into an in-depth discussion on the above questions, which would be somewhat beyond the scope of this book, we would like, in any event, to express our thoughts on the same.

---

[146] We have already come across the notion of the vicious circle that instils itself between the behaviour of social actors and the regularities of the social systems in Chapter 19, where we highlighted the difficulty of recognizing and formulating deterministic laws to describe market dynamics, because such laws are immediately violated by the actors in their constant search for ways of maximizing profits.

A basic characteristic of the human mind is knowing how to make reductions, i.e. knowing how to recognize, in the incessant flow of empirical observations of a continually changing reality, the presence of elements that repeat themselves (recurrences) or elements common to different situations. All scientific thought in the past four centuries has been employed in a continual attempt to make reductions, expressed in the form of recurrences in the varied and many-faceted phenomenology that surrounds us. We could even go as far as saying that science exists because reality appears to be 'algorithmically compressible', that is to say, susceptible of being read in such a way that its description can be expressed, very often at least, in a 'compressed' way, without stating, one by one, all of the facts observed, but by formulating a set of general laws that describe the facts, expressed in a more concise way than would be necessary if we were to describe the phenomena one by one. The mathematical formulae that we use to state recurrences that we call 'laws', from this perspective, are no more than reductions made to economize both the language and the information that the language encodes and transmits, eliminating the superfluous, starting from enormous sequences of data on the world.

If we were to imagine a completely random world, its properties could only be described by means of a very long list of innumerable sequences of phenomena that follow one another over time. When we talk about chaotic phenomenology in real systems, on the other hand, we are imagining that there is an underlying determinism; let us imagine that there is a deterministic law at the origin of the observed data, interpreted as the manifest result of the dynamics generated by the relevant unknown deterministic law. The identification of the deterministic law (if such exists) at the origin of the data and the formulation of a model that reproduces the series of observed data (the dynamics) would enable us to consider those data, and in general the phenomenon to which they refer, not as having a random origin, and, therefore, to economize the quantity of information to be acquired and preserved in order to obtain a complete description of the systems (see Chapter 19).

In practice, however, we are not able to pin down the underlying determinism, i.e. the concise law that describes the phenomena in a deterministic way. In fact, it is extremely difficult to identify the underlying recurrences (the laws) that generate chaotic phenomena unless we know them already through another source. Chaos (determinism is present, but it is hidden) and chance (absence of determinism) both manifest themselves in the same way. Both are characterized by disorder: apparent (deterministic) in chaotic phenomena, real (stochastic) in random phenomena.

If we interpret science as an attempt to algorithmically reduce the world of experience and if the universe is a single entity, we would expect, or rather we would like, the entire universe to be algorithmically compressible; i.e. we would like it to be possible to concentrate the description of all of the sequences of all of the phenomena into a series of formulas, in a sort of 'theory of everything'.[147]

As we have already said, there are numerous classes of phenomena that, for various reasons, cannot be reduced. In addition to those we have already cited relative to chaos, we encounter non-reducible phenomena every time, often without considering the question's methodological and epistemological implications, that we qualify a phenomenon as random. In reality, the concept of randomness is a concept of convenience, that represents either our ignorance of the laws of a phenomenon or the impossibility of observing any recurrences for any excessive length of observation time and/or calculation time for data processing. Strictly speaking, in fact, the fact of not seeing a recurrence does not imply that it doesn't exist, but just that it may be characterized by a period superior to the observation time; to be certain that there is no recurrence, an infinite observation time would be needed.

Alongside the previous situations, in which we seek to identify and model recurrences using mathematics, there are other situations in which, even though not seeking mathematical laws, we use partially mathematical tools in any event, i.e. tools that even though not mathematical in the strict sense, share some characteristics with mathematics. Every time we make a comparison between magnitudes, we take measurements and perform oper-

---

[147] '*Theory of everything*' is the expression that originally indicated all those theories that aimed to unite the four fundamental interactions of physics: the gravitational interaction; the electromagnetic interaction (this, in its turn the result of the unification of electrical and magnetic interactions), whose intensity tends to zero as the distance tends to infinity; the strong interaction; and the weak interaction, whose ranges of action are limited to subatomic distances. The most famous and most recent of these theories is known as *superstring theory*. In its original context, the word 'everything' just referred to the description of elementary particles and their interactions and didn't indicate a theory of everything that is recognizable in the universe. The link between the behaviour of the elementary particles and the phenomena of the universe in their entirety is too indirect to allow one of these theories to be useful, for example, for predicting the trend of the financial market on a given day starting from laws of subatomic physics. A theory of everything is fundamentally and essentially reductionist and as such ineffective for a global understanding of the universe. As J. Cohen and Stewart write: 'The physicist's belief that the mathematical laws of a Theory of Everything really do govern every aspect of the universe is very like a priest's belief that God's laws do. The main difference is that the priest is looking outward while the physicist looks inward. Both are offering an interpretation of nature; neither can tell you how it works' (1994, p. 364–65; on this point see also, for example, Gell-Mann, 1994).

ations on the numerical data that come from said measurements. We make comparisons, however, even when we are not dealing with real magnitudes that can be measured. In these cases, the result of the comparison is not expressed as a number. In this case we talk of qualitative evaluations that often, for no other reason than communications convenience, we nonetheless attempt to express in a quantitative form. This is sometimes carried out by using universally comprehensible conventions; think, for example, of the marks of an exam conducted in the form of an interview, in which the evaluation is the result of the subjective impressions that certain errors leave and not an objective calculation of the errors, as is the case, on the contrary, when the exam is conducted in the form of a written test.

Over the past few decades, processes have been developed for the mathematization of qualitative evaluations based on various hypotheses, aimed at the rigorous treatment of qualitative data. These are called multicriteria analyses (see, amongst others, Nijkamp, Rietveld and Voogd, 1991), decisional techniques that are useful when there are, in part or exclusively, qualitative elements of evaluations in the form of an ordinal number that expresses a scale of priority or preference, and not in the form of a cardinal number, the result of a measurement, i.e. a quantitative judgement. In these situations, due to the convenience of the cultural habit of using mental quantitative tools, we seek, by introducing particular hypotheses, to 'mathematize' sensations, impressions, and judgements, that are only qualitative and not the result of measurements expressed in numerical form, by projecting them onto a quantitative plane. This type of method is based on the assumption that the mind applies schemes of logical-deductive reasoning that we are not aware of but that we know the final outcome of (the qualitative judgement), which, in some way, we attempt to justify and render rigorous by transforming it into a quantitative judgement (Nijkamp, Leitner and Wrigley, 1985). This is accomplished by hypothesizing schemes of reasoning constructed and expressed according to the formal mathematics that we are used to in other fields; however, we know nothing of the ways in which the mind actually works; the mathematization that we perform in these decision-making techniques is merely a working hypothesis (Bertuglia and Rabino, 1990), that has a sense and an efficacy from a practical perspective, but no more than that.

On one hand, the method that involves making reductions and expressing them in a mathematical form, i.e. the application of mathematics to both natural and social sciences, as we have said, is imperfect from various viewpoints. Sometimes, in some contexts, it is extremely effective; other times, above all in several branches of social science, it is ineffective (Israel, 1999). On the other hand, mathematics is a useful tool of analysis, in practice, also when the reduction of data is too difficult, or even when the data are not even numerical.

The conclusion that we reach from various directions is that the mind has developed a powerful tool of thought, mathematics, that can be applied to process data gained from experience. This tool is highly sophisticated, in theory, but in practice the mind does not always know how to manage it adequately or to use it appropriately. It works well in some cases, little and badly in others, and it doesn't work at all in numerous situations; in a word, it is an imperfect tool. The abstraction of schemes from the world of empirical data carried out using this tool is, beyond a certain point, imperfect or even fundamentally impossible; this is the case, for example, of complexity and chaos, in which the study of the consequences of the abstractions made, i.e. the application of the model, is too difficult.[148]

In Chapter 25, we touched on the question of defining what mathematics is in reality, namely what is the mental tool that we apply when we make reductions in the enormous flow of data that comes from the external world and that would end up clogging our mental computation processes with superfluous information if we don't identify the recurrences through some sort of mental algorithm. During the course of this book we have used mathematical methods to describe different phenomenologies, and we have often pointed out the shortcomings and the distortions displayed by commonly applied mathematical methods, in particular, the tendency to apply

---

[148] In the 1960s, a logic different to the formal one, called fuzzy logic, was developed (Zadeh, 1965), as an attempt to reproduce the schemes of thought of the human mind in a less rigid and more effective way, in practice, than the usual formal logic. As Lofti Zadeh writes in the preface to a book by Arnold Kaufmann (1975, p. ix): 'The theory of fuzzy sets is in effect a step toward the rapprochement between the precision of classical mathematics and the pervasive imprecision of the real word: a rapprochement born of the incessant human quest for a better understanding of mental processes and cognition.

At present we are unable to design machines that can compete with humans in the performance of such tasks as recognition of speech, translation of languages, comprehension of meanings, abstraction and generalization, decision-making under uncertainty and, above all, summarization of information.

In large measure, our inability to design such machines stems from a fundamental difference between human intelligence on the one side and the machine on the other.

The difference in question lies in the ability of the human brain, an ability which present day digital computers do not possess to think and reason in imprecise, non-quantitative fuzzy terms.

It is the ability of humans to think and reason in imprecise, non-quantitative terms that makes it possible for humans to decipher sloppy hand-writing, understand distorted speech, and focus on that information that is relevant to a decision. It is the lack of this ability that makes even the most sophisticated computer incapable of communicating with humans in natural – rather than artificially constructed – languages' (see also, for example, Sangalli, 1998).

the methods of linear mathematics to all phenomena. We have also observed how in many branches of science, in particular social sciences, at present it is difficult, if not impossible, to provide a mathematical description, as, on the contrary, it is possible to do in other areas of science, in the last four centuries, particularly the natural sciences.

From what does the difficulty of understanding the phenomenology of complexity in mathematical terms originate, or better why, in this ambit, are we not able to associate an effective mathematical description to a phenomenology? Can a natural law, i.e. a recurrence observed in empirical data, be associated to a mathematical law, i.e. a formulation in a language that expresses abstract concepts in a rigorous way? Are the concepts and the mathematical methods at our disposal sufficient, or can they be improved or replaced by new concepts or new methods? We would now like to reflect on these points, with a view to extending mathematical methods and making their application to complexity issues better and more effective. To do this, we need to further investigate the foundations of mathematics themselves, because a reflection of this nature could possibly generate an idea useful for the formalization of complexity. In Chapter 32, therefore, we will be making a digression into the main interpretations that have been given of the foundations of mathematics.

32 Digression into the main interpretations of the foundations of mathematics

Introduction

In this section, we will be making a digression from the main theme of the book, to take a brief look at the main interpretations that scientific thought has given, at least in the last century and a half, to the foundations of mathematics, and more generally, to the meaning of working with mathematics. A reflection on the actual meaning that is attributed to mathematics and its limitations, in fact, is necessary in order to attempt to clarify the question of the applicability of mathematics (and of which mathematics) to complexity, i.e. if and in what way could it be possible to formulate mathematical laws to describe the complex phenomenology.

The problem of answering the questions 'what is mathematics?', 'from what does it draw its foundations?' is by no means new. Right back to the times of Pythagoras, if not as far back as Taletes, there have been different ideas on 'doing mathematics'. The problem became particularly significant from the mid-nineteenth century onwards, following the important development of real and complex analysis and the breakthrough of non-Euclidean geometry into general culture rather than just academic circles. There have been, and there still are, numerous trends of thought that define mathematics in different ways and, based on such, give different meanings to the 'objects' covered by the same. We will briefly list them in this section, without wanting to enter into detail; interested readers should consult the wealth of literature on the philosophy and the history of mathematics (merely by way of example, of the many, Weyl, 1949; Cellucci, 1967; Boyer, 1968; Wilder, 1973; Stewart, 1981; Benacerraf and Putnam, 1983, Kline; 1990, Maddy, 1990; Shapiro, 2000; Lolli, 2002).

Platonism

How 'real' are the objects in the world of mathematics? From Plato's perspective, the objects of geometry are only approximately realized in terms of the world of actual physical things. The mathematically precise objects of pure geometry inhabited a different world: Plato's ideal world of mathematical concepts, accessible to us only via the intellect. One current of thought, probably the most widespread, sees mathematics from a *Platonic* perspective: mathematics deals with abstract objects, entities that exist in a world of ideas; the images that we draw from experience are no more than an imperfect copy of them. These images stimulate in us the reminiscence of that which exists in an unreal world, independent of our minds, and that we in some way and in some place have known. The Platonic approach, which is clearly derived from the thought of the classical era, is often that which dominates scholastic teaching. For a Platonic mathematician, one can talk about the 'truth' of a mathematical concept, like a mirror of the world of ideas: a proposition is true if it reflects one characteristic of an object of which we have a reminiscence from the world of ideas. For a Platonist, research in mathematics, basically, does not invent, but discovers. Objects and properties are discovered that 'exist' somewhere, even if up until now no one has ever seen them. As, for example, the theoretical physicist Roger Penrose (1989, 1994), a declared Platonist, writes (1989, p. 95): 'The Mandelbrot set is not an invention of the human mind: it was a discovery. Like Mount Everest, the Mandelbrot set is just *there*!' (see p. 341 following, note 162).

Platonist mathematics is substantially a form of realism. The mathematical objects 'exist' even though in an immaterial form, mathematics 'exists', the number indicated by p 'exists', no more and no less than whole numbers, than all of the other real and complex numbers. This is true on the Earth just as it is in any other place in the universe, whether or not it is discovered by mathematicians, almost as if it were a divine creation. Mathematical knowledge assumes, in this framework, a completely different meaning to that which it has in empirical sciences. We enter into contact with the objects of empirical science, belonging to the real world, in a more or less random way, through our senses; on the other hand, according to the Platonic vision, we have a very different, almost mystical, relationship with the 'metaphysical' objects of mathematics, as if they belonged to a transcendental world. The Platonic realists, however, do not provide any answer to the question of 'where' these mathematical objects are assumed to 'exist' and to 'how' we enter into contact with them: mathematical concepts simply 'exist' and are said to be 'intuitive'.

Formalism and 'les Bourbaki'

Towards the end of the nineteenth century, new elements in the development of mathematics, such as the introduction of non-Euclidean geometries, the extensive study of functions in the complex field, and the new notions regarding numbers and sets, progressively led to the rejection of the Galilean vision of a direct connection between empirical objects and mathematical objects (the 'felt experiences' and the 'certain demonstrations' as Galileo Galilei called them[149]). Mathematics ceased to find its foundation in the concrete experience of an empirical reality. As Enrico Giusti observes: 'If mathematics can no longer find its justification in nature, it will find it in freedom (...) The need for mathematics retreats towards the most interior regions of demonstrations, which, however, can start on their way only after the liberal choice of the definitions and the axioms has been fully exerted' (1999, p. 19, our translation).

This is how the *formalist* current of thought was established, in particular thanks to the work of David Hilbert. According to this interpretation, mathematics is seen as a simple manipulation of symbols, made according to a grammar and coherent (at least in its intentions). We are no longer dealing with a realist perspective. There are no objects to discover, hidden somewhere; there are only symbols, postulates, and formal rules of deduction of concepts (propositions) that are already contained in the original postulates. This is a complete antithesis of Platonism. For a formalist, the concept of truth refers to the context of the assumed postulates and is strictly limited to its formal coherence with the latter: 'truth' now means 'coherence'.[150]

---

[149] Galileo Galilei sets 'le sensate esperienze' against 'le certe dimostrazioni' and discusses the different roles they have in the 'new science' in his book *Dialogo sopra i due massimi sistemi del mondo, tolemaico e copernicano*, published in Florence in 1632; the full original Italian text is available on the website *http://www.astrofilitrentini.it/mat/testi/dialogo.html*.

[150] The fundamental criticism of formalism stems from this identification between truth and coherence, and is well-described in a passage from a singular book on logic written by the mathematician Raymond Smullyan:

'The formalist position really strikes me as most strange! Physicists and engineers certainly don't think like that. Suppose a corps of engineers has a bridge built and the next day the army is going to march over it. The engineers want to know whether the bridge will support the weight or whether it will collapse. It certainly won't do them one bit of good to be told: "Well, in some axiom systems it can be proved that the bridge will hold, and in other axiom systems

The ambitions of the formalist school of rewriting all of mathematics in a complete and coherent system of postulates, to which rigorous logical–deductive laws could be applied, underwent a dramatic redimensioning following the demonstration of the two well-known theorems of the formal incompleteness of arithmetics by Kurt Gödel, a staunch Platonist, in 1931. Gödel's theorems are: (1) *every* consistent axiomatic system (i.e. such that a statement and its negation cannot be proven in it) sufficient to express the development of the elementary arithmetic of the integers, *always* contains statements that can neither be proven to be true, nor be proven to be false, using its basic axioms (first theorem of incompleteness); (2) the consistency of the system itself is one of the unprovable statements using its basic axioms (second theorem of incompleteness).

In more approximate terms, simplifying the matter somewhat, Gödel showed that mathematics cannot be demonstrated as 'true' from within it.[151] The collapse of the ancient idea of mathematics as a synonym of 'truth', caused by Gödel's theorems, is well-summarized by John Barrow: 'it has been suggested that if we were to define a religion to be a system of thought which contains unprovable statements, so it contains an element of

it can be proven that the bridge will fall." The engineers want to know whether the bridge will *really* hold or not! And to me (and other Platonists) the situation is the same with the continuum hypothesis: is there a set intermediate in size between N and P(N) or isn't there? If the formalist is going to tell me that in one axiom system there is and in another there isn't, I would reply that that does me no good unless I know which of those two axiom systems is the *correct* one! But to the formalist, the very notion of *correctness*, other than mere consistency, is either meaningless or is itself dependent on which axiom system is taken. Thus the deadlock between the formalist and the Platonist is pretty hopeless! I don't think that either side can budge the other on iota!' (1992, p. 249, italics original).

[151] Gödel's theorems were the answer to the widespread desire, in the era between the nineteenth and twentieth centuries, to define rigorous formal foundations for mathematics, that reached its height on August 8, 1900, at the 2nd International Congress of Mathematics in Paris, when David Hilbert, a well-known professor from the University of Göttingen, held a conference in which, summarizing the current progress of mathematics, he listed 23 unresolved problems, as a subject for research for the new century. The second of those problems asked whether it was possible to demonstrate that a given system of axioms, placed at the foundations of arithmetics, was compatible. The idea, in Hilbert's mind, was that, if the system of axioms is compatible, then with a finite number of logical operations it *cannot* lead to a contradictory result, while, if it is incompatible, a contradiction in the consequences of the axioms, *sooner or later*, can be found. The formalist programme asked for a response: yes, compatible system; no incompatible system. It was actually Gödel, 30 years later, who showed this programme to be unachievable, proving that *any* system of axioms at the basis of arithmetics inevitably leads to propositions that are not true and not false, i.e. *undecidable*. Gödel's omnia works were published in English, a few years ago, in three volumes, edited by Feferman et al. (1986, 1990, 1995). For a general illustration of Gödel's theorems, see Nagel and Newman (1958), Benacerraf (1967), Penrose (1989); an in-depth presentation of the figure and works of Gödel can be found in Guerriero (2001).

faith, then Gödel has taught us that not only is mathematics a religion but it is the only religion able to prove itself to be one' (1992b, p. 19).[152]

On December 10, 1943, in a café in Paris's Latin Quarter, a group of young mathematicians, students of the École Normale Supérieure, formed a club that they named Nicolas Bourbaki;[153] the members of this group, who initially wanted to be considered almost like a secret society of revolutionaries, were often nicknamed 'the Bourbaki' (*les Bourbaki*). The group proposed to resume the formalist approach but on less ambitious bases than those of Hilbert, without giving up the use of an absolute formal rigour and avoiding the use of intuition (Marshaal, 2000). This formulation also entailed the refusal of Poincaré's methods, a dominant figure in French mathematics in the first few decades of the twentieth century, who, according to the Bourbaki, used intuition too often, without rigorously demonstrating his claims. The Bourbaki school (because it can be considered a bona fide school, given the widespread popularity of its ideas, particularly within French mathematical schools) is still today the standard bearer of the formalist approach: mathematics is only made up of formal syntactic rules

---

[152] Another well-known theorem demonstrated by the Polish mathematician Alfred Tarski in 1933 (Tarski, 1956), two years after Gödel's theorems, posing further limitations on axiomatic systems. The essence of Tarski's theorem, generalizing somewhat, can be expressed in the statement: 'no program can decide, for each sentence of number theory, whether the sentence is true or false' (see Hofstadter, 1979). According to this theorem, to decide the *truthfulness* of mathematical propositions in the universe of mathematics, a (meta)theory is needed, of which said universe is just an object, that proves the non-contradictoriness of existing mathematics. Tarski demonstrates that no logically coherent language can be gifted with semantic completeness and that a hierarchy of (meta)languages needs to be conceived, each of which is coherent, but none of which is sufficiently wealthy to express its entire semantic content. The semantic completeness of a given language can only be found with a metalanguage at a higher level. Only in this way does it become possible to distinguish the statements *of* logic from those *on* logic, avoiding problems of self-referencing that emerge in the case of propositions like 'this statement is false'. In the light of Tarski's theorem, talking of mathematical truth is inevitably vague if the language is informal, or leads to an infinite deferment to rigorous metalanguages. If Gödel had highlighted that the notion of *demonstrability* is weaker than that of *truthfulness* (because there are statements than *cannot be demonstrated as true*, but by no means are necessarily false), Tarski makes the very concept of arithmetic truth totter. Gabriele Lolli writes, ironically: 'Each realist mathematician should hang a sign in front of his workplace with the words *Memento Tarski*' (2002, p. 81, our translation). On this point see also Tarski (1967) and Barrow (1992b).

[153] The pseudonym draws its origins from the name of Charles Denis Sauter Bourbaki, who was not even a mathematician, but a general in Napoleon III's regiment, of Greek origin, who distinguished himself in the 1870 Franco-Prussian war. The reasons for the choice of this pseudonym can be found in the aura of folklore and mystery, that was not lacking in studentesque elements, that enveloped the group of young mathematicians in Paris in the early 1930s; these reasons, together with many aspects of the activity of the Bourbaki circle, are extensively illustrated in the original, well-documented and in-depth monograph on the *Société secrète des mathématiciens Bourbaki* edited by Maurice Marshaal (2000).

applied to a set of symbols, starting from a set of axioms. The group compiled a famous monumental treatise in ten volumes (called books) and more than 30 tomes (called files), published at irregular intervals from 1939 onwards and not yet complete, *Éléments de mathématique*[154] (Bourbaki, 1939–1998), in which they wanted to re-present all of mathematics from the same and unique formalist perspective. For formalists, contrary to the opinion of the Platonists, mathematics is not a discovery, but an invention, a creation of the mathematician who acts as a skilful, yet rigid, applier of formal rules.

Today, the majority of mathematicians, probably including those that were educated under the Bourbaki school, lean more towards Platonic inspiration, i.e. they are convinced that their affirmations regard objects that have a real and objective existence somewhere, even though such are not tangible (Giusti, 1999). Nonetheless, the formalist school, and the Bourbaki in particular, made a significant contribution to mathematical thought. First of all, it provided an answer to the difficulties that were being debated in mathematics. Perhaps, more than difficulties, we should talk of a real crisis in the foundations, a crisis that reflected the general crisis in science at the beginning of the twentieth century.

The formalist approach forced mathematicians to thoroughly review the basic axioms and led to a rigorous development of formal and logical techniques. Following this work, from the 1930s onwards, Alan Turing brought to light the fundamental question regarding the concept of computability for the system of real numbers, i.e. the question of whether the decimal representation (or that of any other basis) of a number can be generated by an algorithm[155] (Turing, 1937; see also Cutland, 1980;

---

[154] Note in the title the explicit use of the singular '*de mathématique*', whereas in French the plural form '*des mathématiques*' should be used, to underline the unitary vision founded on rigorous formalism that the Bourbaki group proposed for all branches of mathematics.

[155] Without going into the technical details, we can observe that, if for a given real number we have an algorithm that is able to provide *all* of the decimal places in a *finite* time, then we can say that we are in possession of a rule to know the infinite sequences of its decimal places. This is a computable number. For example, $3.\overline{47}$ is a computable number, because the rule to generate all of the decimal places is: 'repeat 47 continuously', therefore we can immediately say that the one thousandth decimal figure is 7, even without expressing (or calculating) all of the previous figures. Various numerical sets are computable. For example, all rational numbers, but also the roots of integers, whatever index they are; logarithms are computable, and even π and the basis of natural logarithms e (each of these, in fact, can be expressed as the sum of a particular numerical series). A number like $0.12112211122211112222\ldots$ is also computable, because the rule that generates the decimal places is immediately identifiable. On the other hand, a number, even though defined in a logical and non-ambiguous way, is not computable, if its definition is, *in practice, inexecutable*. To better clarify the question, we have illustrated below an example taken from Cutland (1980). Let us imagine we have to

Herken, 1983; Penrose, 1989). It was realized that non-computable numbers, i.e. the numbers for which there is no algorithm to calculate the sequence of decimal places, represent the great majority of real numbers (technically, it is said that they represent a non-numerable infinity in the continuum of real numbers). The consequences of this, from a methodological and epistemological point of view, are surprising, to say the least, insofar as computable numbers, all of those that we ordinarily use in all kinds of measurement and in calculations made by machines, turn out to be nothing more than 'rare exceptions' in the 'background' of the continuum of real numbers that was presented by Cantor, a background that now appears to be 'filled' with non-computable numbers.

The awareness of the existence of non-computable numbers, i.e. of calculations that do not lead to a result in finite time, that thus surfaced into the scientific conscience, represented, alongside Gödel's theorem, the negative response to Hilbert's tenth problem (see p. 332, note 151), the famous *Entscheidungsproblem*, the problem of decidability, i.e. the problem of whether there is a general algorithmic procedure that can, in principle, decide on the validity or falsity of *any mathematical proposition*. Given that any mathematical problem translates into a series of propositions, the *Entscheidungsproblem* translated into the question of whether, from a conceptual point of view, an algorithmic procedure existed that was able to solve *any mathematical problem*, i.e. a procedure that once started to be processed on an appropriately defined abstract calculating machine,

---

write a certain positive number A that is less than 1, whose successive decimal places are all 0 or 1. A is, therefore, in the form $0.a_1 a_2 \ldots a_n \ldots$, where the generic place a_n can only be 0 or 1. The rule (algorithm) to write A is as follows: to decide if the decimal place a_n is 0 or 1, let us consider the number π and examine its decimal places in sequence from left to right. If in the infinite decimal places of π we find, somewhere, a sequence of the number 7 that repeats itself n times in succession, then we put $a_n = 1$; if, on the other hand, we *never* find (and this is the crucial point) a sequence of this type, then we put $a_n = 0$. The definition of the procedure is clear and unambiguous, but in operational terms, it does not work. By applying this rule, in fact, when we find the sequence of n number 7s we were looking for, then we can assume that $a_n = 1$ without any problem. The problem, on the other hand, lies in the fact that, while we are looking for this sequence, we do not know whether it *exists*, somewhere, in the decimal representation of π and we haven't found it yet, but sooner or later we will find it, or if we never find it, simply because it *doesn't exist*. We can *never* know it, unless we find it: the search that is made on the infinite decimal places of π can end *only* with a positive answer and *never* with a negative one, as the latter case would only be the never-achievable outcome of an infinite search. In other words, if for a given value of n, we do not find a sequence of n number 7s, we cannot decide to interrupt the search: the fact that up until now we haven't found it, doesn't authorize us to conclude that such a sequence doesn't exists, therefore, in reality, we will *never* be able to assume that $a_n = 0$. This makes the number A uncomputable. This is the basis for the link between uncomputability and logical undecidability.

called a Turing machine (Turing, 1937), gave a result in a finite time. The negative response suggested by the existence of non-computable numbers, signified, in other words, the awareness that not all mathematical problems can be resolved in a finite time.

Constructivism

Towards the end of the nineteenth century, following the attempt to establish the entire framework of mathematics on a rigorous logical basis starting from the little-developed theory of sets, an attempt led by figures such as Gottlob Frege and Giuseppe Peano, the inevitability of internal contradictions came to light, the so-called antinomies (often, although inappropriately called logical paradoxes), such as Russell's (1903) famous antinomy.[156]

---

[156] Bertrand Russell told the mathematician Gottlob Frege of his antinomy in a letter dated 1902, just before the latter published the result of his lengthy work of logical formalization of arithmetics on set bases in a book. The antinomy was then published and discussed in Chapter 10 of Russell's *The Principles of Mathematics* (1903). Russell's antinomy is formulated as follows. Let us imagine dividing all sets into two large groups: (1) the sets that are not elements of themselves, called *ordinary* sets (for example, the set of cats is not a cat and therefore is not an element of the set of cats), and (2) sets that are elements of themselves, called *non-ordinary* sets (for example the set of all the sets of cats grouped according to the various breeds, is still a set of cats, and therefore belongs to the set of all of the sets of cats, i.e. is an element in itself; in the same way also the set of all abstract concepts is itself an abstract concept and therefore and element of itself). Let us now consider the set N of all the (infinite) ordinary sets (*the set of the sets that do not belong to themselves*) and let us ask ourselves: is N an element of itself or not? Let us try to answer (Devlin, 1988, 1998):

1. If we respond yes, i.e. if we say that N belongs to itself, this means that, by the same definition of N, N has the property of not having itself as an element; if this is true, it cannot belong to itself, and therefore we fall into contradiction.
2. If we respond no, i.e. if we say that N does not belong to itself, this means that, by the same definition of N, N has the property of having itself among its elements, and therefore, we fall into contradiction again.

What made Russell's antinomy so profound was its absolute simplicity. It only used fundamental concepts on which practically all mathematics is based. Russell's antinomy, in reality, was a simplified version, so as to speak, of a previous antinomy, which, in 1897, Cantor himself, the creator of the theory of the sets, had found.

Setting aside the technical aspects that link it to Cantor's, Russell's antinomy is not new, also because it is none other than a reformulation of the old antinomy known as the paradox of Epimenides the Cretan (who lived in the sixth century BC) according to which: 'Epimenides, Cretan, claims that all Cretans lie', translated more simply as 'I always lie'. The paradox is attributed to St. Paul, who involuntarily fell into the trap citing Epimenides in a

One of the answers given to the ferment of uncertainty caused by logical paradoxes at the beginning of the century was *constructivism* (also indicated as *intuitionism*) that derived from a vision that had already been proposed several decades earlier by Leopold Kronecker, in open and polemic antithesis to Cantor and Dedekind's development of the theory of infinite sets, which he considered to be metaphysics and not science. The non-realistic vision of constructivism holds that mathematical entities that we are unable to tangibly experience, such as infinite sets, cannot be taken into consideration (Heyting, 1956). In other words, mathematical constructivism asserts that it is necessary to find ('construct') a mathematical object to prove that it exists: when we assume that an object does not exist, and derive a contradiction from that assumption, we still have not found it, and therefore not proved its existence.

For constructivists, mathematics is none other that a series of propositions that can be construed by means of a finite number of deductive operations on natural numbers only, the only acceptable starting point, that are assumed almost as if they were the subject of a revelation. In this regard, Kronecker made the famous remark 'God created the integers, all else is the work of man' (quoted in Cajori, 1919).[157] Constructivists refuse, retaining them beyond the scope of mathematics, all proofs *ab absurdo* (the so-called *reductio ad absurdum*): a proposition can be true or false if it can be demonstrated as such in a number of *finite* operations, but it can also be neither true nor false, if neither its truthfulness or its falsity can be demonstrated in a finite number of operations. From this it follows that the non-truthfulness of a statement does not guarantee its falsity. Falsity must be demonstrated as such, not as a 'non-truth'. Constructivism did not win a great deal of popularity with mathematicians,[158] nor with other scientists

---

letter to Titus, in which he wrote: 'Even one of their own prophets has said, "Cretans are always liars, evil brutes, lazy gluttons. This testimony is true. Therefore, rebuke them sharply, so that they will be sound in the faith" ' (Titus, 1, 12–13, Bible's New International Version). See Weyl (1949), on Russell's antinomy and on other philosophical matters connected to it.

[157] 'Die ganze Zahl schuf der liebe Gott, alles Ubrige ist Menschenwerk' (Cajori, 1919). To be precise, this position is known as finitism, an extreme form of constructivism, according to which a mathematical object does not exist unless it can be constructed from natural numbers in a *finite* number of steps, whereas most constructivists, in contrast, allow a *countably infinite* number of steps.

[158] Here, we will restrict ourselves to remembering the Dutch mathematician Luitzen Egbertus Jan Brouwer, one of the most dogmatic supporters of constructivism (Brouwer, 1913; see also Heyting, 1956; Van Sigt, 1990), editor of the very important German periodical *Mathematische Annalen*. For the entire time he held this position, Brouwer stubbornly refused to accept any article that used proofs *ab absurdo*. After a long struggle in often very harsh tones, Hilbert, who incidentally had helped Brouwer in 1912 to obtain the chair of the

that make extensive use of mathematics, such as physicists. Physical theories, such as quantum theory and general relativity, in reality, made wide use of a non-constructivist way of reasoning.

In complete antithesis to Platonic realism, constructivism is totally anthropocentric. There is no mathematics outside of natural numbers and everything that can be deduced from them must be done so in a finite number of operations. But a constructivist could never say whether the very intuition of natural numbers is the same for all people and at all times.[159]

Note that, although the constructivist vision was developed in the nineteenth century, some of the elements that qualify it are not that distant from the characteristics that geometry was thought to have, in its most remote pre-Greek origins. Pre-Greek geometry, for example, also does not use the metaphysical entities typical of Platonism, and it is not concerned with a rigorous deductive–logical formalism, nonexistent before Euclid, while it focuses, instead, on resolving tangible problems such as, for example, land surveying.

The fundamental message of constructivisim is that there will *always* be undecidable propositions. This concept is strictly linked to the question that we mentioned a little earlier, regarding the meaning of non-computable real numbers, brought to light by Turing in the 1830s. At present, we do not know whether the nature of the world we are immersed in is such that it includes non-computable elements; in other words, we do not know whether the laws of nature are all computable, i.e. whether they provide a result in finite time, or whether they contain non-computable elements, in a framework compliant to that of the constructivist perspective. In the first case, we will be able to 'manage' the numbers obtained; in the second case, on the other hand, we find ourselves unable to treat non-computable numbers that evade an algorithmic definition and therefore an 'understanding' in terms of logic, which would render a complete description of the world impossible.

Finally, it is important to observe that the essence of Gödel's theorems of incompleteness, that had managed to shatter Hilbert's formalist pro-

---

theory of sets at the University of Amsterdam, managed to get the editorial committee disbanded and reconstituted without Brouwer, a person whose behaviour was considered tormented, unbalanced and even paranoiac by his contemporaries (Barrow, 1992ab, Van Dalen, 1999).

[159] It should be noted that constructivism is often confused with mathematical intuitionism, but, in fact, intuitionism is only one kind of constructivism. Intuitionism maintains that the foundations of mathematics lie in the individual mathematician's intuition, thereby making mathematics into an intrinsically subjective activity. Constructivism does not, and is entirely consonant with, an objective view of mathematics.

gramme, lies in the self-referential nature of the problem: the arithmetic that wants to find its foundations in itself, justifying itself in arithmetic form. More generally, we encounter self-referential situations every time that the symbol of an object gets confused with the object and becomes an object itself, making the difference between the *use* of a word and the *mention* of a word disappear. In these cases, our logic *inevitably* finds elements of undecidability. This is what is found at the origin of Russell's and Cantor's antinomies, in which the self-referencing originates from the introduction of the idea of the 'set of sets', and of Epimenides the Cretan's antinomy (paradox) (see p. 336, note 156), in which the self-referencing lies in the fact that 'I am talking about what I am saying'. The situations without a logical outlet that self-referencing leads to[160] are the consequence of an intrinsic limitation of our minds and of our way of thinking.[161] It is

---

[160] The tippler that the little prince meets in the story of the same name by Saint-Exupéry finds himself in a situation of this type (see the quote at the beginning of this part).
[161] As regards this last point, we quote Smullyan again:
' "Now, the Russell and Cantor paradoxes are more serious and disturbing because they show that there is something basically wrong with our way of thinking. What I have in mind is this: Doesn't it seem obvious that given any property, there exists the set of all things having that property?"
"It would certainly seem so!" said Annabelle.
"That seems obvious enough," said Alexander.
"That's the whole trouble!" said the Sorcerer. "This principle – called the *unlimited abstraction principle* – the principle that every property determines the set of all things having that property – that principle indeed seems self-evident, yet it leads to a logical contradiction!"
"How so?" asked Annabelle.
"It leads to the Russell paradox as well as the Cantor paradox. Suppose that it were really true that for any property, there exists the set of all things having that property. Well, take the property of being an ordinary set. Then there must exist the set B of all ordinary sets, and we get the Russell paradox: the set B can be neither a member of itself nor not a member of itself without presenting a contradiction. Thus the unlimited abstraction principle leads to Russell's paradox. It also leads to the Cantor paradox, since we can consider the property of being a set, and so there is then the set of all sets, and this set is on the one hand as large as any set can be, but on the other hand, for any set there is a larger set (by Cantor's theorem), and so there must be a set larger than the set of all sets, which is absurd. Thus the fallacy of the Cantor paradox is that there is such a thing as the set of all sets, and the fallacy of Russell's paradox is that there is such a thing as the set of all ordinary sets. Yet, the unlimited abstraction principle, which seems so obvious, leads to these paradoxes, and hence cannot be true. The fact that it *seems* true is what I meant when I said that there was something basically wrong with our naïve way of thinking about sets.
The discovery of these paradoxes was at first very disturbing, since it seemed to threaten that mathematics might be inconsistent with logic. A reconstruction of the foundations of mathematics was necessary which I will tell you about next time." ' (1992, p. 238–239, italics original).

precisely in this limitation that the origins of Godel's theorems can be found.

So self-referencing implies that there are situations of undecidability and, therefore, non-computable operations; this means that by adopting the constructivist viewpoint, we cannot achieve complete knowledge and an effective description of the world. This is true if the time needed for the calculation is infinite, but even if the time needed is finite but 'too' long, for example longer than the life of the universe; in practice there is no difference: neither case leads to a result.

Experimental mathematics

In recent years, especially following the significant development of information technology and, in particular, of automatic iterative calculation methods, a vision of mathematics has been developed that we can call *experimental* (Barrow, 1992ab). More than a philosophical definition, and all the more reason, a school of thought, perhaps it is more appropriate to speak about, pragmatically, research ambit and methodology in a sector that is totally new with respect to the past, open and made practicable by new technological means. Research is conducted essentially on 'objects' constructed using automatic calculation procedures, iterations of formulae and graphic visualizations on a computer screen. The graphic visualizations in particular are fundamental. They represent the only tool useful to highlight the properties of certain figures that are created from the iterative formulae that they generate and that are not intuitable or predictable by any other means.

A large part of the development of fractal geometry and of its applications is only possible in this new experimental formulation of mathematics, which, in a certain sense, brings together the investigations conducted by a mathematician with the research work of a scientist in a laboratory. We are no longer dealing with just demonstrating theorems, but also of observing, in a strict sense, the properties of formulae, equations, etc. This is particularly important in the mathematical study of nonlinear dynamics, of chaos, and of complexity. For example, the Feigenbaum tree (Figures 22.2–22.4) and the Mandelbrot set are fractal objects constructed using algorithmic procedures that we would never have been able to visualize, describe, and examine, or even conceive, even though in possession of the elementary formulae that are at their origin and of the techniques to process them, if we had not had the technological means at our disposal for the creation of the images that derived from these formulae and that, in a certain sense, make them tangible.[162]

The paradigm of the cosmic computer in the vision of experimental mathematics

Up until the end of the Renaissance era and Aristotelianism, the universe was considered to be a large organism, which displayed Plato-inspired geometric harmony. At the end of the seventeenth century, in Newton's time, when the pendulum clock had been significantly perfected and was finding widespread application, the mechanistic perspective (or paradigm, as we would call it in contemporary scientific language) was dominant, according to which the universe works like a huge clock (a 'cosmic clock'),

---

[162] One of the possible extensions of the logistic map to the complex field, the simplest and certainly the most well-known case, is the following map:

$$z_{n+1} = z_n^2 + c \tag{32.1}$$

where z is a complex variable and c is a complex parameter. By separating the real ($x_n = \mathrm{Re}\, z_n$) and imaginary ($y_n = \mathrm{Im}\, z_n$) components, the map (32.1) is transformed into the following two-dimensional map with real variables:

$$x_{n+1} = x_n^2 - y_n^2 + \mathrm{Re}\, c \quad y_{n+1} = 2x_n y_n + \mathrm{Im}\, c \tag{32.2}$$

Equation (32.1) can be understood, and therefore iterated in two different ways:

1. A complex value is established for parameter c and the region of the complex plane (or 'Argand plane') in which the initial values z_1 of (32.1) are examined, which generate a sequence of z_n (an orbit) that does not tend towards infinity, but towards an attractor. The z_1 from which such orbits originate define the basin of attraction of the attractor, the boundary of which is called the Julia set for that given value of c.
2. In (32.1) we assume that $\mathrm{Re}\, z_1 = \mathrm{Im}\, z_1 = 0$ and we examine which values of c generate a sequence of z_n (an orbit) that does not tend towards infinity, but remains *limited* in the complex plane; the set of such points c constitutes the Mandelbrot set (Figure 32.1).

The Mandelbrot set re-proposes geometric properties in a different form that correspond to some of those described by the Feigenbaum tree generated by the logistic map: in this regard, this is the most famous case interpretable as a chaotic dynamic in the complex field. The Mandelbrot set is characterized by very particular properties that we will not be discussing here. They are illustrated in a wide number of books and articles, both specifically dedicated to chaos (for example, Mandelbrot, 1986; Peitgen and Richter, 1986; Gleick, 1987; Devaney, 1989, 1992; Stewart, 1989; Falconer, 1990; Pickover, 1990; Schroeder, 1991; Casti, 1994; Cohen and Stewart, 1994), and that mention the subject marginally with respect to other topics (for example Penrose, 1989; Bertuglia and Vaio, 1997; Puu, 2000); interested readers are invited to consult these texts.

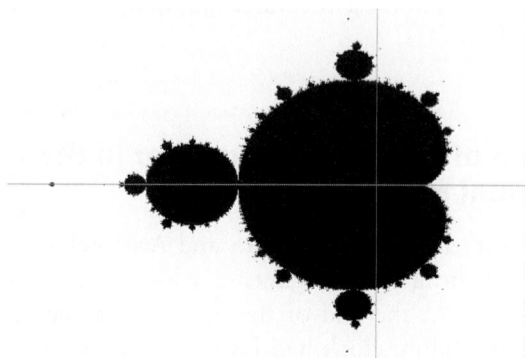

Figure 32.1 The Mandelbrot set$(-2 \leq \text{Re}z \leq +0.8; -1.25 \leq \text{Im}z \leq +1.25)$.

in itself a metaphor of determinism. In the second half of the nineteenth century, however, in the midst of the second industrial revolution, the dominant metaphor was that of the steam engine. The in-depth studies that had been conducted in the field of gases and steam engines, aimed at perfecting the latter, had led to a significant development in thermodynamics theory and had established the idea of an arrow of time, i.e. of a temporal asymmetry, that had not been previously conceived, that surpassed Newton and Kant's Euclidean and isotropic spatial–temporal schema. In that cultural era, the industrialized West, in analogy to the metaphor of the steam engine, had matured a concept of the universe as a huge machine dominated by exchanges and flows of all types of energy (Dioguardi, 2001).

Today's technology leads us to consider the treatment of information as the dominant metaphor and, in relation to this, to assume the computer, obviously meant in a very general sense of 'machine that runs a calculation program', as a paradigm in our attempt to understand the world (Barrow, 1992ab). The computer is not only a paradigm, but also the real foundation and the engine of the enormous development in experimental mathematics over the past few decades, a development that has taken place by adopting a perspective that in practice is not that distant from the realist one, therefore in antithesis with the formalist one.[163]

---

[163] James Gleick (1987) tells how Benoît Mandelbrot, the inventor (according to the constructivist perspective) or the discoverer (according to the Platonic one) of fractals, abandoned France, the country he had grown up in and had been educated in, but where mathematical research was dominated by the Bourbaki group, to travel to the United States to IBM. There Mandelbrot found a cultural environment that was less rigid and more open to the new way of 'doing mathematics', that which we have called experimental, which he adopted.

In the evolution of scientific thought, the perspective of experimental mathematics appeared as one of the most important steps, an authentic new category of thought, a manifestation of the culture of this era. By adopting the vision of experimental mathematics (sometimes called *computational* mathematics) and by taking the computer as a paradigm of the world ('the cosmic computer'), as the watch had been to the mechanistic vision that sought deterministic laws at the origin of the observed dynamics, we could ask ourselves whether the evolution and the structure of the universe itself should not be considered as the manifestation of a recursive calculation. We could ask ourselves, that is, whether the laws of mathematics and those of society should be sought (or conceived) in a recursive form, as if they were a calculation program; in other words, if the world functions and evolves due to the effect of a sort of large and universal calculation program.

The vision that imposes the laws of nature in a symmetrical and invariant form, such as those provided by classical dynamics, is in line with the Platonic vision: the laws within which nature mirrors itself and that control it 'exist' somewhere. It is the task of the scientist to discover them and mathematics is one of the tools used to do this. According to the formalist vision, however, the image of the world is that of an all-comprehensive mathematical law, logically necessary and absolutely general, able to account for any sequence of observations in the form of pure logical connections. The vision that is founded on the image of a cosmic computer is closer to the notions of constructivism. There are neither transcendent nor immanent deterministic laws, i.e. that 'exist' at the origins of the observed dynamic. On the contrary it is just a calculation program, whose results, i.e. the general dynamics of the world, we (also) describe by using a tool such as mathematics.

A comparison between Platonism, formalism, and constructivism in mathematics

The study of mathematical objects such as fractals, conducted using the methodologies of experimental mathematics, leads to the adoption of elements of all three of the above interpretations of mathematics. The most relevant of all is the Platonic realistic vision, in which the images of the fractals represent 'existing' mathematical objects that are simply visualized by means of an approximation of a purely technical nature. In the non-realistic vision of formalism, on the other hand, fractal figures are taken into consideration insofar as they are based on rigorously defined formulae.

Lastly, the antirealist constructivist vision gives fundamental significance to the method by means of which such objects are produced, but does not fill the gap between a simple calculation procedure iterated by a machine a finite number of times, and the abstract idea that we have of the object, which goes beyond the simple graphic image produced by the machine, but refers to the outcome (or rather the limit) of an iteration repeated an infinite number of times.[164]

Let us look further into the diversities and the limitations of the contributions of the three approaches, still with the aim of better clarifying the meaning of 'doing mathematics', by referring, for the sake of convenience, to a particular case of a fractal: the Mandelbrot set (see p. 341, note 162).

The constructivist vision gives its important contribution when it warns us against the hasty and superficial use of certain concepts and definitions, such as, for example, the concept of the boundary line of a geometric figure that we would take from elementary geometry, a boundary line between an 'inside' and an 'outside' that are not clearly defined in specific figures such as, for example, the Mandelbrot set. This, in actual fact, does not have an 'inside' and an 'outside' separated by a line in the usual sense of elementary Euclidean geometry, which has a realistic and Platonic perspective and according to which the lines 'exist' and we conceive them 'intuitively' as unidimensional objects.[165]

According to Platonism, the Mandelbrot set 'was already there', i.e. it 'existed' even before it was seen (Penrose, 1989). However, we would never have been able to find it without the use of mental processes and methods that, in reality, are unrelated to the Platonic vision. We would never have even been able to sense it intuitively, as we 'intuitively sense', for example, a sphere or a straight line, i.e. simply by *abstracting* from experience, by closing our eyes, so to say, and 'by imagining' a sphere or a straight line that will never exist as such in the experience of our senses. Our senses, in fact, perceive objects that are only approximately round or straight. In these objects, according to Platonism, we *recall* the image of a sphere or of a

---

[164] We must be very careful to keep the object that we talk about and its image on the computer screen separate. The image is merely the result of a *finite* number of operations iterated automatically, and furthermore the same image is inevitably reproduced with an approximation in the resolution. We, on the other hand, see a different object behind it: what we imagine to be obtained (if we can say that) with an *infinite* number of iterations of the calculation formula, of which the image displayed is nothing but an approximation, as long as, obviously, the iterated formula converges towards a finite limit; otherwise the image created would have no meaning, but would be an intermediate step in a construction process that leads to nothing.

[165] Mitsuhiro Shishikura (1994) demonstrated that the border of the Mandelbrot set is a fractal object with a Hausdorff–Besicovich dimension of 2.

mathematical line with which we have already entered into contact in a world of ideas. A process of abstraction from experience, alone, however, can never lead us to 'discover' the Mandelbrot set, even conceding that it 'exists' in a world of ideas, and, in particular, will never lead us to 'encounter' an object such as its boundary, simply on the basis of its definition, which, remember, is all we possess. To discover it, therefore, we must start from natural numbers, extend the concept of number to the complex field and apply the definition, operating according to the rules that define the properties of the numbers, as constructivist mathematics does.

However, the constructivist vision of mathematics alone is not sufficient. It replaces the concept of fractal, the limit of an *infinite* number of iterated calculations, with the image of a geometric object that is not fractal, because it has been obtained with a *finite* number of operations, as we have said, but that we can bring close to the real fractal that cannot be represented as such. Visual experience, thus enhanced with respect to what it would be without the use of constructivist mathematics and of the computer, at this point, is able to abstract the Mandelbrot set in the Platonic sense, starting from various approximations and without ever having *seen* it as such. Constructivist mathematics, in fact, shows its intrinsic limitations every time it refuses to take into account any situation that cannot be treated in algorithmic terms, as being, in their view, *tout court* lacking in meaning; it refuses, that is, any question that cannot be resolved in finite time and which requires the use of mental tools such as abstraction and intuition, without which we would never be able to understand the properties (we would say the 'essence'), for example, of objects such as the Mandelbrot set.

The formalist vision of mathematics, for its part, fails when it encounters the logical difficulties that derive from the formal definition of the Mandelbrot set. The fact that a line (intended in the standard meaning of the term, i.e. as a unidimensional object) that separates an 'inside' from an 'outside' of the Mandelbrot set cannot be conceived Platonically, is a situation that is totally analogous to that of the undecidability of a proposition; just like 'neither true nor false', in the same way 'neither in nor out'.

In more general terms, the formalist vision fails because it inevitably comes up against the intrinsic impossibility for mathematics to demonstrate the coherence of its entire system from within, as proved by Gödel's theorems (see p. 332, note 151). In this regard, the aphorism with which Russell, a realist like Gödel, ironically expressed his critical position with regard to formalism is well known: according to this approach 'pure mathematics may be defined as the subject in which we never know what we are talking about, not whether what we are saying is true' (Russell, 1917, chapter 4, see also Nagel and Newman, 1958). And that's not all, we could also add that formalists isolate mathematics into a sterile mental exercise that is not related to any reality, and, in any case, gives no consideration whatsoever to the fact that mathematical methodologies can actually be applied successfully to descriptions of the world.

Undoubtedly, the attention that the formalist approach pays to the internal aspects of mathematics and to the importance of rigour in the arguments with which Platonic intuitions are sustained, has been and still is a very useful contribution to the modern development of the discipline, as we have already seen; however, we must not forget the risk implicit to formalism of falling into unfruitful self-contemplation. As the mathematician Bruno de Finetti wrote: 'The ivory tower of the "end in itself" is full of pitfalls. Would it not be an aberration – by way of an analogy – if a linguist, to purify the language by suppressing that which regards uses and purposes beyond it, wanted to reduce it to only the words and structures necessary to exclusively express the rules of its grammar and syntax? Or if someone placed a red triangle in the middle of the road with the sole purpose of signalling the inconvenience caused by the presence of the same red triangle?' (1963, p. 5, our translation).

To conclude this section, we observe how it is difficult, in any event, to respond in a general way to the question 'what is mathematics?'.[166] The excerpt that follows, taken from the book *La serva padrona* (*The bossy maid*) (Boncinelli and Bottazzini, 2000, p. 117–118, our translation),

---

[166] In reality a vast number of varied answers have been given to this question in the history of Western thought, by philosophers and mathematician alike. However, as Lolli (2002) observes, each answer is partial and the result reflects both the general culture of the era it was formulated in, and the state of advancement of mathematics at that time. It should be borne in mind, in any event, that the question '*what* is mathematics?', similar to the question 'what does doing mathematics mean?', is different to the question 'what are the foundations of mathematics?'. The first two questions are more general and still actual, while the third, concerning the philosophy of mathematics, has perhaps become less actual from the mid-twentieth century onwards. As Lolli himself writes: 'It is worth remembering that set theory is not the last metamorphosis of mathematics; abstract mathematics continues to be important, above all in close symbiosis with physics theories, but today computational mathematics, with the problems set by the computer, is preferred and prominent. In the second half of the twentieth century, the advent of the computer was the only really new phenomenon in mathematics and its environs. The problems that it poses, however, do not seem to be dramatic, do not necessitate any conceptual revisions – except for (. . .) certain empiricists. They may be worrying for other humanistic or spiritualistic reasons. It is possible that, as at the present time, no dramatic challenges are envisaged, not only have we lost interest in the problem of its foundations, but also that there is no need for interpretation and reconstruction, and, in definitive, that philosophy doesn't have any (new) problems to deal with. If the philosophy of mathematics languishes, perhaps this is a sign of the vitality of mathematics' (2002, p. 64, our translation).

provides a good summary of the main visions (in addition to this paragraph, see also Chapters 2 and 3). The two scientists that wrote the book, Edoardo Bonicelli, a scholar of natural science, and Umberto Bottazzini, a scholar of so-called 'pure' mathematics, discuss their own different opinions on mathematics:

Edoardo – Human beings observe the world and conceive questions that are generally tangible and specific. To respond to some of these, modelling and at times formalization is required. This often implies a translation of the tangible problem into the abstract and ultra-rarefied language of mathematics. The resulting mathematical problem can then be treated with the tools that mathematicians have perfected and that are applied according to their principles. This application can also be very long and laborious, but at a certain point it all has to be retranslated into the language of the real world to be studied and compared with the data of the observation. In my mind, this fairly well clarifies what mathematics represents for a physicist or a biologist with respect to what it represents for a mathematician. For the mathematician, it is a stand-alone plan of the real situation, autonomous, self-sufficient and (almost) always complete. For the experimental scientist, mathematics is this subterranean navigation that leads from certain things to other certain things thanks to the rigorous use of an impressive number of mental constructs.

Umberto – I prefer to think that mathematics is a very complex, beautiful mental construction and, at the same time, the most efficient tool that mankind has equipped itself with to identify regularities in natural phenomena that are apparently chaotic.

33 The need for a mathematics of (or for) complexity

The problem of formulating mathematical laws for complexity

As we have said on a number of occasions, sciences exist as such because the world often appears algorithmically compressible into laws, often expressed in mathematical forms, that are none other than reductions carried out starting from enormous sequences of data. Reductions are possible when the phenomena display recurrences; in a completely chaotic world, the latter would not appear and reductions would be impossible. The chaos in the measured data means precisely that the deterministic law that is assumed to be behind the phenomena and that should appear as a result of the reduction made on the observations, is not easily visible, so much so that it makes the same phenomena appear to be random, i.e. not due to a deterministic law. But even a complex world is difficult to describe in a deterministic way. The phenomenology of complex systems, in fact, is characterized by properties such as emergence, self-organization, etc. (see Chapters 27 and 28) and evades (at least for now) the possibility of a formulation that enables, a priori, the manifestation of such to be predicted. The situations of complex systems are not repeated in the same way, maybe similar, but never exactly the same, and it is precisely because of this that it is difficult to recognize the laws constructed following reductions. The phenomenology that we observe in nature and in society often appears to be characterized by chaos and/or complexity, and therefore, in reality, the operation of mathematically reducing the data, effective in other situations, in these cases, ceases to be so. Nonetheless, we want to find effective laws, expressed in a mathematical form, to describe chaotic and/or complex phenomenology.

When we attempt to describe a chaotic and/or complex phenomenology in mathematical terms, we find ourselves in difficulty especially if we adopt the formalist interpretation of mathematics, as we run aground in the undecidability that evades the claim of any omni-comprehensive law. The condition we find ourselves in with chaos, in this sense, is analogous to that which leads us to state undecidable propositions of mathematical logic such as Russell's antinomy or that of Epimenides (Agnes and Rasetti, 1998).

If the observed dynamics are very intricate, the 'compression' of the phenomenology into laws is difficult to perform: the data on which the reductions are to be made, in this case, must be very numerous, because the laws that are obtained from them are often too detailed and they are extremely difficult to analyse.

On occasion, the laws are known and they are perhaps relatively simple in their formulation, but nevertheless they are ineffective because we do not know how to use the equations written to describe the phenomenology observed, and, in some way, make predictions. For example, a two-body system in gravitational interaction is relatively simple to describe and its evolution is predictable with arbitrary precision, starting from the law of universal gravitation, which, so to speak, 'works' very well. Up until now no interaction in an isolated two-body system has ever been observed that violates this law,[167] such as, for example, by a sort of antigravity that distances one body from the other. Already with three bodies, however, as we know, the description of the system's dynamics presents a significant level of difficulty: the law of universal gravitation is still the same, but it is not sufficient for predictions. The formal laws do not change with respect to the two-body case, but we are not able to use them adequately. Long-term prediction is impossible and, in practice, it is as if the law of universal gravitation were not effective. In other words, if the system is made up of three bodies, the phenomenology that is displayed is too rich and varied, and the reductions operated on the data are not sufficient to lead to a better formal law, that is really effective, even though the system is subjected to gravitational forces, the functioning of which is well known. In the absence of theoretical results obtained from laws, the numerical simulations conducted by applying iterative formulae, in accordance with the notions of modern experimental mathematics, shows that the motions present dynamic situations that are extremely hard to imagine, a priori, on the basis

---

[167] In reality, it could be argued that neither has a completely isolated system ever been seen, as it is a totally abstract concept, nonexistent in the real universe. Nonetheless, however, with good approximation, we can consider a double star system, i.e. a system made up of two stars that are close to one another, and that both orbit around the common barycentre, very distant with respect to the other stars, to be 'sufficiently' isolated from the rest of the universe. The phenomenology displayed by these systems, which are very common in our galaxy, is described well by the law of universal gravitation.

of intuition. If the system is made up of more than three bodies, the phenomenology is even more intricate and the description even more complex. In certain cases, it may even occur that gravity, although exclusively attractive, gives rise to a dynamics such that one or more interacting bodies are literally expelled from the system (Saari and Xia, 1995).

Just as for systems that display chaotic dynamics, such as, for example, many-body systems in gravitational interaction with one another, even with complex systems we have to tackle the question of the formulation (if possible) of a formal law on mathematical bases that manages to describe the rich and many-faceted phenomenology that they display. We have to tackle the question, that is, of a mathematical theory of complexity. But in doing this we encounter several problems.

As we know, mathematics is a descriptive and predictive tool that finds it difficult to 'work' with chaotic dynamics, for which, even if the laws are known, the predictability can only be extended for a short distance. Can the same type of mathematics be applied to describe complex systems? Or what type of mathematics would be able to? More specifically, is differential and integral calculus, that has achieved so many brilliant successes in numerous fields (especially in several natural sciences), but that has also had its fair share of failures, such as, for example, the case of the study of many-body systems or that of turbulence, or in the case of the various failed attempts to apply it to social sciences, still an effective mathematical tool for complexity? Are the concepts of real numbers and continuum, on which mathematical analysis is based, and the very concept of differential equation still adaptable to complex systems, without being changed? What mathematical objects should we refer to, to describe the evolution of a complex system? Do we need to review the postulates and the foundations of mathematics, for example like Robinson's (1966) non-standard analysis?[168] Could

---

[168] Non-standard analysis represents one of the more successful attempts to re-establish infinitesimal calculus on new conceptual bases. It is developed starting from axioms that are weaker than those of classical mathematical analysis. Basically it is characterized by the introduction of the set of hyperreal numbers, considered as an extension of the, so to say, 'usual' set of real numbers. In the set of hyperreal numbers we assume (using for the sake of simplicity a rather inappropriate but straightforward expression) the 'existence' of infinite numbers and infinitesimal numbers. The former have the property of being larger than any natural number, the latter that of being smaller than the inverse of all natural numbers, 1/2, 1/3, 1/4, ..., but larger than zero. Non-standard analysis, in reality, is portrayed as a real theory of infinitesimals; it has had various applications, amongst others, in mathematical physics (Albeverio et al., 1986; Albeverio, Luxemburg and Wolff, eds., 1995; Honig, 1995) and in economic mathematics (Anderson, 1991; van den Berg, 2000). In the preface to the second edition of 1973 of his book entitled *Non-Standard Analysis*, first published in 1966, Abraham Robinson writes: 'A non-standard proof can always be replaced by a standard one,

further extensions of the very concept of number possibly be usefully applied? Is the mathematics that we use effective as a tool for investigation, in particular mathematical analysis, with the concept of the continuity and of the infinity on which it is based, or do we need to seek new forms? We would like to remind readers that these are the fundamental questions that need to be tackled. Lastly, we would like to make some reflections on these points, by referring to the different interpretations of the foundations of mathematics illustrated in Chapter 32.

The description of complexity linked to a better understanding of the concept of mathematical infinity: some reflections

When attempting to define the laws of the phenomenology of complexity, we cannot ignore the fact that, in practice, the models of the processes observed, even if written as sets of mathematical formulae, have been made to evolve only after having been translated into the form of computer programs, with the exception of a few simple cases. This inevitably recalls the paradigm of the world as a cosmic computer and re-proposes the perspective of experimental mathematics. The fact that, in practice, what we use is simply the numerical result of a program, leads us to assume numbers as the only knowable element of reality: not only the *measured* number, i.e. the figure obtained from the measurements, but also the *calculated* number, i.e. the number resulting from the descriptions and predictions of the system's evolutions.

Note, however, that we are by no means returning to a Pythagorean perspective, according to which numbers have an almost mystical significance, the most profound essence of reality. We simply find it somewhat

<hr/>

even though the latter may be more complicated and less intuitive (...) The application of non-standard analysis to a particular mathematical discipline is a matter of choice: it is natural for the actual decision of an individual to depend on his early training' (p. xv). Again, the same foreword also illustrates Kurt Gödel's opinion: 'There are good reasons to believe that non-standard analysis, in some version or other, will be the analysis of the future (...) I think, in coming centuries it will be considered a great oddity in the history of mathematics that the first exact theory of infinitesimals was developed 300 years after the invention of the differential calculus' (p. xvi). An axiomatic approach to non-standard analysis, different from Robinson's constructivist approach has been given by Edward Nelson (1977), whereas, in a celebrated article, Imre Lakatos (1978) discussed a few philosophical and historical questions relating to non-standard analysis; presentations of ordinary calculus at college level from a non-standard standpoint are in Keisler (1976), Deledicq and Diener (1989); for a general introduction to non-standard analysis see, e.g., Diener and Diener (1995), Goldblatt (1998).

difficult to adopt a Platonic vision of complexity referred to a transcendent object, that exists in the world of ideas, and that expresses an 'exact' law. Basically, laws wouldn't be abstract entities that express the most profound expression (the quintessence, to put it in Platonic terms) of a phenomenon in its entirety, but merely a program that generates numbers which describe a sequence of states over time. The logical game of formalist mathematics is even less relevant to a vision of this nature, in which everything depends on the choice of a set of initial postulates. Even constructivism, while less remote from the Platonism and formalism of the paradigm of the cosmic computer, in reality, does not appear to be wholly sufficient in this context, because rather than a positive proposal of an effective method to describe evolutive processes, it places restrictions on what is permitted, like a track that forces us only to give values to numbers and what can be obtained from them.

The experimental mathematics approach does not use the symmetry of laws as a primary characteristic, but processes 'small pieces' of information, due to its substantially discrete vision. The search for a 'law for everything', that includes all of the symmetrical laws established to date (and even those that we don't know, because we haven't been able to determine them, but that we imagine to exist), not only in natural sciences but also in social sciences, is based on the basic assumption that the laws possess symmetrical properties in time and/or space (for example translation symmetry in periodic phenomena, symmetry by reflection in mechanical and classical phenomena), and that time and space are continuous, because this is the intuition we have of them. However, up until now this has only worked partially. Whole branches of science, for example, biology, sociology, and economics, as we have observed on a number of occasions, make poor use of laws, intended in the way that they are generally conceived, i.e. expressed in a mathematical form, or are completely lacking in them. And this generally happens even if the cultural setting is reductionist and at times, in practice, when adopting approaches similar to the so-called hard sciences, such as physics and chemistry, are adopted.

And that's not all; the same mathematical laws that exist in a disciplinary field that make wide use of them, such as physics, sometimes display imperfections, and above all, inadequacies, when the systems are very ('too') complex. We can say that the more complex the systems considered are, as often happens when passing from the scope of physics to that of chemistry, and then biology, and then further to the ambits of social sciences, the greater the difficulty in recognizing and applying laws expressed in a mathematical form, the current nature of which is essentially reductionist.

The interpretation of complexity according to the vision proposed by experimental mathematics, i.e. the laws are considered as calculation programs, leads us back to the concepts of numbers, sets, and numerical systems. But, inevitably, this leads us to encounter the same concepts of the continuity of the set of real numbers and, in general, of numerical infinity, imposing an in-depth analysis and (possibly) a review of the same.[169] From this perspective, infinity would be considered a human intuition in a reality that is substantially discrete, in which the very existence of tiny elements that constitute the 'minimum' element, absent if the reality was continuous, may be at the origin of the effects that are increasingly displayed during the course of the evolution of a system, giving rise to all those characteristics of complexity that we call surprise, emergence, self-organization, and to the very phenomenology of chaos. In this way, infinity would change the role that it has laboriously attained through interminable and harsh conceptual, philosophical, and even technological debate (Suber, 1998).

We would like to better explain what we mean. All things considered, we can state that the set of real numbers, which is at the basis of the intuitive concept that we have of continuity, in a certain sense is *only* a representation that we create (or that we 'intuitively sense') and that we use to construct differential equations. However, we mustn't forget that the results of any experimental measurement are only represented by rational numbers, far from the continuity that characterizes the set of real numbers, which leads us to think that a description in non-continuous terms, i.e. in discrete terms, is more relevant. Thinking in terms of spatial–temporal continuity, at least for many processes, is instinctive. In the same way, we all instinctively think of the objects of ordinary geometry as continuous bodies as this is how the real situation appears to us in first approximation (or on first sight).

A sudden change in the evolutive dynamics of a system (a 'surprise') can emerge, apparently violating a symmetrical law that was formulated by making a reduction on some (or many) finite sequences of numerical data. This is the crucial point. As we have said on a number of occasions, complexity emerges as a breakdown of symmetry (a system that, by evolving with continuity, suddenly passes from one attractor to another) in laws which, expressed in mathematical form, are symmetrical. Nonetheless, this breakdown happens. It is the surprise, the paradox, a sort of butterfly effect that can highlight small differences between numbers that are very close to one another in the continuum of real numbers; differences that may evade the experimental interpretation of data, but that may increasingly amplify in the system's dynamics. However, we can *never* disregard numerical approximations, both when taking measurements of any magnitude, and

---

[169] For the sake of clarity, we would like to specify that in this concept we are referring to the actual infinite and not the potential infinite (see p. 318, note 142).

when processing the data contained in a computer, the only tool which, in practice, enables us to make the laws 'work' and make systems evolve. Non-rational numbers, in practice, are always and in any event approximated with numbers belonging to the infinite, discrete and countable set of rational numbers. In particular, in applications with automatic calculation, approximations are even made with integers, almost completely losing sight of the non-countable infinity of the continuum.

In this perspective, the mathematical techniques used have to be reconsidered on new bases, possibly starting from a substantial critical review of the concept of continuum, maybe resuming some of the ideas of constructivism, as in Weyl (1918); see also Freiling (1986) and Maddy (1998). The tool of investigation provided by mathematical analysis, and that originates from intuition, is based on the intuitive concept of continuity, which proves difficult and evasive to formalize: numerical continuity (non-computable numbers included), on one hand, in its conceptual foundation, spatiotemporal continuity, on the other hand, insofar as its applications are concerned. In practice, however, we operate with rational numbers, i.e. with a countable infinity of an order that is no different to that of the more familiar whole numbers. When we use the tools of mathematical analysis, we oscillate, so to say, between the discrete perspective of rational numbers, which is more immediate and easier in terms of practical calculations, and the continuous perspective of real numbers, which, at least on occasion, is more immediately intuited, at least in some aspects, but is often more difficult in terms of calculation techniques, even though, where calculation techniques permit, it leads to effective conclusions. One moment we use this vision, the next we use the other, and often we use the first as an image of the second (this is also how, for example, fractals are represented in computational mathematics).

However, by working in this way, we use concepts that partially belong to the technical perspective, but that possess several aspects that refer to foundations that are not wholly clear. For example, we do not have any idea of how to bridge the gap – by means of a numerical construction, a theorem or otherwise – between the countable infinity and the non-countable infinity of the continuum of a higher order (see p. 157, note 71), nor do we know if there is an intermediate order of infinity (the assumption that there is not is known as Cantor's hypotheses of the continuum). And we are lacking this idea, both as reminiscence, according to the Platonic vision, and as the logical deduction of a set of postulates, according to formalism, and further still as a construction starting from the natural numbers, according to the constructivist approach. We use the concept of going to the limit, in mathematical analysis, to jump between two orders of infinity, but this does not bridge the gap between an experimental reality, made up of measurements necessarily expressed using numbers with finite decimal places (countable infinity), and an abstraction that lives in a different ambit, the numerical continuum, the vast majority of which is made up of non-computable numbers and that intrinsically leads to situations similar to those of logical undecidability, as we saw in Chapter 32.

We do not know if the dynamics of nature and of society are intrinsically continuous or discrete, even if we instinctively picture them as continuous and we portray them in continuous spaces. We have no proof one way or the other, with the sole exception of the fact that the energy is transmitted in discrete quantities that are multiples of a basic magnitude. It is interesting to observe, in this regard, that such basic magnitude, the so-called energy quantum, was introduced in a daring hypothesis of Max Planck in 1900, initially as a simple calculation artifice: a step backwards with respect to the use of classical mathematical analysis, made by replacing an integral (i.e. a sum of infinitesimals in the continuum of real numbers) with a summation of finite quantities, the energy quanta. This was done with the aim of providing a description, i.e. a law, for certain experimental facts that evaded understanding (see, in this regard, the detailed presentation by the physicist Max Born, one of the fathers of quantum physics in his well-known book published in 1935). Later, the daring hypothesis of quanta revealed itself to be extremely fecund to significant development of all of modern physics, but that hypothesis arose from a disconnection between the technique of mathematical analysis, incidentally still widely, continuously, and very successfully used in quantum physics, and real experimental facts that didn't fit the idea of the continuum which is at the basis of mathematical analysis and its techniques.

Probably, complexity, although applicable to numerous scientific disciplines, is in a situation similar to that of physics at the end of the nineteenth century. Perhaps some audacious hypothesis, such as Planck's, is needed, which, by revolutionizing the foundations of mathematics, and not simply mathematical technique, would provide new and more fertile notions and a mathematical tool to describe a system's dynamics more effectively as well as being able to account for the emergent phenomena that characterize complexity. For this reason, it will probably be necessary to clarify concepts such as the numerical continuum, and the different orders of infinity in more depth, continuing the research work that started in the seventeenth century. The objective is to develop a theoretical–mathematical framework that is able to describe the dynamics characterized by complexity and by chaos, in the same way as Newton's and Leibniz's mathematical analyses made their contribution to the field of mechanics in the seventeenth century.

References

Abarbanel H.D.I. (1996) *Analysis of Observed Chaotic Data*, Springer Verlag, Heidelberg.
Accardi L. (1997) *Urne e camaleonti*, Il Saggiatore, Milan.
Agnes C., Rasetti M. (1988) Chaos and Undecidability: A Group Theoretical View, in Solomon A.I. (ed.) *Advances in Statistical Mechanics*, World Scientific, Singapore, 57–73.
Albeverio S., Fenstad J., Hoegh-Krohn R., Lindstrøm T. (1986) *Nonstandard Methods in Stochastic Analysis and Mathematical Physics*, Academic Press, New York.
Albeverio S., Luxemburg W.A.J., Wolff R.P.H. (eds.) (1995) *Advances in Analysis, Probability and Mathematical Physics*, Mathematics and its Application, Vol. 314, Kluwer, Dordrecht.
Albin P.S. (1998) *Barriers and Bounds to Rationality*, Princeton University Press, Princeton, New Jersey.
Allen P.M. (1997a) *Cities and Regions as Self-organizing Systems. Models of Complexity*, Gordon and Breach, London.
Allen P.M. (1997b) Le città come sistemi complessi autoorganizzativi, in Bertuglia C.S., Vaio F. (eds.) *La città e le sue scienze*, Volume 1, *La città come entità altamente complessa*, Franco Angeli, Milan, 1–60 [English edition: (1998) Cities as Self-Organising Complex Systems, in Bertuglia C.S., Bianchi G., Mela A. (eds.) *The City and Its Sciences*, Physica-Verlag, Heidelberg, 95–144].
Allen P.M., Deneubourg J.L., Sanglier M., Boon F., de Palma A. (1978) The Dynamics of Urban Evolution, Final Report of the U.S. Department of Transportation, Washington D.C.
Allen R.G.D. (1959) *Mathematical Economics*, Macmillan, London.
Alligood K., Sauer T., Yorke J.A. (1996) *Chaos, An Introduction to Dynamical Systems*, Springer Verlag, Heidelberg.
Alvesson M. (1995) *Management of Knowledge-intensive Companies*, Walter De Gruyter, Berlin.
Anderson Ph.W., Arrow K.J., Pines D. (eds.) (1988) *The Economy as an Evolving Complex System*, Santa Fe Institute Studies in the Sciences of Complexity, Proceedings Volume 5, Perseus Books, Reading, Massachusetts.
Anderson R.M. (1991) Nonstandard Analysis with Applications to Economics, in Hildenbrand W., Sonnenschein H. (eds.) *Handbook of Mathematical Economics*, Volume 4, 2145–2208, North Holland, Amsterdam.
Andrecut M. (1998) Logistic map as a random number generator, *International Journal of Modern Physics B*, **12**, 921–930.
Anosov D.V., Arnold V.I. (eds.) (1988) *Dynamical Systems I, Ordinary Differential Equations and Smooth Dynamical Systems*, Springer Verlag, Berlin.
Arditi R., Ginzburg L.R. (1989) Coupling in Predator–Prey Dynamics: Ratio-Dependence, *Journal of Theoretical Biology*, **139**, 311–326.
Arnold V.I. (1983) *Geometrical Methods in the Theory of Ordinary Differential Equations*, Springer Verlag, New York.
Arnold V.I. (ed.) (1987) *Dynamical Systems III, Mathematical Aspects of Classical and Celestial Mechanics*, Springer Verlag, Berlin.

Arnold V.I. (1992) *Ordinary Differential Equations*, Springer Verlag, Berlin.
Arsenin V.Ya. (1987) Problems on Computer Diagnostics in Medicine, in Tikhonov A.N., Goncharsky A.V. (eds.) *Ill-Posed Problems in the Natural Sciences*, Mir, Moscow, 206–219.
Arthur W.B. (1994) On the Evolution of Complexity, in Cowan G.A., Pines D., Meltzer D. (eds.) *Complexity – Metaphors, Models and Reality*, Santa Fe Institute Studies in the Sciences of Complexity, Proceedings Volume 19, Addison-Wesley, Reading, Massachusetts, 65–81.
Arthur W.B., Durlauf S.N., Lane D.A. (eds.) (1997) *The Economy as an Evolving Complex System*, Santa Fe Institute Studies in the Sciences of Complexity, Proceedings Volume 27, Perseus Books, Reading, Massachusetts.
Arthur W.B., Holland J.H., Le Baron B., Palmer R.G., Tayler P. (1997) Asset Pricing Under Endogenous Expectations in an Artificial Stock Market, in Arthur W.B., Durlauf S.N., Lane D.A. (eds.) *The Economy as an Evolving Complex System*, Santa Fe Institute Studies in the Sciences of Complexity, Proceedings Volume 27, Perseus Books, Reading, Massachusetts, 15–44.
Ashby W.R. (1957) *An Introduction to Cybernetics*, Chapman & Hall, London.
Atlan H. (1994) Ideali e limiti della conoscenza e della libertà, in Ceruti M., Fabbri P., Giorello G., Preta L. (eds.) *Il caso e la libertà*, Laterza, Rome, 38–51.
Axelrod R. (1984) *The Evolution of Co-operation*, Basic Books, New York.
Axelrod R. (1997) *The Complexity of Co-operation. Agent-Based Models of Competition and Collaboration*, Princeton University Press, Princeton, New Jersey.
Bachgus R., Van Nes P., Van der Zaan A. (1985) Stedelijke Dinamiek, een Schatting van het Prooi-roofdier Model voor Rotterdam, Erasmus University, Department of Theoretical Spatial Economy, Working Paper.
Bak P. (1996) *How Nature Works*, Springer Verlag, New York.
Barber C.L. (1964) *The Story of Language*, Pan Books, London [revised edition: 1972].
Barkley Rosser J. (1991) *From Catastrophe to Chaos, A General Theory of Economic Discontinuities*, Kluwer Academic Publishers, Boston.
Barrow J.D. (1992a) *Perché il mondo è matematico?*, Lezioni italiane, Laterza, Rome.
Barrow J.D. (1992b) *Pi in the Sky. Counting, Thinking and Being*, Oxford University Press, Oxford.
Barrow J.D. (1994) Caso e determinismo, in Ceruti M., Fabbri P., Giorello G., Preta L. (eds.) *Il caso e la libertà*, Laterza, Rome, 15–27.
Barrow J.D., Tipler F.J. (1986) *The Anthropic Cosmological Principle*, Oxford University Press, New York.
Barrow-Green J. (1997) *Poincaré and the Three Body Problem*, American Mathematical Society, Providence, Rhode Island.
Batten D.F. (2000a) *Discovering Artificial Economics*, Westview Press, Boulder, Colorado.
Batten D.F. (2000b) Emergence and Co-Evolutionary Learning in Self-Organised Urban Development, in Batten D.F., Bertuglia C.S., Martellato D., Occelli S. (eds.) *Learning, Innovation and Urban Evolution*, Kluwer Academic Publishers, Boston, 45–74.
Batten D.F., Bertuglia C.S., Martellato D., Occelli S. (eds.) (2000) *Learning, Innovation and Urban Evolution*, Kluwer Academic Publishers, Boston.
Bayly B. (1990) Complexity in Fluids, in Nadel L., Stein D.L. (eds.) *1990 Lectures in Complex Systems*, Santa Fe Institute Studies in the Sciences of Complexity, Lecture Volume 3, Addison Wesley, Redwood, California, 53–97.
Bechhoeffer J. (1996) The Birth of Period 3, Revisited, *Mathematics Magazine*, **69**, 115–118.

References

Bellacicco A. (1990) *La rappresentazione frattale degli eventi*, La Nuova Italia Scientifica, Rome.

Bellomo N. (1982) *Modelli matematici e sistemi con parametri aleatori*, Levrotto e Bella, Turin.

Bellomo N., Preziosi L. (1995) *Modelling Mathematical Methods and Scientific Computation*, CRC Press, Boca Raton, Florida.

Beltrami E. (1987) *Mathematics for Dynamic Modelling*, Academic Press, London.

Benacerraf P. (1967) God, the Devil and Gödel, *The Monist*, 51, 9–32.

Benacerraf P., Putnam H. (eds.) (1983) *Philosophy of Mathematics*, Cambridge University Press, Cambridge.

Ben-Akiva M., Lerman S.R. (1979) Disaggregate Travel and Mobility Choice Models and Measures of Accessibility, in Hensher D.A., Stopher P.R., *Behaviour Travel Modelling*, Croom Helm, London, 654–679.

Bender W., Orszag S. (1978) *Advanced Mathematical Methods for Scientists and Engineers*, McGraw-Hill, New York.

Benhabib J. (ed.) (1992) *Cycles and Chaos in Economic Equilibrium*, Princeton University Press, Princeton, New Jersey.

Berkeley G. (1734) *The Analyst; Or, A Discourse Addressed To An Infidel Mathematician*, J. Tonson, London [available on the website http://www.maths.tcd.ie/~dwilkins/Berkeley].

Bertuglia C.S., Bianchi G., Mela A. (eds.) (1998) *The City and its Sciences*, Physica Verlag, a Springer Verlag Company, Heidelberg.

Bertuglia C.S., Clarke G.P., Wilson A.G. (1994) *Modelling the City*, Routledge, London.

Bertuglia C.S., Fischer M.M., Preto G. (eds.) (1995) *Technological Change, Economic Development and Space*, Springer Verlag, Berlin.

Bertuglia C.S., Leonardi G., Occelli S., Rabino G.A., Tadei R., Wilson A.G. (eds.) (1987) *Urban Systems: Contemporary Approach to Modelling*, Croom Helm, London.

Bertuglia C.S., Leonardi G., Wilson A.G. (eds.) (1990) *Urban Dynamics: Designing an Integrated Model*, Routledge, London.

Bertuglia C.S., Lombardo S., Nijkamp P. (eds.) (1997) *Innovative Behaviour in Space and Time*, Springer Verlag, Berlin.

Bertuglia C.S., Rabino G.A. (1990) The Use of Mathematical Models in the Evaluation of Actions in Urban Planning: Conceptual Premises and Operative Problems, *Sistemi urbani*, 12, 121–132.

Bertuglia C.S., Staricco L. (2000) *Complessità, autoorganizzazione, città*, Franco Angeli, Milan.

Bertuglia C.S., Vaio F. (eds.) (1997a) *La città e le sue scienze*, 4 volumes, Franco Angeli, Milan.

Bertuglia C.S., Vaio F. (1997b) Introduzione, in Bertuglia C.S., Vaio F. (eds.) *La città e le sue scienze*, Volume 1, *La città come entità altamente complessa*, Franco Angeli, Milan, I–XCIII.

Besicovich A.S. (1934) On Rational Approximation to Real Numbers, *Journal of the London Mathematical Society*, 9, 126–131.

Besicovich A.S. (1935) On the Sum of Digits of Real Numbers Represented in the Dyadic System, *Mathematische Annalen*, 110, 321–330.

Birkhoff G.D. (1927) *Dynamical Systems*, American Mathematical Society, Providence, Rhode Island, [revised edition: (1966) available on the website http://www.ams.org/online_bks/coll9/].

Black F., Scholes M. (1973) The Pricing of Options and Corporate Liabilities, *Journal of Political Economics*, 81, 637–659.

Blair D., Pollack R. (1999) La logique du choix collectif, *Pour la science, Dossier: Les mathématiques sociales*, July, 82–89.

Blum L. (1989) Lectures on a Theory of Computation and Complexity over the Reals (or an Arbitrary Ring), in Jen E. (ed.) *1989 Lectures in Complex Systems*, Santa Fe Institute Studies in the Sciences of Complexity, Lectures Volume 2, Addison-Wesley, Redwood City, California, 1–47.

Bohm D. (1957) *Casualty and Chance in Modern Physics*, Routledge, London.

Boncinelli E., Bottazzini U. (2000) *La serva padrona. Fascino e potere della matematica*, Raffaello Cortina Editore, Milan.

Born M. (1935) *Atomic Physics*, Blackie and Son Ltd., London.

Borwein J., Bailey D. (2004) *Mathematics by Experiment: Plausible Reasoning in the 21st Century*, A.K. Peters, Natick, Massuchusetts.

Bottazzini U. (1999) Poincaré, il cervello delle scienze razionali, *Le scienze, I grandi della scienza*, 2, 7.

Bourbaki N. (1939–1998) *Éléments de mathématiques*, Hermann & C. Éditeurs, Paris [English edition: (1987–1999) *Elements of Mathematics*, 10 volumes, Springer-Verlag, Heidelberg].

Boyce D.E., Nijkamp P., Shefer D. (eds.) (1991) *Regional Science, Retrospect and Prospect*, Springer Verlag, Berlin.

Boyer C.B. (1968) *A History of Mathematics*, Wiley, New York [Italian edition: (1980) *Storia della matematica*, Mondadori, Milan].

Briggs K.M. (1991) A Precise Calculation of the Feigenbaum Constants, *Mathematics of Computation*, 57, 435–439.

Briggs K.M., Quispel G.R.W., Thompson C.J. (1991) Feigenvalues for Mandelsets, *Journal of Physics A*, 24, 3363–3368.

Brock W., Hsieh D., Le Baron B. (1991) *Nonlinear Dynamics, Chaos and Instability: Statistical Theory and Economic Evidence*, MIT Press, Cambridge, Massachusetts.

Brouwer L.E.J. (1913) Intuitionism and Formalism, *Bulletin of the American Mathematical Society*, 20, 81–96 [re-edited: (1999) in *Bulletin (new series) of the American Mathematical Society*, 37, 1, 55–64].

Bruner J.S. (1994) Modelli del mondo, modelli di menti, in Ceruti M., Fabbri P., Giorello G., Preta L. (eds.) *Il caso e la libertà*, Laterza, Rome, 67–94.

Bruschini Vincenzi L. (1998) *Storia della Borsa*, Il sapere, Newton, Rome.

Burke Hubbard B., Hubbard J. (1994) Legge e ordine nell'universo: il teorema KAM, *Quaderni di Le Scienze*, 81, 8–16.

Byrne D. (1998) *Complexity Theory and the Social Sciences*, Routledge, London.

Cajori F. (1919) *History of Mathematics* (2nd edition), Macmillan, New York.

Camagni R. (1985) Spatial Diffusion of Pervasive Process Innovations, *Papers and Proceedings of the Regional Science Association*, 58, 83–95.

Çambel A.B. (1993) *Applied Chaos Theory*, Academic Press, San Diego, California.

Campisi D. (1991) I fondamenti della modellistica urbana, in Bertuglia C.S., La Bella A. (eds.) *I sistemi urbani*, Franco Angeli, Milan, 509–551.

Cantor G. (1915) *Contributions to the Founding of the Theory of Transfinite Numbers* (edited by Jourdain E.B.), Open Court, Chicago [new edition: (1955), Dover, New York].

Carey H.C. (1858) *Principles of Social Science*, J.B. Lippincot, Philadelphia [available on the website http://socserv2.mcmaster.ca/~econ/ugcm/3ll3/].

Casti J.L. (1986) On System Complexity: Identification, Measurement and Management, in Casti J.L., Karlqvist A. (eds.) *Complexity, Language and Life: Mathematical Approaches*, Springer Verlag, Berlin, 146–173.

Casti J.L. (1987) *Alternate Results. Mathematical Models of Nature and Man*, Wiley, New York.

Casti J.L. (1994) *Complexification*, Abacus, London.

Casti J.L. (2000) *Five More Golden Rules*, Wiley, New York.

Cauchy A.L. (1882–1970) *Oeuvres complètes*, 27 volumes, Gauthier-Villars, Paris.

Cellucci C. (eds.) (1967) *La filosofia della matematica*, Laterza, Bari.

Changeux J.-P., Connes A. (1989) *Matière à pensée*, Éditions Odile Jacob, Paris [English edition: (1995) *Conversations on Mind, Matter and Mathematics*, Princeton University Press].

Chavel I. (1994) *Riemanian Geometry: A Modern Introduction*, Cambridge University Press, New York.

Chiarella C. (1990) *The Elements of a Nonlinear Theory of Economic Dynamics*, Springer Verlag, Berlin.

Chossat P., Iooss G. (1994) *The Couette–Taylor Problem*, Applied Mathematical Sciences, Volume 102, Springer Verlag, Heidelberg.

Christensen K.S. (1999) *Cities and Complexity, Making Intergovernmental Decisions*, Sage Publications, London.

Cohen I.B. (1993) *The Natural Sciences and the Social Sciences: Some Historical and Critical Perspectives*, Boston Studies in the Philosophy of Science, Boston, Massachusetts.

Cohen J., Stewart I. (1994) *The Collapse of Chaos*, Penguin, New York.

Collatz L. (1966) *Functional Analysis and Numerical Mathematics*, Academic Press, New York.

Condorcet M.J.A.N. (1785) Essai sur l'application de l'analyse à la probabilité des décisions prises dans une pluralité de voix, in *Arithmétique politique. Textes rares ou inédits* (1994) INED, Paris.

Coveney P., Highfield R. (1995) *Frontiers of Complexity. The Search for Order in a Chaotic World*, Fawcett Columbine Books, New York.

Crépel P. (1999) La naissance des mathématiques sociales, *Pour la science, Dossier: Les mathématiques sociales*, July, 8–13.

Crilly A.J., Earnshaw R.A., Jones H. (eds.) (1991) *Fractals and Chaos*, Springer Verlag, New York.

Crutchfield J.P., Farmer J.D., Packard N.H., Shaw R.S. (1986) Chaos, *Scientific American*, 255, 38–49.

Cutland N.J. (1980) *Computability*, Cambridge University Press, Cambridge.

Czyz J. (1994) *Paradoxes of Measures and Dimensions Originating in Felix Hausdorff's Ideas*, World Scientific, Singapore.

Daganzo C. (1979) *Multinomial Probit*, Academic Press, New York.

Darwin C. (1859) *On the Origin of Species by Means of Natural Selection or the Preservation of Favored Races in the Struggle for Life*, Murray, London.

Davies P.C.W. (1989) The Physics of Complex Organization, in Goodwin B, Saunders P. (eds.) *Theoretical Biology, Epigenetic and Evolutionary Order from Complex Systems*, Edinburgh University Press, Edinburgh, 101–111.

Davies S. (1979) *The Diffusion of Process Innovation*, Cambridge University Press, Cambridge.

Dedekind R. (1901) *Essays on the Theory of Numbers* (English translation ed. Beman W.W.), Open Court, Chicago [new edition: (1963) Dover, New York].

de Finetti B. (1963) *L'apporto della matematica nell'evoluzione del pensiero economico*, Edizione Ricerche, Rome.

Deledicq A., Diener M. (1989) *Lesons de calcul infinitésimal*, Editions Armand Colin. Paris.

Dendrinos D. (1980) A Basic Model of Urban Dynamics Expressed as a Set of Volterra–Lotka Equations, Catastrophe Theory in Urban and Transport Analysis, Department of Transportation, Washington D.C.

Dendrinos D. (1984a) Madrid's Aggregate Growth Pattern: A Note on the Evidence Regarding the Urban Volterra–Lotka Model, *Sistemi urbani*, 6, 237–246.

Dendrinos D. (1984b) The Decline of the US Economy: A Perspective from Mathematical Ecology, *Environment and Planning A*, 16, 651–662.

Dendrinos D. (1984c) The Structural Stability of the US Regions: Evidence and Theoretical Underpinnings, *Environment and Planning A*, 16, 1433–1443.

Dendrinos D. (1991a) Methods in Quantum Mechanics and the Socio-Spatial World, *Socio-Spatial Dynamics*, 2, 81–110.

Dendrinos D. (1991b) Quasi-Periodicity in Spatial Population Dynamics, *Socio-Spatial Dynamics*, 2, 31–59.

Dendrinos D. (1992) *The Dynamics of Cities*, Routledge, London.

Dendrinos D. (1997) Sui fondamenti della dinamica sociale. Efficiente formulazione matematica di un quadro generale alla base del determinismo sociale complesso non lineare: sovraosservatore e sovrastruttura, in Bertuglia C.S., Vaio F. (eds.) *La città e le sue scienze*, Volume 1, *La città come entità altamente complessa*, Franco Angeli, Milan, 129–153 [English edition: (1998) On the Foundation of Social Dynamics: An Efficient Mathematical Statement of a General Framework Underlying a Complex Nonlinear Social Determinism, Incorporating a Supra-Observer and a Suprastructure, in Bertuglia C.S., Bianchi G., Mela A. (eds.) *The City and Its Sciences*, Physica-Verlag, Heidelberg, 203–223].

Dendrinos D. (2000) Nonlinear Dynamics, Innovation and Metropolitan Development, in Batten D.F., Bertuglia C.S., Martellato D., Occelli S. (eds.) (2000) *Learning, Innovation and Urban Evolution*, Kluwer Academic Publishers, Boston, 75–106.

Dendrinos D., Mullally H. (1981) Evolutionary Patterns of Urban Populations, *Geographical Analysis*, 13, 328–344.

Dendrinos D., Mullally H. (1985) *Urban Evolution: Studies in the Mathematical Ecology of Cities*, Oxford University Press, Oxford.

Dendrinos D., Sonis M. (1986) Variational Principles and Conservation Conditions in Volterra's Ecology and in Urban Relative Dynamics, *Journal of Regional Science*, 26, 359–372.

Dendrinos D., Sonis M. (1990) *Chaos and Socio-Spatial Dynamics*, Springer Verlag, Berlin.

Devaney R.L. (1989) *An Introduction to Chaotic Dynamical Systems*, Perseus Books, Reading, Massachusetts.

Devaney R.L. (1992) *A First Course in Chaotic Dynamical Systems: Theory and Experiment*, Perseus Books, Reading, Massachusetts.

Devlin K. (1980) *Fundamentals of Contemporary Set Theory*, Springer Verlag, New York.

Devlin K. (1988) *Mathematics. The New Golden Age*, Penguin Books, New York.

Diener F., Diener M. (1989) *Nonstandard Analysis in Practice*, Springer Verlag, Heidelberg.

Dioguardi G. (1995) *Dossier Diderot*, Sellerio editore, Palermo.

Dioguardi G. (2000) Al di là del disordine, discorso sulla complessità e sull'impresa, CUEN, Naples.

Dioguardi G. (2001) *Sui sentieri della scienza*, Sellerio editore, Palermo.

Dirac P.A.M. (1982) Pretty Mathematics, *International Journal of Theoretical Physics*, **21**, 603–605.

Ditto W.L., Spano M.L., Savage H.T., Rauseo S.N., Heagy J., Ott E. (1990) Experimental Observation of a Strange Non-chaotic Attractor, *Physics Review A*, **39**, 5, 2593–2598.

Dobb M. (1973) *Theories of Value and Distribution since Adam Smith – Ideology and Economic Theory*, Cambridge University Press, Cambridge.

Drazin P.S. (1992) Nonlinear Systems, Cambridge University Press, Cambridge.

Duffing G. (1918) *Erzwungene Schwingungen bei Veränderlicher Eigenfrequenz*, F. Vieweg und Sohn, Braunschweig.

Eckmann J.-P., Oliffson Kamphorst S., Ruelle D., Ciliberto S. (1986) Lyapunov Exponents from Time Series, *Physics Review, A*, **34**, 4971–4979.

Eckmann J.-P., Oliffson Kamphorst S., Ruelle D., Scheinkman J. (1988) Lyapunov Exponents for Stock Returns, in Anderson Ph.W., Arrow K.J., Pines D. (eds.) *The Economy as an Evolving Complex System*, Santa Fe Institute Studies in the Sciences of Complexity, Proceedings Volume 5, Addison-Wesley, Reading, Massachusetts, 301–304.

Eckmann J.-P., Ruelle D. (1985) Ergodic Theory of Chaos and Strange Attractors, *Review of Modern Physics*, **57**, 617–686.

Edmonds B. (1999) What is Complexity? – The Philosophy of Complexity per se With Application to Some Examples in Evolution, in Heylighen F., Bollen J., Riegler A. (eds.) *The Evolution of Complexity, The Violet Book of 'Einstein Meets Magritte'*, Kluwer Academic Publishers, Dordrecht [article available on the website http://bruce.edmonds.name/evolcomp].

Einstein A. (1922) *The Meaning of Relativity*, Methuen & Co., London.

Einstein A., Podolski B., Rosen N. (1935) Can Quantum-Mechanical Description of Physical Reality Be Considered Complete?, *Physical Review*, **47**, 777.

Ekeland I. (1995) *Le chaos*, Flammarion, Paris.

Elliott R.N. (1946) *Nature's Law: The Secret of the Universe* [new edition in: Prechter R.J. (ed.) (1994) *Elliott's Masterworks: The Definitive Collection*, Lambert-Gann, Gainesville, Georgia].

Elsgolts L. (1977) *Differential Equations and the Calculus of Variations*, Mir, Moscow.

Enriques F., de Santillana G. (1936) *Compendio di storia del pensiero scientifico*, Zanichelli, Bologna.

Epstein I.R. (1989) Chemical Oscillators and Nonlinear Chemical Dynamics, in Jen E. (ed.) *1989 Lectures in Complex Systems*, Santa Fe Institute Studies in the Sciences of Complexity, Lectures Volume 2, Addison-Wesley, Redwood City, California, 213–269.

Epstein J.M. (1997) *Nonlinear Dynamics, Mathematical Biology, and Social Science*, Addison Wesley, Reading, Massachusetts.

Eubank S., Farmer J.D. (1989) An Introduction to Chaos and Randomness, in Jen E. (ed.) *1989 Lectures in Complex Systems*, Santa Fe Institute Studies in the Sciences of Complexity, Lectures Volume 2, Addison-Wesley, Redwood City, California, 75–190.

Falconer K.J. (1985) *The Geometry of Fractal Sets*, Cambridge University Press, Cambridge.

Falconer K.J. (1990) *Fractal Geometry: Mathematical Foundations and Applications*, Wiley, New York.

Fang F., Sanglier M. (eds.) (1997) *Complexity and Self-Organization in Social and Economic Systems*, Springer Verlag, Berlin.

Farmer J.D., Ott E., Yorke J.A. (1983) The Dimension of Chaotic Attractors, *Physica*, **7D**, 153–170.

Feigenbaum M. (1978) Quantitative Universality for a Class of Nonlinear Transformation, *Journal of Statistical Physics*, **19**, 25–52.
Feigenbaum M. (1980) Universal Behavior in Nonlinear Systems, *Los Alamos Science*, Summer, 4–27.
Fermi E., Pasta J., Ulam S. (1955) Studies on Nonlinear Problems, Document LA-1940, Los Alamos, May.
Fischer M.M. (1989) Innovation, Diffusion and Regions, in Andersson Å.E., Batten D.F., Karlsson C. (eds.) *Knowledge and Industrial Organization*, Springer Verlag, Berlin, 47–61.
Flood R.L., Carson E.R. (1986) *Dealing with Complexity*, Plenum Press, London.
Forrester J.W. (1969) *Urban Dynamics*, MIT Press, Cambridge, Massachusetts.
Forrester J.W. (1971) *World Dynamics*, Wight-Allen Press, New York.
Fotheringham A.S., O'Kelly M.E. (1989) *Spatial Interaction Models: Formulations and Applications*, Kluwer Academic Publishers, Dordrecht.
Frankhauser P. (1994) *La fractalité des structures urbaines*, Anthropos, Paris.
Franses P.H., Van Dijk D. (2000) Non-Linear Time Series Models in Empirical Finance, Cambridge University Press, Cambridge.
Freiling Ch. (1986) Axioms of Symmetry: Throwing Darts at the Real Number Line, *Journal of Symbolic Logic*, **51**, 1, 190–200.
Frenkel A., Shefer D. (1997) Technological Innovation and Diffusion Models: A Review, in Bertuglia C.S., Lombardo S., Nijkamp P. (eds.) *Innovative Behaviour in Space and Time*, Springer Verlag, Berlin, 41–63.
Frost A.J., Prechter R.J. (1990) *Elliott Wave Principle: Key to Stock Market Profits*, Lambert-Gann, Gainesville, Georgia.
Fujita M., Krugman P., Venables A.J. (2000) *The Spatial Economy*, MIT Press, Cambridge, Massachusetts.
Gandolfi A. (1999) *Formicai, imperi, cervelli*, Bollati Boringhieri, Turin.
Gaudry M.J.I., Dagenais M.G. (1979) The Dogit Model, *Transportation Research*, **13B**, 105–111.
Gause G.F. (1934) *The Struggle for Existence*, Williams and Wilkins, Baltimore, Maryland [available in Russian and in English on the website http://www.ggause.com].
Gause G.F. (1935) Experimental Demonstrations of Volterra's Periodic Oscillations in the Number of Animals, *British Journal of Experimental Biology*, **12**, 44–48.
Gell-Mann M. (1994) *The Quark and the Jaguar*, Little, Brown and Company, London.
Giusti E. (1999) *Ipotesi sulla natura degli oggetti matematici*, Bollati Boringhieri, Turin.
Glasko V.B., Mudretsova E.A., Strakhov V.N. (1987) Inverse Problems in Gravitometry and Magnetometry, in Tikhonov A.N., Goncharsky A.V. (eds.) *Ill-Posed Problems in the Natural Sciences*, Mir, Moscow, 21–52.
Gleick J. (1987) *Chaos*, Penguin, New York [Italian edition: (1989) *Caos*, Rizzoli, Milan].
Gödel K. (1931) Über Formal Unentscheidbare Sätze der Principia Mathematica und Verwandter Systeme, *Monatschefte der Mathematik und Physik*, 38, 173–198.
Gödel K (1986) *Collected Works*, Volume 1, *Publications 1929–1936*, edited by Feferman S., Dawson J.W., Goldfarb W., Parsons Ch., Solovay R.N., Oxford University Press, Oxford.
Gödel K. (1990) *Collected Works*, Volume II, *Publications 1938–1974*, edited by Feferman S., Dawson J.W., Goldfarb W., Parsons Ch., Solovay R.N., Oxford University Press, Oxford.
Gödel K. (1995) *Collected Works*, Volume III, *Unpublished Essays and Lectures*, edited by Feferman S., Dawson J.W., Goldfarb W., Parsons Ch., Solovay R.N., Oxford University Press, Oxford.

Goldberg A.L., Rigney D.R., West B.J. (1994) Caos e frattali in fisiologia umana, *Quaderni di Le Scienze*, **81**, 45–50.

Goldblatt R. (1998) *An Introduction to Nonstandard Analysis*, Springer Verlag, Heidelberg.

Goncharsky A.V. (1987) Ill-Posed Problems and Their Solution Methods, in Tikhonov A.N., Goncharsky A.V. (eds.) (1987) *Ill-Posed Problems in the Natural Sciences*, Mir, Moscow, 21–52.

Goodwin R.M. (1951) The Non-Linear Accelerator and the Persistence of Business Cycles, *Econometrica*, **19**, 1–17.

Goodwin R.M. (1991) Economic Evolution, Chaotic Dynamics and the Marx–Keynes–Schumpeter System, in Hodgson G.M., Screpanti E. (eds.) *Rethinking Economics*, Edward Elgar, Aldershot, 138–152.

Gordon W.B. (1996) Period Three Trajectories of the Logistic Map, *Mathematics Magazine*, **69**, 118–120.

Granger G.G. (1993) *La science et les sciences*, Presses Universitaires de France, Paris.

Grassberger P. (1981) On the Hausdorff Dimension of Fractal Attractors, *Journal of Statistical Physics*, **26**, 173–179.

Grassberger P., Procaccia I. (1983a) Measuring the Strangeness of Strange Attractors, *Physica*, **9D**, 189–208.

Grassberger P., Procaccia I. (1983b) Characterization of Strange Attractors, *Physics Review Letters*, **50**, 346–349.

Grebogi C., Ott E., Pelikan S., Yorke J.A. (1984) Strange Attractors that are not Chaotic, *Physica*, **13D**, 261–268.

Grebogi C., Ott E., Yorke J.A. (1987) Chaos, Strange Attractors and Fractal Basin Boundaries in Nonlinear Dynamics, *Science*, **238**, 632–638.

Guerrien B. (1999) La société, objet complexe et changeant, *Pour la science, Dossier: Les mathématiques sociales*, July, 20–21.

Guerriero G. (2001) Kurt Gödel, paradossi logici e verità matematica, *Le scienze, I grandi della scienza*, **4**, 19.

Gullen M. (1995) *Five Equations that Changed the World*, Hyperion, New York.

Haag G., Muntz M., Pumain D., Saint-Julien T., Sanders L. (1992) Interurban Migration and the Dynamics of a System of Cities, *Environment and Planning A*, **24**, 181–198.

Hadamard J. (1932) *Le problème de Cauchy et les équations aux dérivées partielles linéaires hyperboliques*, Hermann, Paris.

Haken H. (1977) *Synergetics: An Introduction*, Springer Verlag, Berlin.

Haken H. (1983) *Advanced Synergetics*, Springer Verlag, Berlin.

Haken H. (1985) *Complex Systems – Operational Approach*, Springer Verlag, Berlin.

Haken H. (1990) *Synergetics of Cognition*, Springer Verlag, Berlin.

Hallett M. (1984) *Cantorian Set Theory and Limitation of Size*, Oxford University Press, Oxford.

Hannon B., Ruth M. (1997a) *Modelling Dynamic Biological Systems*, Springer Verlag, New York.

Hannon B., Ruth M. (1997b) *Modelling Dynamic Economic Systems*, Springer Verlag, New York.

Hausdorff F. (1919) Dimension und äusseres Mass, *Mathematische Annalen*, **79**, 157–179.

Havel I. (1995) Scale Dimensions in Nature, *International Journal of General Systems*, **23**, 2, 303–332.

Heisenberg W. (1927) Über den anschaulichen Inhalt der quantentheoretischen Kinematik und Mechanik, *Zeitschrift für Physik*, **43**, 172–198.

Hénon M.A. (1976) A Two-Dimensional Mapping with Strange Attractors, *Communications in Mathematical Physics*, **50**, 69–70.

Hensher D.A., Johnson L.W. (1981) *Applied Discrete Choice Modelling*, Croom Helm, London.

Herken R. (ed.) (1983) *The Universal Turing Machine: A Half-Century Survey*, Oxford University Press, Oxford.

Herrmann H. (1998) *From Sociobiology to Sociopolitics*, Yale University Press, New Haven, Connecticut.

Hewings G.J.D., Sonis M., Madden M., Kimura Y. (eds.) (1999) *Understanding and Interpreting Economic Structure*, Springer Verlag, New York.

Heylighen F. (1990) Relational Closure: A Mathematical Concept for Distinction-making and Complexity Analysis, in Trappl R. (ed.) *Cybernetics and Systems '90*, World Science, Singapore, 335–342.

Heyting A. (1956) *Intuitionism, an Introduction*, North-Holland, Amsterdam.

Hicks J.R. (1950) *A Contribution to the Theory of the Trade Cycle*, Oxford University Press, Oxford.

Higgs H. (1897) *The Physiocrats*, Langland Press, London [available on the website http://socserv2.mcmaster.ca/~econ/ugcm/3ll3/].

Hirsch M., Smale S. (1974) *Differential Equations, Dynamical Systems, and Linear Algebra*, Academic Press, New York.

Hodgson G.M., Screpanti E. (eds.) *Rethinking Economics*, Edward Elgar, Aldershot.

Hofstadter D. (1979) *Gödel, Escher, Bach, The Eternal Golden Braid*, Harper and Row, New York.

Holland J. (1996) *Hidden Order: How Adaptation Builds Complexity*, Perseus Books, New York.

Holland J. (1999) *Emergence, from Chaos to Order*, Perseus Books, New York.

Holden A.V., Muhamad M.A. (1986) A Graphical Zoo of Strange and Peculiar Attractors, in Holden A.V. (ed.) *Chaos*, Manchester University Press, Manchester, 15–35.

Holling C.S. (1959) Some Characteristics of Simple Types of Predation and Parasitism, *Canadian Entomology*, **91**, 385–398.

Honig W.R. (1959) *Nonstandard Logics and Nonstandard Metrics in Physics*, World Scientific, Singapore .

Horgan P. (1995) From Complexity to Perplexity, *Scientific American*, June, 104–109.

Horrocks G. (1987) *Generative Grammar*, Longman, London.

Householder A.S. (1974) *Principles of Numerical Analysis*, Dover, New York.

Isard W. (1960) *Methods of Regional Analysis*, MIT Press, Cambridge, Massachusetts.

Isard W. (1999) Regional Science: Parallels from Physics and Chemistry, *Papers in Regional Science*, **78**, 5–20.

Israel G. (1986) *Modelli matematici*, Editori Riuniti, Rome.

Israel G. (1994) Immagini matematiche della realtà, *Quaderni di Le Scienze*, **81**, 3–7.

Israel G. (1996) *La mathématisation du réel. Essai sur la modélisation mathématique*, Éditions du Seuil, Paris [Italian edition: (1997) *La visione matematica della realtà. Introduzione ai temi e alla storia della modellistica matematica*, Laterza, Rome].

Israel G. (1999) *Scienza e storia: la convivenza difficile*, Di Renzo Editore, Rome.

Jones H. (1991) Fractals before Mandelbrot – A Selected History, in Crilly A.J., Earnshaw R.A., Jones H. (eds.) *Fractals and Chaos*, Springer Verlag, New York, 7–33.

Jordan D.W., Smith P. (1977) *Nonlinear Ordinary Differential Equations*, Oxford University Press, New York.

Jost Ch., Arino O., Arditi R. (1999) About Deterministic Extinction in Ratio-dependent Predator Prey Models, *Bulletin of Mathematical Biology*, **61**, 19–32.

Kalecki M. (1954) *Theory of Economic Dynamics*, Allen and Unwin, London.

Katok A., Hasselblatt B. (1995) *Introduction to the Modern Theory of Dynamical Systems*, Cambridge University Press, Cambridge.

Kaufmann A. (1975) *Introduction to the Theory of Fuzzy Subsets*, Academic Press, New York.

Kauffman S. (1986) Autocatalytic Sets of Proteins, *Journal of Theoretical Biology*, **119**, 1–24.

Kauffman S. (1994) Whispers from Carnot: The Origins of Order and Principles of Adaptation in Complex Nonequilibrium Systems, in Cowan G.A., Pines D., Meltzer D. (eds.) *Complexity – Metaphors, Models and Reality*, Santa Fe Institute Studies in the Sciences of Complexity, Proceedings Volume 19, Addison-Wesley, Reading, Massachusetts, 83–160.

Kauffman S. (1995) *At Home in the Universe*, Oxford University Press, New York.

Keisler H.J. (1986) *Elementary Calculus: An Approach Using Infinitesimals* (2nd ed) Prindle, Weber and Schmidt, Boston, Massachusetts [available on the website http://www.math.wisc.edn/~keisler/calc.html]

Kline M. (1990) *Mathematical Thought from Ancient to Modern Times*, Oxford University Press, Oxford.

Koyré A. (1961) *La révolution astronomique*, Herman, Paris.

Kraft R.L. (1999) Chaos, Cantor Sets, and Hyperbolicity for the Logistic Maps, *American Mathematical Monthly*, **106**, 400–408.

Krugman P. (1996) *The Self-Organizing Economy*, Blackwell, Malden, Massachusetts.

Lakatos I. (1978) Cauchy and the continuum: The Significance of Non-Standard Analysis for the History and Philosophy of Mathematics. *The Mathematical lutelligencer*, **3**, 151–161.

Landau L.D., Lifshitz E.M. (1959) *Fluid Mechanics*, Mir, Moscow [2nd edition: (1987), reprinted (1989), Reed Educational and Professional Publishing, Oxford].

Landau L.D., Lifshitz E.M. (1970) *Field Theory*, Mir, Moscow [4th edition: (1975), reprinted (1987), Reed Educational and Professional Publishing, Oxford].

Langton C.G. (1990) Computation at the Edge of Chaos: Phase Transition and Emergent Computation, *Physica*, **42D**, 12–37.

Laplace P.S. (1814) *Essai philosophique sur les probabilités*, Courcier, Paris.

Laskar J. (1989) A Numerical Experiment on the Chaotic Behaviour of the Solar System, *Nature*, **338**, 237–238.

Le Bras H. (1999) Des modèles d'occupation de l'espace, *Pour la science, Dossier, Les mathématiques sociales*, July, 28–31.

Lekachman R. (1959) *A History of Economic Ideas*, Harper & Row, New York [Italian edition: (1993) *Storia del pensiero economico*, Franco Angeli, Milan].

Leonardi G. (1985) Equivalenza asintotica tra la teoria delle utilità casuali e la massimizzazione dell'entropia, in Reggiani A. (ed.) *Territorio e trasporti*, Franco Angeli, Milan, 29–66.

Leonardi G. (1987) The Choice-Theoretic Approach: Population Mobility as an Example, in Bertuglia C.S., Leonardi G., Occelli S., Rabino G.A., Tadei R., Wilson A.G. (eds.) *Urban Systems: Contemporary Approach to Modelling*, Croom Helm, London, 136–188.

Leven R.W., Pompe B., Wilke C., Koch B.P. (1985) Experiments on Periodic and Chaotic Motions of a Parametrically Forced Pendulum, *Physica*, **16D**, 371.

Levy D. (1994) Chaos Theory and Strategy, Theory, Application and Managerial Implications, *Strategic Management Journal*, **15**, 167–178.

Lewin R. (1992) *Complexity, Life at the Edge of Chaos*, Macmillan, New York.

Li T.Y., Yorke J.A. (1975) Period Three Implies Chaos, *American Mathematical Monthly*, **82**, 985–992.

Lloyd S. (1989) Physical Measures of Complexity, in Jen E. (ed.) *1989 Lectures in Complex Systems*, Santa Fe Institute Studies in the Sciences of Complexity, Lectures Volume 2, Addison-Wesley, Redwood City, California, 67–73.

Lorenz E.N. (1963) Deterministic Non-Periodic Flows, *Journal of Atmospheric Science*, **20**, 134–141.

Lotka A.J. (1925) *Elements of Physical Biology*, Williams and Wilkins Co., Baltimore, Maryland.

Lowry I.S. (1964) *A Model of Metropolis*, Memorandum RM-4035–RC, Rand Corporation, Santa Monica, California.

Lyapunov A.M. (1954) *Collected Works*, Akademia Nauk, Moscow.

Lyons J. (1968) *Introduction to Theoretical Linguistics*, Cambridge University Press, London.

Maddy P. (1988a) Believing the Axioms I, *Journal of Symbolic Logic*, **53**, 2, 481–511.

Maddy P. (1988b) Believing the Axioms II, *Journal of Symbolic Logic*, **53**, 3, 736–764.

Maddy P. (1990) *Realism in Mathematics*, Oxford University Press, Oxford.

Majorana E. (1942) Il valore delle leggi statistiche nella fisica e nelle scienze sociali, *Scientia*, **36**, 58–66.

Malthus Th.R. (1798) *Essay on the Principle of Population as it Affects the Future Improvement of Society with Remarks on the Speculation of Mr. Godwin, M. Condorcet, and Other Writers*, Johnson, London [available on line at the site http://socserv2.mcmaster.ca/~econ/ugcm/3ll3/]

Mandelbrot B. (1975) *Les objets fractals*, Flammarion, Paris.

Mandelbrot B. (1986) Fractals and the Rebirth of Iteration Theory, in Peitgen H.O., Richter P.H. *The Beauty of Fractals: Images of Complex Dynamical Systems*, Springer-Verlag, Berlin, 151–160.

Mandelbrot B. (1997a) *Fractales, hasard et finance*, Flammarion, Paris.

Mandelbrot B. (1997b) *Fractals and Scaling in Finance: Discontinuity, Concentration, Risk*, Springer Verlag, Berlin.

Mandelbrot B. (1999a) *Multifractals and 1/f Noise: Wild Self-affinity in Physics*, Springer Verlag, Berlin.

Mandelbrot B. (1999b) A Multifractal Walk Down Wall Street, *Scientific American*, February, 50–53.

Manneville P., Pomeau Y. (1980) Different Ways to Turbulence in Dissipative Dynamical Systems, *Physica*, **1D**, 219–226.

Mantegna R.N. (2001) Fisica e mercati finanziari, *Le scienze*, **394**, June, 92–97.

Mantegna R.N., Stanley H.E. (2000) *An Introduction to Econophysics: Correlations and Complexity in Finance*, Cambridge University Press, New York.

Marshaal M. (ed.) (2000) Bourbaki, une société secrète de mathématiciens, *Pour la science*, Dossier, 2, Paris.

Martinet A. (1960) *Éléments de linguistique générale*, Librairie Armand Colin, Paris.

Marx K. (1864) *Das Kapital*, Verlag Otto Meissner, Hamburg [English edition: (1942) *Capital*, Penguin, London; available in German on the website http://www.mlwerke.de/me/default.htm; available in English on the website http://www.marxists.org/archive/marx/*works*/1867-c1].

May R.M. (1973) *Stability and Complexity in Model Ecosystems*, Princeton University Press, Princeton, New Jersey.

May R.M. (1976) Simple Mathematical Models with Complicated Dynamics, *Nature*, **261**, 459–467.

May R.M., Leonard W.J. (1975) Nonlinear Aspects of Competition between Three Species, *Siam Journal of Applied Mathematics*, **29**, 243–253.

Medio A. (1992) *Chaotic Dynamics. Theory and Applications to Economics*, Cambridge University Press, Cambridge.

Medio A., Lines M. (2001) *Nonlinear Dynamics: A Primer*, Cambridge University Press, Cambridge.

Meyer-Spasche R. (1999) *Pattern Formation in Viscous Flows: The Taylor Couette Problem and Rayleigh-Bénard Convection*, International Series in Numerical Mathematics, Springer Verlag, Heidelberg.

Moon F.C. (1992) *Chaotic and Fractal Dynamics*, Wiley, New York.

Mullin T. (1991) Chaos in Physical Systems, in Crilly A.J., Earnshaw R.A., Jones H. (eds.) *Fractals and Chaos*, Springer Verlag, New York, 237–245.

Nagel E., Newman J.R. (1958) *Gödel's Proof*, New York University Press, New York.

Neftci S. (2000) *Introduction to the Mathematics of Financial Derivatives*, Academic Press, New York.

Nelson E. (1977) Internal set theory: A New Approach to Nonstandard Analysis, *Bulletin American Mathematical Society*, **83**, 1165–1198 [available on the website www.math.princeton.edu/~nelson/books.html].

Newton I. (1687) *Philosophiae naturalis principia mathematica*, Royal Society, London [reprinted in Latin: (1871) *Philosophiae naturalis principia mathematica*, Thomson Blackburn, Glasgow; English edition: (1968) *The Mathematical Principles of Natural Philosophy*, Dawson, London; English translation by Andrew Motte (1729), partly available on the website http://gravitee.tripod.com/PRINCIPIA.HTM].

Nicolis G., Prigogine I. (1987) *Exploring Complexity. An Introduction*, Freeman, New York.

Nijkamp P., Leitner H., Wrigley N. (eds.) (1985) *Measuring the Unmeasurable. Analysis of Qualitative Spatial Data*, Martinus Nijhoff, The Hague.

Nijkamp P., Reggiani A. (1990) An Evolutionary Approach to the Analysis of Dynamic Systems with Special Reference to Spatial Interaction Model, *Sistemi urbani*, **12**, 95–112.

Nijkamp P., Reggiani A. (1991) Chaos Theory and Spatial Dynamics, *Journal of Transport Economics and Policy*, **25**, 81–96.

Nijkamp P., Reggiani A. (1992a) *Interaction, Evolution and Chaos in Space*, Springer Verlag, Berlin.

Nijkamp P., Reggiani A. (1992b) Spatial Competition and Ecologically Based Socio-Economic Models, *Socio-Spatial Dynamics*, **3**, 89–109.

Nijkamp P., Reggiani A. (eds.) (1993) *Nonlinear Evolution of Spatial Economic Systems*, Springer Verlag, Berlin.

Nijkamp P., Reggiani A. (1998) *The Economics of Complex Spatial Systems*, Elsevier, Amsterdam.

Nijkamp P., Rietveld P., Voogd H. (1991) *The Use of Multicriteria Analysis in Physical Planning*, Elsevier, Amsterdam.

Orishimo I. (1987) An Approach to Urban Dynamics, *Geographical Analysis*, **19**, 200–210.

Orléan A. (1999) À quoi servent les économistes?, *Pour la science, Dossier: Les mathématiques sociales*, July, 4–6.

Osborne A.R., Provenzale A. (1989) Finite Correlation Dimension for Stochastic Systems with Power-law Spectra, *Physica*, **35D**, 357.

Ott E. (1993) *Chaos in Dynamical Systems*, Cambridge University Press, Cambridge.
Overman E. (1996) The New Science of Management, Chaos and Quantum Theory and Method, *Journal of Public Administration Research and Theory*, **6**, 75–89.
Paelinck J. (1992) De l'économétrie spatiale aux nouvelles dynamiques spatiales, in Derycke P.-H. (sous la direction de) *Espace et dynamiques territoriales*, Economica, Paris, 137–174.
Paelinck J. (1996) Économétrie urbaine dynamique, in Huriot J.-M., Pumain D. (sous la direction de) *Penser la ville*, Anthropos, Paris, 91–106.
Palmer R.G., Arthur W.B., Holland J.H., Le Baron B., Tayler P. (1994) Artificial Economic Life: A Simple Model of Stockmarket, *Physica*, **75D**, 264–274.
Pareto V. (1896) *Cours d'économie politique*, Geneva.
Parisi G. (1998) *Statistical Field Theory*, Perseus Books, Reading, Massachusetts.
Parks P.C. (1992) A.M. Lyapunov Stability Theory – 100 years on, *IMA Journal of Mathematics Control Information*, **9**, 4, 275–303.
Pearl R. (1925) *The Biology of Population Growth*, Knopf, New York.
Pearl R., Reed L.J. (1920) On the Rate of Growth of the Population of the United States since 1790 and its Mathematical Representation, *Proceedings of the National Academy of Sciences, USA*, **6**, 275–288.
Peitgen H.O., Richter P.H. (1986) *The Beauty of Fractals: Images of Complex Dynamical Systems*, Springer Verlag, Berlin.
Peitgen H.O., Saupe D., Jürgens H. (1992) *Chaos and Fractals: New Frontiers of Science*, Springer Verlag, Heidelberg.
Penrose R. (1989) *The Emperor's New Mind*, Oxford University Press, Oxford [Italian edition: (1992) *La mente nuova dell'imperatore*, Rizzoli, Milan].
Penrose R. (1994) *Shadows of the Mind*, Oxford University Press, Oxford.
Perko L. (1991) *Differential Equations and Dynamical Systems*, Springer Verlag, New York.
Pesenti Cambursano O. (ed.) (1967) *Opere di Pierre Simon Laplace*, UTET, Turin.
Peters E.E. (1994) *Fractal Market Analysis*, Wiley, New York.
Peters E.E. (1996) *Chaos and Order in the Capital Markets*, Wiley, New York.
Peterson I. (1993) *Newton's Clock*, Freeman, New York.
Phillips A.W. (1954) Stabilisation Policy in a Closed Economy, *Economic Journal*, **64**, 290–323.
Phillips A.W. (1957) Stabilisation Policy and the Time-form of Lagged Responses, *Economic Journal*, **67**, 265–277.
Pickover C. (1990) *Computers, Chaos, Pattern and Beauty – Graphics from an Unseen World*, St Martin's Press, New York [new edition (2001) Dover, New York].
Pickover C. (1994) *Chaos in Wonderland*, St. Martin's Press, New York.
Pitfield D.E. (ed.) (1984) *Discrete Choice Models in Regional Science*, Pion, London.
Poincaré H. (1899) *Les méthodes nouvelles de la mécanique céleste*, Gauthier-Villars, Paris.
Poincaré H. (1908) *Science et méthode*, Flammarion, Paris.
Pomeau Y., Manneville P. (1980) Intermittent Transition to Turbulence in Dissipative Dynamical Systems, *Communications in Mathematical Physics*, **74**, 189–197.
Portugali J. (2000) *Self-Organization and the City*, Springer Verlag, Berlin.
Prigogine I. (1962) *Non-Equilibrium Statistical Mechanics*, Wiley, New York.
Prigogine I. (1980) *From Being to Becoming*, Freeman, S. Francisco.
Prigogine I. (1993) *Le leggi del caos*, Lezioni italiane, Laterza, Rome.
Prigogine I., Stengers I. (1979) *La nouvelle alliance*, Gallimard, Paris.
Prigogine I., Stengers I. (1984) *Order out of Chaos*, Bantam Books, New York.
Prigogine I., Stengers I. (1988) *Entre le temps et l'éternité*, Librairie Arthème Fayard, Paris.

Pumain D. (1997) Ricerca urbana e complessità, in Bertuglia C.S., Vaio F. (eds.) *La città e le sue scienze*, Volume 2, *Le scienze della città*, Franco Angeli, Milan, 1–45 [English edition: (1998) Urban Research and Complexity, in Bertuglia C.S., Bianchi G., Mela A. (eds.) *The City and its Sciences*, Physica-Verlag, Heidelberg, 323–361].

Pumain D., Sanders L., Saint-Julien Th. (1989) *Villes et auto-organisation*, Economica, Paris.

Puu T. (1990) A Chaotic Model of the Business Cycle, *Occasional Paper Series on Socio-Spatial Dynamics*, 1, 1–19.

Puu T. (2000) *Attractors, Bifurcations and Chaos, Nonlinear Phenomena in Economics*, Springer Verlag, Berlin.

Quetelet A. (1835) *Sur l'homme et le développement de ses facultés, ou essai de physique sociale*, Bachelier, Paris.

Rabino G. (1991) Teoria e modelli di interazione spaziale, in Bertuglia C.S., La Bella A. (eds.) *I sistemi urbani*, Franco Angeli, Milan, 485–508.

Rasband S.N. (1990) *Chaotic Dynamics of Nonlinear Systems*, Wiley, New York.

Rasetti M. (1986) *Modern Methods in Equilibrium Statistical Mechanics*, World Scientific, Singapore.

Ravenstein E.G. (1885) The Laws of Migration, *Journal of the Royal Statistical Society*, 48, 167–235.

Ray Th.S. (1991) An Approach to the Synthesis of Life, in Langton C.G., Taylor C., Farmer J.D., Rasmussen S. (eds.) *Artificial Life II*, Santa Fe Institute Studies in the Science of Complexity, Volume 11, Addison-Wesley, Redwood City, California, 371–408.

Ray Th.S. (1994) Evolution and Complexity, in Cowan G.A., Pines D., Meltzer D. (eds.) *Complexity – Metaphors, Models and Reality*, Santa Fe Institute Studies in the Sciences of Complexity, Proceedings Volume 19, Addison-Wesley, Reading, Massachusetts, 161–176.

Rayleigh, Lord (1902) *Scientific Papers*, Volume 3, Cambridge University Press, Cambridge, 1–14.

Regge T. (1999) *L'universo senza fine*, Mondadori, Milan.

Reggiani A. (1997) Verso la città complessa: approcci e sperimentazioni in economia spaziale, in Bertuglia C.S., Vaio F. (eds.) *La città e le sue scienze*, Volume 4, *Le metodologie delle scienze della città*, Franco Angeli, Milan, 129–162 [English edition: (1998) Towards the Complex City: Approaches and Experiments in Spatial Economics, in Bertuglia C.S., Bianchi G., Mela A. (eds.) *The City and its Sciences*, Physica-Verlag, Heidelberg, 797–824].

Reggiani A. (ed.) (2000) *Spatial Economic Science*, Springer Verlag, Berlin.

Reilly W.J. (1931) *The Law of Retail Gravitation*, G.P. Putnam and Sons, New York.

Ricardo D. (1815) *Essay on the Influence of a Low Price of Corn on the Profits of Stock*, John Murray, London. [available on several websites, e.g.: http://www.oswego.edu/~kane/eco322.html].

Riganti R. (2000) *Biforcazioni e caos*, Levrotto e Bella, Turin.

Robbins P., Coulter M. (1998) *Management*, Prentice Hall, Upper Saddle River, New Jersey.

Robert R. (2001) L'effet papillon n'existe plus!, *Pour la science*, 283, May, 28–35.

Robinson A. (1966) *Non-standard Analysis*, North-Holland, Amsterdam [revised edition: (1996) Princeton University Press, Princeton, New Jersey].

Robinson C. (1999) *Dynamical Systems: Stability, Symbolic Dynamics and Chaos*, CRC Press, Boca Raton, Florida.

Rössler O.E. (1976a) An Equation for Continuous Chaos, *Physics Letters A*, 57, 397.

Rössler O.E. (1976b) Chemical Turbulence: Chaos in a Small Reaction-Diffusion System, *Zeitschrift für Naturforschung*, 31, 1168–1172.

Ruelle D. (1991) *Hasard et chaos*, Éditions Odile Jacob, Paris.
Ruelle D., Takens F. (1971) On the Nature of Turbulence, *Communications in Mathematical Physics*, 20, 167–192.
Russell B. (1903) *The Principles of Mathematics*, Norton, New York.
Saari D.G., Xia Zh. (1995) Off to Infinity in Finite Time, *Notices of American Mathematical Society*, 42, 5, 538–546 [available on the website http://www.ams.org/notices/199505/saari-2.pdf].
Saha P., Strogatz S.H. (1995) The Birth of Period Three, *Mathematics Magazine*, 68, 42–47.
Samuelson P.A. (1939) Interactions Between the Multiplier Analysis and the Principle of Acceleration, *Review of Economics and Statistics*, 21, 75–78.
Samuelson P.A. (1947) *Foundations of Economic Analysis*, Harvard University Press, Cambridge, Massachusetts.
Sanders L. (1992) *Systèmes de villes et synergétique*, Anthropos, Paris.
Sangalli A. (1998) *The Importance of Being Fuzzy*, Princeton University Press, Princeton New Jersey.
Sharkovsky, A.N. (1964) Co-Existence of Cycles of a Continuous Mapping of a Line onto Itself, *Ukrainian Mathematical Journal*, 16, 61–71.
Schroeder M. (1991) *Fractals, Chaos, Power Laws: Minutes from an Infinite Paradise*, Freeman, New York.
Schumpeter J.A. (1934) *The Theory of Economic Development*, Harvard University Press, Cambridge, Massachusetts.
Schumpeter J.A. (1939) *Business Cycles*, McGraw-Hill, New York.
Schuster H.G. (1984) *Deterministic Chaos. An Introduction*, Physik-Verlag, Weinheim.
Segrè E. (1971) *Enrico Fermi, fisico*, Zanichelli, Bologna [English edition: (1995) *Enico Fermi, Physicist*, University of Chicago Press, Chicago].
Sen A., Smith T.E. (1995) *Gravity Models of Spatial Interaction Behavior*, Springer Verlag, Heidelberg.
Serra R., Zanarini G. (1986) *Tra ordine e caos*, CLUEB, Bologna.
Serra R., Zanarini G. (1990) *Complex Systems and Cognitive Processes*, Springer Verlag, Berlin.
Shapiro S. (2000) *Thinking about Mathematics*, Oxford University Press. Oxford.
Shishikura M. (1994) The Boundary of the Mandelbrot Set has Hausdorff Dimension Two, *Astérisque*, 7, 222, 389–405.
Siegler L.E. (2002) *Fibonacci's Liber Abaci. A Translation into Modern English of Leonardo Pisano's Book of Calculation*, Springer Verlag, Heidelberg.
Simmons G.F. (1963) *Introduction to Topology and Modern Analysis*, McGraw-Hill International Student Edition, Singapore.
Simó C. (2000) New Families of Solutions in N-Body Problems, Preprints 3rd European Congress of Mathematics, Barcelona, 10–14 July.
Smale S. (1967) Differentiable Dynamical Systems, *Bulletin of the American Mathematical Society*, 73, 747–817.
Smith D.E. (1958) *History of Mathematics*, Dover, New York.
Smith M.J. (1974) *Models in Ecology*, Cambridge University Press, London.
Smullyan R. (1992) *Satan, Cantor, and Infinity, and Other Mind-Boggling Puzzles*, Alfred A. Knopf, New York.
Sokal A., Bricmont J. (1997) *Impostures intellectuelles*, Éditions Odile Jacob, Paris [English edition (1998): *Fashionable Nonsense: Postmodern Intellectuals' Abuse of Science*, Picador, New York.

Sonis M. (1983a) Competition and Environment: A Theory of Temporal Innovation Diffusion, in Griffith D.A., Lea A.C. (eds.) *Evolving Geographical Structures*, Martinus Nijhoff, The Hague, 99–129.

Sonis M. (1983b) Spatio-Temporal Spread of Competitive Innovations: An Ecological Approach, *Papers of the Regional Science Association*, **52**, 159–174.

Sonis M. (1984) Dynamic Choice of Alternatives, Innovation Diffusion and Ecological Dynamics of the Volterra–Lotka Model, *London Papers in Regional Science*, **14**, 29–43.

Sonis M. (1988) Discrete Time Choice Models Arising from Innovation Diffusion Dynamics, *Sistemi urbani*, **10**, 93–107.

Sonis M. (1990a) From 'Homo Economicus' to 'Homo Socialis' in Innovative Diffusion and Dynamic Choice Processes, *Horizons in Geography*, **28–29**, 19–27.

Sonis, M. (1990b) Logistic Growth: Models and Applications to Multiregional Relative Dynamics and Competition of Socio-Economic Stocks, *Sistemi urbani*, **1**, 17–46.

Sonis M. (1991) A Territorial Socio-ecological Approach in Innovation Diffusion Theory: Socio-ecological and Economic Interventions of Active Environment into Territorial Diffusion of Competitive Innovation, *Sistemi urbani*, **13**, 29–59.

Sonis M. (1995) A Territorial Socio-ecological Approach to Innovation Diffusion, Schumpeterian Competition and Dynamic Choice, in Bertuglia C.S., Fischer M.M., Preto G. (eds.) (1995) *Technological Change, Economic Development and Space*, Springer Verlag, Berlin, 34–74.

Sonis M., Dendrinos D. (1990) Multi-stock, Multi-location Relative Volterra–Lotka Dynamics Are Degenerate, *Sistemi urbani*, **12**, 7–15.

Sprott J.C. (1993) *Strange Attractors: Creating Patterns in Chaos*, Henry Holt, New York.

Stacey R. (1996) *Complexity and Creativity in Organizations*, Berret Koehler, San Francisco, California.

Staricco L., Vaio F. (2003) Cooperazione dalla competizione: l'approccio della complessità e l'emergenza dalla cooperazione tra gli attori urbani, in Fusco Girard L., Forte B., Cerreta M., De Toro P., Forte F. (eds.) *L'uomo e la città, Verso uno sviluppo umano sostenibile*, Proceedings of the International Congress 'The Human Being and the City: Towards a Human and Sustainable Development', Naples, 6–8 September 2000, Franco Angeli, Milan, 238–254.

Stephenson J.W., Wang Y. (1990) Numerical Solution of Feigenbaum's Equation, *Applied Mathematics Notes*, **15**, 68–78.

Stephenson J.W., Wang Y. (1991a) Relationships between Eigenfunctions Associated with Solutions of Feigenbaum's Equation, *Applied Mathematics Letters*, **4**, 53–56.

Stephenson J.W., Wang Y. (1991b) Relationships between the Solutions of Feigenbaum's Equation, *Applied Mathematics Letters*, **4**, 37–39.

Stewart I. (1981) *Concepts of Modern Mathematics*, Penguin, London.

Stewart I. (1989) *Does God Play Dice? The New Mathematics of Chaos*, Penguin, London.

Strogatz S.H. (1994) *Nonlinear Dynamics and Chaos*, Addison Wesley, Reading, Massachusetts.

Suber P. (1998) Infinite Reflections, *St John's Review*, **44**, 2, 1–59.

Swinney H.L. (1983) Observations of Order and Chaos in Nonlinear Systems, in Campbell D.K., Rose H.A. (eds.) *Order and Chaos*, North-Holland, Amsterdam, 3–15.

Swinney H.L. (1985) Observations of Complex Dynamics and Chaos, in Cohen E.G.D. (ed.) *Fundamental Problems in Statistical Mechanics*, VI, Elsevier, New York, 253–289.

Tabor M. (1989) *Chaos and Integrability in Nonlinear Dynamics: An Introduction*, Wiley, New York.

Tanner J.T. (1975) The Stability and the Intrinsic Growth Rates of Prey and Predator Populations, *Ecology*, **56**, 855–867.

Tarski A. (1967) Truth and Proof, *Scientific American*, **220**, 63–77.

Tarski A. (1975) *Logic Semantics Metamathematics. Papers from 1923 to 1938*, Clarendon Press, Oxford.

Taylor G.I. (1923) Stability of a Viscous Liquid Contained between Two Rotating Cylinders, *Philosophical Transactions of the Royal Society A*, **223**, 289.

Thiel D. (sous la direction de) (1998) *La dynamique des systèmes*, Hermès, Paris.

Thom R. (1972) *Stabilité structurelle et morphogénèse*, W.A. Benjamin Inc., Massachusetts [English edition: (1975) *Structural Stability and Morphogenesis*, Westview Press, Reading, Massachusetts].

Tikhonov A.N., Arsenin V.Ya. (1979) *The Methods for Solving Ill-Posed Problems*, Nauka, Moscow.

Tikhonov A.N., Goncharsky A.V. (eds.) (1987) *Ill-Posed Problems in the Natural Sciences*, Mir, Moscow.

Tricomi F.G. (1965) *Lezioni di analisi matematica*, Volume 1, CEDAM, Paduva.

Tufillaro N.B., Reilly J., Abbott T. (1992) *An Experimental Approach to Nonlinear Dynamics and Chaos*, Addison-Wesley, Reading, Massachusetts [available on the website http://www.drchaos.net/drchaos/Book/node2.html].

Turing A.M. (1937) On Computable Numbers, with an Application to the Entscheidungs-problem, *Proceedings of the London Mathematical Society*, **42**, 230–265.

Van Dalen D. (1999) *Mystic, Geometer and Intuitionist: the Life of L.E.J. Brouwer*, Oxford University Press, Oxford.

Van den Bery I. (1990) *Principles of Infinitesimal Stochastic and Financial Analysis*, Word Scientific, Singapore.

Van der Pol B.L. (1927) Forced Oscillations in a Circuit with Non-linear Resistance, *Philosophical Magazine*, **3**, 65.

Van der Pol B.L., Van der Mark J. (1928) The Heartbeat Considered as a Relaxation Oscillation, and an Electrical Model of the Heart, *The London, Edinburgh and Dublin Philosophical Magazine and Journal of Science*, **6**, 763–775.

Van Sigt W.P. (1990) *Brouwer's Intuitionism*, Elsevier-North-Holland, Amsterdam.

Verhulst P.F. (1838) Notice sur la loi que la population suit dans son accroissement, *Correspondence Mathématique et Physique*, **10**, 113–121.

Verhulst P.F. (1850) À Quetelet, Pierre Francois Verhulst, *Annuaire de l'Académie royale des science de Belgique*, **16**, 97–124.

Volterra V. (1926) Variazioni e fluttuazioni del numero d'individui in specie animali conviventi, *Memorie della Regia Accademia Nazionale dei Lincei*, Serie VI, Volume 2, 31–113.

Von Haeseler F., Peitgen H.O. (1988) *Newton's Method and Complex Dynamical Systems*, Acta Applied Mathematics, **13**, 3–58 [also in: Peitgen H.O. (ed.) (1989) *Newton's Method and Dynamical Systems*, Kluwer, Dordrecht].

Vulpiani A. (1994) *Determinismo e caos*, La Nuova Italia Scientifica, Rome.

Waldrop W.M. (1992) *Complexity, The Emerging Science at the Edge of Order and Chaos*, Viking, London.

Walras L. (1874) *Élements d'économie pure*, L. Corbaz, Geneva.

Weick K. (1995) *Sensemaking in Organizations*, Sage, Thousand Oaks, California.

Weidlich W. (1991) Physics and Social Science, the Approach of Synergetics, *Physics Reports*, **204**, 1–163.

Weidlich W., Haag G. (1983) *Concepts and Models of a Quantitative Sociology*, Springer Verlag, Berlin.
Weidlich W., Haag G. (1988) *Interregional Migration, Dynamic Theory and Comparative Results*, Springer Verlag, Berlin.
Weyl H. (1918) *Kontinuum* [edizione in English, trans. Pollard S., Bole Th. (1987): *The Continuum, a Critical Examination of the Foundation of Analysis*, The Thomas Jefferson University Press, Kirksville, Missouri; new edition: (1994) Dover, New York].
Weyl H. (1949) *Philosophy of Mathematics and Natural Science*, Princeton University Press, Princeton, New Jersey.
Wigner E.P. (1960) The Unreasonable Effectiveness of Mathematics in the Natural Sciences, *Communications on Pure and Applied Mathematics*, **13**, 1–14 [available on several websites, e.g.: http://nedwww.ipac.caltech.edu/level5/March02/Wigner/Wigner.html].
Wilder R.L. (1973) *Evolution of Mathematical Concepts*, Wiley, New York.
Wilmott P., Howison S., Dewynne J. (1995) *The Mathematics of Financial Derivatives*, Cambridge University Press, Cambridge.
Wilson A.G. (1967) A Statistical Theory of Spatial Distribution Models, *Transportation Research*, **1**, 253–269.
Wilson A.G. (1970) *Entropy in Urban and Regional Modelling*, Pion, London.
Wilson A.G. (1974) *Urban and Regional Models in Geography and Planning*, Wiley, London.
Wilson A.G. (1981) *Catastrophe Theory and Bifurcation*, Croom Helm, London.
Wolf A. (1984) Quantifying Chaos with Lyapunov Exponents, in Holden A.V. (ed.) *Non-linear Scientific Theory and Applications*, Manchester University Press, Manchester.
Wolf A., Swift J.B., Swinney H.L., Vasano J.A. (1985) Determining Lyapunov Exponents from a Time Series, *Physica*, **16D**, 285–317.
Wolfram S. (1994) *Cellular Automata and Complexity*, Addison-Wesley, Reading, Massachusetts.
Zadeh L.A. (1965) Fuzzy Sets, *Information and Control*, **8**, 338–356.
Zanarini G. (1996) Complessità come modo di pensare il mondo, in Cerrato S. (ed.) *Caos e complessità*, CUEN, Naples, 7–29.
Zeldovich Ya.B., Myškis A.D. (1976) *Elements of Applied Mathematics*, Mir, Moscow.
Zermelo E. (1908) Untersuchungen über die Grundlagen der Mengenlehre I, *Mathematische Annalen*, **65**, 261–281 [English edition, in Van Heijenoort J. (1967) *From Frege to Gödel: A Source Book in Mathematical Logic, 1879–1931*, Harvard University Press, Cambridge, Massachusetts].
Zhang W.-B. (1991) *Synergetic Economics*, Springer Verlag, Berlin.
Zhang W.-B. (1997) The Complexity of Economic Evolution, in Fang F., Sanglier M. (eds.) *Complexity and Self-Organization in Social and Economic Systems*, Springer Verlag, Berlin, 57–75.
Zurek W.H. (1989) Algorithmic Information Content, Church–Turing Thesis, Physical Entropy, and Maxwell's Demon, in Jen E. (ed.) *1989 Lectures in Complex Systems*, Santa Fe Institute Studies in the Sciences of Complexity, Lectures Volume 2, Addison-Wesley, Redwood City, California, 49–65.

Subject index

Age of Enlightenment; 4; 200; 203; 206; 208;
   classical science; 13–15; 203–209; 211
   mechanism of; 200
analysis
   non-standard; 279
anthropic principle; 198
antimatter; 30
artificial stock market of Santa Fe; 249
astronomy; 16; 110; 200; 202; 203; 204; 252
attractor(s)
   chaotic; 103; 112–122; 126; 128; 147; 148; 150; 171; 240; 245
   dimension; 147–150;
   Lorenz; 116; 117; 128; 147; 148
   Rössler; 118; 245
   strange; 100; 111–119; 123; 126; 146; 245
   vs. chaotic; 112; 147; 179
autonomous equation/system/model
   definition; 27; 37; 145
average man; 4
axiom of choice (Zermelo's); 254

baker's map; 119; 120; 121; 122
Black and Scholes formula; 247
Bourbaki (les), 265, 272; 288; 296
butterfly effect; 17; 103; 207; 281

Cambrian explosion; 233
capturing software; 234
carrying capacity; 80; 81; 82; 87; 89; 92; 94; 97; 154; 155; 156; 157; 159
cat map (Arnold's); 112; 113; 120
cellular automata; 37; 195; 222
Charles Sturt University (CSU); 220
Complexity International; 220
complex systems
   emergence in; 212; 221; 224; 226; 227; 277; 281
   properties; 219–230
   self-organization in; 212; 215–227; 232; 237; 239; 240; 243; 244; 277; 281
control complexity; 227; 251
Cours d'analyse de l'École Polytechnique (Cauchy); 318
Cours d'économie politique (Pareto); 62
cybernetics; 212

Das Kapital (Marx); 322
design complexity; 227; 251
Dialogo sopra i due massimi sistemi del mondo, tolemaico e copernicano (Galilei); 263
dimension
   correlation; 148; 149
      method of Grassberger and Procaccia; 183
   Hausdorff–Besicovich; 112; 147–148; 274
double star system; 278

economics
   classical; 259–261; 302
   marginalist; 61
   mathematical; 60–64
   neoclassical; 15; 62–64; 261
econophysics; 151; 248

Éléments de mathématique (Bourbaki); 265
Elliott waves; 50
Encyclopédie; 208
Entscheidungsproblem; 267; 301
equation
  Duffing; 79; 104
  Schrödinger; 214
  Van der Pol; 79; 301
  Verhulst; 102; 191–198
Essay on the Influence of a Low Price of Corn on the Profit of Stock (Ricardo); 322
Essay on the Principle of Population (Malthus); 153
Essai philosophique sur les probabilités (Laplace); 203–204
Essai sur l'application de l'analyse (Condorcet); 203

feedback; 11; 12; 33; 74; 75; 87; 129; 155; 156; 209–226; 239; 240; 247; 249;251; 253
Feigenbaum
  numbers (Feigenvalues); 177–178
  tree; 172–180
financial instruments; 246; 247; 248
first integral; 41; 42
fixed point(s)
  centre; 67
  focal point
    stable; 70; 89
    unstable; 70; 89
  nodal point
    stable; 60; 70
    unstable; 62; 64; 70
  of a function; 162–164
  saddle point; 63; 64; 65; 85
functional response of the predator; 82; 83; 85
futures; 50; 246–248
fuzzy logic/sets; 260

holism; 15; 18; 218; 219; 225; 229; 241

homo oeconomicus; 244
homo socialis; 244
human language; 233

independence of irrelevant alternatives; 184
infinity
  countable/uncountable; 125; 254; 281; 282
  actual infinite; 253; 280
  potential infinite; 253; 280
inflation; 11; 12
Interjournal; 220
invisible hand (A. Smith's); 209; 210; 240
iteration method; 162–164

Jupiter and Saturn (resonance between the planets' revolution periods); 137; 155

Kolmogorov–Sinai entropy; 140
Kondratiev waves; 241

La Monadologie (Leibniz); 206
Larsen effect; 129
law(s) of
  gravitation of the moral world (Durkheim); 7
  motion; 7; 19; 37; 39; 40; 41; 42; 111
  Pearl: see logistic function
  planets' orbits (Kepler); 201
  retail gravitation (Reilly); 5
  tendential fall in the profit rate (Marx); 322
  universal gravitation (Newton); 4; 5; 14; 16; 34; 182; 201; 277; 278
Les méthodes nouvelles de la mécanique céleste (Poincaré); 18; 131
life appearance/evolution; 247; 262; 269; 270; 291; 295; 297; 298
limit cycle; 77; 78; 79; 82; 101; 102; 103; 114; 140; 145

linear
  approximation; 12; 29; 33; 44; 45; 127; 132; 139; 211
  causality; 210
  differential equation; 19–22
  map; 104; 105; 106; 115; 121
  model of two interacting populations
    definition and qualitative properties; 54–57
    characteristic equation; 58; 60; 65
    solutions; 57–72
logistic
  function (growth); 5; 8; 98; 100; 156–159; 186
  map
    definition and qualitative characteristics; 12; 45; 48; 102; 126; 129; 145; 152; 278
    dynamics of; 134; 135; 193–219
    example of application to modelling in economics; 308; 310
    example of application to spatial interaction modelling; 224–230
    extended to the complex field; 341
    Lyapunov exponent of; 176
Lyapunov
  exponent(s); 137–145; 151; 178–180; 187; 221
  function; 134
  stability
    definition; 134
    first criterion; 134
    second criterion; 134

Mandelbrot set; 262; 271; 272; 274; 275
market indicators; 250
mathematical truth; 196; 262–264; 268
mathematics (interpretation of)
  computational; 273; 276; 282
  constructivism; 268; 269; 273; 274; 280; 282

finitism; 268
formalism; 263; 265; 277
intuitionism; 268
Platonism; 263; 269; 274; 280
realism; 262; 269
mechanism; 200
metaphor of the world
  cosmic clock; 31; 200; 272
  cosmic computer; 272; 273; 279; 280
  steam engine; 272
model(s)
  Allen (spatial self-organization); 244
  Arthur et al. (financial markets); 249
  Dendrinos (urban area of Madrid); 92
  Dendrinos (urban areas of the United States); 92
  demographic of logistic growth; 196; 197
  diffusion of technological innovation; 197–198; 307
  discrete choice; 183
  evolutionistic Tierra; 235
  Goodwin; 245
  Goodwin–Kalecki–Phillips; 48
  harmonic oscillator; 24; 41–45; 48; 51; 68; 79; 86; 114
  Holling (two interacting prey–predator populations); 82–84; 294
  Holling–Tanner (two interacting prey–predator populations); 84
  joint population–income dynamics; 85
  two populations interacting linearly: see linear model of two interacting populations
  logistic growth: see logistic function
  logit; 184
  Lowry (fluxes in an urban system); 183
  May (two interacting prey–predator populations); 82; 84

Malthus (exponential growth of a population); 29; 30; 73; 162; 168–171; 177; 190–192; 233; 367
niche formation; 121–122
Puu (economic cycle); 245
pendulum
　linear; 23–25
　linear in presence of friction; 25–27
　nonlinear; 34–35
　predictability horizon of; 135; 136; 159; 176–178
Reilly (retail gravitation); 5
Samuelson and Hicks (economic cycle); 48; 245
Sanders (the system of French cities); 244
Sonis (diffusion of innovation); 244
spatial interaction; 183
symmetrical competition between two interacting populations; 94
Verhulst (population growth with limited carrying capacity); 102; 191–198
Volterra–Lotka (prey–predator)
　basic model; 73–76; 241
　variants applied to ecology; 80–85; 96; 97; 101
　variants applied to urban and regional science; 85–94; 186;
modelling
　general properties; 7–10
　linear hypothesis; 193
　role of mathematics in; 195–202
Monash University; 275
multicriteria analysis; 259

neodarwinism; 210
New England Complex System Institute (NECSI); 220
non-countable infinity; 282
numbers
　computable; 266; 269
　continuum; 252–254; 263; 266; 278; 281–283

hyperreal; 279
numerical response of the predator; 82; 83

options; 246; 247

paradox (antinomy) of
　Achilles and the tortoise; 253
　Cantor; 270
　Condorcet; 3
　Epimenides; 268; 269; 277
　Russell; 267
　Zeno of Elea; 253
Permian extinction; 234
phase space
　definition; 37–45
Philosophiae naturalis principia mathematica (Newton); 13; 14; 15; 206; 292; 296
physicalism; 3; 6; 46; 240
physiocracy; 208; 209
Poincaré
　recurrence; 114
　stability criterion; 130
political arithmetics; 6; 259
principle of
　energy conservation; 28; 42; 47
　Gause (competitive exclusion); 95
　Heisenberg (quantum uncertainty); 15; 107; 108; 205
　least action; 207
　second principle of dynamics; 24; 201
　second principle of thermodynamics; 213; 214
　sufficient reason; 18; 206; 207; 211; 212; 222
　superposition of states; 20; 205
Principles of Social Science (Carey); 182
prisoner's dilemma; 237
problem
　direct in modelling; 8–10
　ill-posed; 10
　inverse in modelling; 8–10

many-body (three bodies); 15; 16; 18; 199; 278
well-posed; 10

reductio ad absurdum; 268
reductionism; 15; 18; 192; 193; 222
Regional Science Association International (RSAI); 108; 226

Santa Fe Institute (SFI); 219; 248
Science et méthode (Poincaré); 124
social mathematics; 4; 255
spatial interaction; 5; 6; 9; 14; 182; 183; 185; 186
stability of the solar system; 109–111; 202
state variable
  definition; 8; 36–45
synergetics; 304–307
system
  autonomous; 27; 51; 115; 133
  ergodic; 230
  linear; 54–72

technological innovation; 87; 159; 160; 199; 256
The Analyst; Or, A Discourse Addressed To An Infidel Mathematician (Berkeley); 252
The Laws of Migration (Ravenstein); 225
The Law of Retail Gravitation (Reilly); 8

The Physiocrats (Higgs); 208
The Principles of Mathematics (Russell); 336
The Struggle for Existence (Gause); 120
The Unreasonable Effectiveness of Mathematics in the Natural Science (Wigner); 196
theorem(s) of
  Cantor; 254; 125
  Cohen; 254
  conservation of mechanical energy; 42
  existence and uniqueness of the solutions of a differential equation; 27; 57; 114; 115; 156
  Fourier; 193
  Gödel (incompleteness); 254; 264; 267; 269; 270; 275
  KAM; 110
  Li and Yorke (*Period Three Implies Chaos*); 171
  Liouville; 101
  Poincaré–Bendixon; 114
  Sharkovsky; 171
  Tarski; 264; 265
theory of random utility; 227
trading on line; 248
turbulence in fluids; 32; 33; 111; 215; 240; 278
Turing machine; 267

undecidability; 264; 266; 269; 270; 275; 277; 282

Name index

Abarbanel H.D.I.; 112; 141; 144; 285
Abbott T.; 171; 172; 301
Accardi L.; 251; 285
Agnes C.; 277; 285
Albeverio S.; 279; 285
Albin P.S.; 220; 285
Allen P.M.; 159; 160; 216; 244; 285
Allen R.G.D.; 48; 240; 285
Alligood K.; 53; 285
Alvesson M.; 223; 285
Anderson C.D.; 30
Anderson Ph.W.; 245; 285; 290
Anderson R.M.; 279; 285
Andrecut M.; 179; 285
Anosov D.V.; 53; 54; 132; 285
Arditi R.; 73; 85; 285; 294
Arino O.; 73; 85; 294
Aristotle; 201
Arnold V.I.; 15; 37; 53; 54; 74; 110; 112; 113; 120; 132; 285; 286
Arrow K.J.; 245; 285; 290
Arsenin V.Ya.; 9; 10; 286; 300
Arthur W.B.; 232; 233; 234; 245; 248; 249; 286; 297
Ashby W.R.; 212; 286
Atlan H.; 207; 286
Axelrod R.; 216; 237; 286

Bachgus R.; 87; 286
Bailey D.; 177; 288
Bak P.; 192; 286
Barber C.L.; 233; 235; 286
Barkley Rosser J.; 33; 172; 239; 245; 286
Barrow J.D.; 197; 199; 255; 264; 265; 269; 270; 272; 286

Barrow-Green J.; 16; 286
Batten D.F.; 49; 192; 199; 239; 249; 286; 290; 291
Bayly B.; 32; 286
Bechhoeffer J.; 171; 286
Bellacicco A.; 172; 286
Bellomo N.; 7; 10; 53; 54; 114; 128; 132; 190; 287
Beltrami E.; 53; 54; 74; 80; 81; 82; 114; 132; 172; 287
Beman W.W.; 289
Benacerraf P.; 262; 264; 287
Ben-Akiva M.; 184; 287
Bender C.M.; 35; 54; 287
Bendixon I.O.; 114
Benhabib J.; 150; 239; 245; 287
Berkeley G.; 14; 252; 287
Bernoulli D.; 4; 32; 204
Bertuglia C.S.; 5; 96; 116; 183; 192; 199; 228; 244; 260; 271; 285; 286; 287; 288; 290; 291; 295; 297; 298; 300
Besicovich A.S.; 112; 147; 148; 274; 287
Bianchi G.; 244; 285; 287; 290; 297; 298
Birkhoff G.D.; 15; 287
Black F.; 247; 287
Blair D.; 4; 287
Blum L.; 228; 287
Bohm D.; 14; 210; 288
Bohr N.; 15
Bole Th.; 301
Bollen J.; 291
Boltzmann L.; 15; 213
Boncinelli E.; 31; 128; 197; 276; 288

Name index

Boole G.; 46
Boon F.; 285
Born M.; 282; 288
Borwein J.; 177; 288
Bottazzini U.; 16; 18; 31; 110; 128; 131; 132; 197; 276; 288
Bourbaki N.; 265; 272; 288; 295
Boyce D.E.; 183; 288
Boyer C.B.; 20; 190; 255; 317; 329; 359
Bricmont J.; 6; 299
Briggs K.M.; 177; 288
Brock W.; 150; 239; 288
Brouwer L.E.J.; 268; 288
Bruner J.S.; 192; 288
Bruschini Vincenzi L.; 248; 288
Burke Hubbard B.; 110; 288
Byrne D.; 223; 288

Cajori F.; 268; 288
Camagni R.; 87; 288
Çambel A.B.; 74; 141; 148; 172; 180; 288
Campbell D.K.; 300
Campisi D.; 9; 183; 185; 288
Cantor G.; 125; 170; 253; 254; 266; 268; 269; 270; 282; 288
Carey H.C.; 182; 288
Carson E.R.; 226; 228; 291
Carter B.; 198
Casti J.L.; 7; 11; 37; 111; 148; 192; 227; 271; 288
Catherine of Russia; 208
Cauchy A.L.; 253; 288
Cavalieri B.; 252
Cellucci C.; 262; 288
Cerrato S.; 302
Cerreta M.; 300
Ceruti M.; 286; 288
Changeux J.-P.; 31; 197; 288
Chavel I.; 101; 289
Chiarella C.; 245; 289
Chossat P.; 32; 289
Christensen K.S.; 223; 289
Ciliberto S.; 290

Clarke G.P.; 183; 287
Clausius R.; 15; 213
Cohen E.G.D.; 300
Cohen I.B.; 2; 4; 14; 46; 47; 228; 289
Cohen J.; 192; 199; 211; 258; 271; 289
Cohen P.; 254
Collatz L.; 164; 289
Condorcet J.-A. de Caritat; 3; 4; 154; 289
Connes A.; 31; 197; 288
Couette M.; 32
Coulter M.; 210; 298
Cournot A.A.; 4
Coveney P.; 192; 289
Cowan G.A.; 286; 294; 298
Craig J.; 14
Crépel P.; 3; 4; 208; 289
Crilly A.J.; 112; 289; 294; 296
Crutchfield J.P.; 114; 116; 122; 289
Curie P.; 188
Cutland N.J.; 266; 289
Czyz J.; 112; 289

D'Ancona U.; 73
Daganzo C.; 184; 289
Dagenais M.G.; 184; 292
Darwin Ch.; 132; 210; 211; 235; 237; 289
Darwin G.; 132
Davies P.C.W.; 222; 289
Davies S.; 160; 289
Dawson J.W.; 292
de Finetti B.; 275; 289
de Palma A.; 285
de Santillana G.; 207; 291
De Toro P.; 300
Dedekind R.; 253, 268, 289
Deledicq A.; 351; 361
Dendrinos D.; 6; 85; 87; 89; 90; 91; 92; 93; 95; 97; 200; 215; 231; 244; 245; 247; 289; 290; 300
Deneubourg J.L.; 285
Derycke P.-H.; 296
Descartes R.; 21; 317

Devaney R.L.; 164; 171; 172; 271; 290
Devlin K.; 4; 254; 267; 290
Dewynne J.; 247; 301
Dicke R.; 198
Diderot D.; 208
Diener F.; 351; 361
Diener M.; 351; 361
Dioguardi G.; 192; 208; 224; 228; 272; 290
Dirac P.A.M.; 15; 30; 198; 290
Ditto W.L.; 112; 290
Dobb M.; 209; 290
Dostoevsky F.M.; 99; 189
Dow Ch.; 49
Drazin P.S.; 172; 290
Duffing G.; 79; 104; 290
Durkheim E.; 4; 14
Durlauf S.N.; 245; 286

Earnshaw R.A.; 112; 289; 294; 296
Eckmann J.-P.; 140; 151; 290
Eco U.; 1
Edmonds B.; 226; 227; 228; 290
Einstein A.; 207; 214; 251; 291
Ekeland I.; 197; 203; 207; 291
Elliott R.N.; 49; 50; 291
Elsgolts L.; 54; 132; 291
Enriques F.; 207; 291
Epimenides the Cretan; 268; 269; 277
Epstein I.R.; 73; 291
Epstein J.M.; 54; 74; 291
Eubank S.; 118; 141; 172; 291
Euclid; 254; 269
Euler L.; 15; 16; 29; 32; 66

Fabbri P.; 286; 288
Falconer K.J.; 112; 170; 172; 271; 291
Fang F.; 245; 291; 302
Farmer J.D.; 118; 120; 141; 148; 172; 289; 291; 298
Feferman S.; 264; 292
Feigenbaum M.; 33; 161; 172; 174; 175; 176; 177; 178; 179; 180; 181; 187; 245; 271; 291

Fels R.; 299
Fenstad J.; 285
Fermat P. de; 207; 252
Fermi E.; 18; 31; 103; 291
Fibonacci (Leonardo Pisano); 153
Fischer M.M.; 159; 199; 287; 291; 300
Fisher I.; 45; 46; 47
Flood R.L.; 226; 228; 291
Forrester J.W.; 223; 291
Forte B.; 300
Forte F.; 300
Fotheringham A.S.; 183; 291
Fourier J.-B.; 4; 10; 32; 110; 193
Frankhauser P.; 112; 291
Franses P.H.; 248; 291
Frege G.; 267
Freiling Ch.; 281; 291
Frenkel A.; 159; 291
Frost A.J.; 50; 291
Fujita M.; 183; 292
Fusco Girard L.; 300

Galilei G.; 1; 2; 24; 252; 263
Gandolfi A.; 192; 220; 229; 292
Gaudry M.J.I.; 184; 292
Gause G.F.; 76; 85; 95; 292
Gell-Mann M.; 192; 228; 259; 292
Gentile G. jr.; 5
Gibbs J.W.; 46
Ginzburg L.R.; 285
Giorello G.; 286; 288
Giusti E.; 31; 252; 263; 265; 292
Glasko V.B.; 9; 292
Gleick J.; 18; 31; 128; 143; 172; 271; 272; 292
Gödel K.; 254; 264; 267; 269; 275; 279; 292
Godwin W.; 153
Goldberg A.L.; 79; 292
Goldblatt R.; 351; 364
Goldfarb W.; 292
Goncharsky A.V.; 9; 10; 286; 292; 300
Goodwin R.M.; 48; 49; 241; 245; 246; 289; 292

Gordon W.B.; 171; 292
Granger G.G.; 31; 292
Grassberger P.; 147; 148; 149; 292
Grebogi C.; 112; 292; 293
Griffith D.A.; 299
Guerrien B.; 256; 293
Guerriero G.; 264; 293
Gullen M.; 14; 293

Haag G.; 244; 293; 301
Hadamard J.; 9; 10; 293
Haken H.; 241; 243; 293
Hallett M.; 253; 254; 293
Hamilton W.R.; 207
Hannon B.; 8; 54; 293
Hasselblatt B.; 141; 294
Hausdorff F.; 112; 147; 148; 274; 293
Havel I.; 293
Heagy J.; 290
Heisenberg W.; 15; 32; 107; 205; 251; 293
Hénon M.A.; 181; 293
Hensher D.A.; 183; 287; 293
Herken R.; 266; 293
Herrmann H.; 223; 293
Hewings G.J.D.; 245; 293
Heylighen F.; 231; 291; 293
Heyting A.; 268; 293
Hicks J.R.; 48; 245; 293
Higgs H.; 208; 293
Highfield R.; 192; 289
Hilbert D.; 263; 264; 265; 267; 268; 269
Hildenbrand W.; 285
Hirsch M.; 114; 293
Hodgson G.M.; 245; 292; 293
Hoegh-Krohn R.; 285
Hofstadter D.; 264; 293
Holden A.V.; 112; 293; 301
Holland J.H.; 216; 223; 237; 286; 293; 297
Holling C.S.; 82; 83; 84; 294
Honig W.M.; 350; 365
Horgan P.; 228; 294

Horrocks G.; 233; 294
Householder A.S.; 164; 294
Howison S.; 247; 301
Hsieh D.; 150; 239; 288
Hubbard J.; 110; 288
Hume D.; 14
Huriot J.-M.; 297

Iooss G.; 32; 289
Isard W.; 8; 108, 226; 365
Israel G.; 8; 21; 73; 74; 77; 79; 128; 197; 294; 326

Jeans J.; 132
Jen E.; 287; 291; 295; 302
Jevons W.S.; 45; 46; 47
Johnson L.W.; 183; 293
Jones H.; 112; 289; 294; 296
Jordan D.W.; 35; 53; 54; 131; 132; 294
Jost Ch.; 73; 85; 294
Jourdain E.B.; 288
Jürgens H.; 112; 148; 297

Kalecki M.; 48; 49; 294
Kant I.; 272
Karlqvist A.; 288
Karlsson C.; 291
Katok A.; 141; 294
Kauffman S.; 216; 226; 233; 234; 235; 237; 238; 239; 294
Kaufmann A.; 260; 294
Keisler J.; 351; 366
Kelvin, Lord (William Thomson); 15; 213
Kepler J.; 201
Keynes J.M.; 240; 242; 245
Kimura Y.; 293
Kline M.; 252; 262; 294
Koch B.P.; 295
Kolmogorov A.N.; 110; 140; 172
Kondratiev N.D.; 241
Koyré A.; 201; 294
Kraft R.L.; 179; 294

Kronecker N.D.; 268
Krugman P.; 183; 239; 245; 292; 294

La Bella A.; 288; 298
Lagrange L.; 15; 16; 109; 110
Lakatos I.; 351; 366
Landau L.D.; 31; 32; 207; 294
Lane D.A.; 245; 286
Langton C.G.; 216; 294; 298
Laplace P.S.; 4; 13; 15; 18; 108; 109; 110; 203; 204; 205; 206; 294
Larsen S.; 129
Laskar J.; 111; 202; 294
Le Baron B.; 150; 239; 286; 288; 297
Le Bras H.; 157; 158; 294
Lea A.C.; 299
Leibniz G.W.; 15; 18; 206; 207; 211; 222; 252; 283
Leitner H.; 259; 296
Lekachman R.; 208; 209; 294
Leonard W.J.; 97; 296
Leonardi G.; 183; 184; 185; 287; 294; 295
Lerman S.R.; 184; 287
Leven R.W.; 52; 295
Levy D.; 223; 224; 295
Lewin R.; 216; 295
Li T.Y.; 171; 178; 295
Lifshitz E.M.; 32; 207; 294
Lindstrøm T.; 285
Lines M.; 141; 296
Liouville J.; 101
Lloyd S.; 228; 295
Lolli G.; 31; 262; 265; 275
Lombardo S.; 199; 287; 291
Lorenz E.N.; 15; 18; 103; 116; 117; 128; 143; 147; 148; 194; 295
Lotka A.; 6; 72; 73; 76; 80; 81; 82; 84; 85; 87; 88; 89; 90; 92; 94; 96; 97; 101; 152; 158; 186; 236; 241; 295
Lowry I.S.; 183; 295
Luxemburg W.A.J.; 350; 356
Lyapunov A.M.; 127; 132; 133; 134; 135; 136; 137; 138; 139; 140; 141; 142; 143; 144; 145; 151; 178; 179; 180; 187; 221; 295
Lyons J.; 233; 295

Madden M.; 293
Maddy P.; 262; 281; 295
Majorana E.; 5; 295
Malthus Th. R.; 21; 22; 55; 129; 134; 142; 153; 154; 155; 188; 295
Mandelbrot B.; 33; 50; 112; 151; 181; 248; 262; 271; 272; 274; 275; 295
Manneville P.; 32; 172; 295; 297
Mantegna R.N.; 151; 248; 295
Marshaal M.; 265; 295
Martellato D.; 286; 290
Martinet A.; 233; 295
Marx K.; 240; 242; 245; 256; 367
Maupertuis P.L.M. de; 207
May R.; 33; 81; 82; 84; 97; 154; 160; 171; 172; 295; 296
Medio A.; 141; 296
Mela A.; 244; 285; 287; 290; 297; 298
Meltzer D.; 286; 294; 298
Mendel G.; 196; 197
Menger C.; 47
Menten M.; 83
Mercier de la Rivière P.-P.; 208
Meyer-Spasche R.; 32; 296
Michaelis L.; 83
Monge G.; 4
Moon F.C.; 18; 53; 54; 109; 120; 122; 132; 141; 148; 172; 201; 296
Moser J.; 110
Mudretsova E.A.; 9; 292
Muhamad M.A.; 112; 293
Mullally H.; 85; 290
Mullin T.; 32; 296
Muntz M.; 293
Myškis A.D.; 54; 164; 302

Nadel L.; 286
Nagel E.; 264; 275; 296
Napier J.; 252
Neftci S.; 247; 296

Nelson E.; 351; 368
Newman J.R.; 264; 275; 296
Newton I.; 4; 5; 13; 14; 15; 16; 18; 24; 25; 26; 27; 34; 47; 111; 128; 182; 200; 201; 202; 203; 208; 209; 210; 211; 213; 214; 252; 272; 283; 288; 296
Nicolis G.; 30; 120; 122; 146; 150; 191; 216; 296
Nijkamp P.; 87; 94; 97; 172; 182; 183; 185; 186; 199; 239; 244; 245; 259; 287; 288; 291; 296

Occelli S.; 286; 287; 290; 295
O'Kelly M.E.; 183; 291
Oliffson Kamphorst S.; 290
Orishimo I.; 87; 296
Orléan A.; 255; 256; 296
Orszag S.; 35; 54; 287
Osborne A.R.; 149; 296
Oscar II, king of Sweden and Norway; 18; 110
Ott E.; 112; 120; 148; 172; 290; 291; 292; 293; 296
Overman E.; 224; 296

Packard N.H.; 289
Paelinck J.; 86; 296; 297
Palmer R.G.; 249; 286; 297
Pareto V.; 45; 46; 47; 297
Parisi G.; 230; 297
Parks P.C.; 132; 297
Parsons Ch.; 292
Pasta J.; 18; 291
Peano G.; 116; 267
Pearl R.; 158; 159; 297
Peitgen H.O.; 112; 148; 164; 271; 295; 297; 301
Pelikan S.; 292
Penrose R.; 199; 262; 264; 266; 271; 274; 297
Perko L.; 53; 54; 297
Pesenti Cambursano O.; 204; 297
Peters E.E.; 248; 297

Peterson I.; 110; 297
Phillips A.W.; 48; 49; 297
Pickover C.; 112; 172; 177; 271; 297
Pines D.; 245; 285; 286; 290; 294; 298
Pitfield D.E.; 183; 297
Planck M.; 199; 282
Plato; 262
Podolsky B.; 251; 291
Poincaré H.; 15; 18; 30; 32; 103; 110; 111; 114; 124; 127; 130; 131; 132; 133; 134; 265; 297
Poisson S.D.; 130
Pollack R.; 4; 287
Pollard S.; 301
Pomeau Y.; 32; 172; 295; 297
Pompe B.; 295
Portugali J.; 244; 297
Prechter R.J.; 50; 291
Preta L.; 286; 288
Preto G.; 199; 287; 300
Preziosi L.; 7; 10; 53; 54; 114; 128; 132; 287
Prigogine I.; 15; 30; 120; 122; 146; 150; 191; 216; 296; 297
Procaccia I.; 148; 149; 292
Provenzale A.; 149; 296
Pumain D.; 73; 89; 95; 243; 293; 297
Putnam H.; 262; 287
Puu T.; 132; 141; 172; 245; 271; 297; 298
Pythagoras; 261

Quesnay F.; 208; 209
Quetelet A.; 4; 298
Quispel G.R.W.; 288

Rabino G.; 5; 9; 183; 185; 260; 287; 295; 298
Rasband S.N.; 141; 148; 172; 177; 298
Rasetti M.; 207; 277; 285; 298
Rasmussen S.; 298
Rauseo S.N.; 290
Ravenstein E.G.; 183; 298
Ray Th.S.; 235; 298

Rayleigh, Lord (John William Strutt); 52; 298
Reed L.J.; 158; 297
Regge T.; 199; 298
Reggiani A.; 87; 94; 97; 172; 182; 183; 185; 186; 239; 244; 245; 295; 296; 298
Reilly J.; 171; 172; 301
Reilly W.J.; 5; 183; 298
Ricardo D.; 256; 370
Richter P.H.; 112; 271; 295; 297
Riegler A.; 291
Rietveld P.; 259; 296
Riganti R.; 54; 132; 141; 148; 172; 298
Rigney D.R.; 79; 292
Robbins P.; 210; 298
Robert R.; 203; 298
Robinson A.; 279; 298
Robinson C.; 141; 298
Rose H.A.; 300
Rosen N.; 251; 291
Rössler O.E.; 118; 245; 298
Ruelle D.; 32; 103; 111; 140; 172; 240; 290; 298
Russell B.; 254; 267; 268; 269; 270; 275; 277; 298
Ruth M.; 8; 54; 293

Saari D.G.; 278; 298
Saha P.; 171; 298
Saint-Exupéry A. de; 191; 270
Saint-Julien Th.; 73; 95; 293; 297
Samuelson P.A.; 3; 47; 48; 49; 240; 245; 298; 299
Sanders L.; 73; 95; 243; 244; 293; 297; 299
Sangalli A.; 260; 299
Sanglier M.; 245; 285; 291; 302
Sauer T.; 53; 285
Saunders P.; 289
Saupe D.; 112; 148; 297
Savage H.T.; 290
Scheinkman J.; 290
Scholes M.; 247; 287

Schrödinger E.; 15; 214; 251
Schroeder M.; 112; 116; 120; 148; 170; 172; 175; 271; 299
Schumpeter J.A.; 47; 48; 240; 241; 242; 245; 299
Schuster H.G.; 172; 179; 299
Screpanti E.; 245; 292; 293
Segrè E.; 18; 299
Sen A.; 14; 183; 299
Serra R.; 172; 299
Shapiro S.; 262; 299
Sharkovsky O.M.; 171; 299
Shaw R.S.; 289
Shefer D.; 159; 183; 288; 291
Shishikura M.; 274; 299
Siegler L.E.; 190; 371
Simmons G.F.; 116; 299
Simò C.; 16; 17; 299
Sinai Ya.G.; 140
Smale S.; 15; 103; 114; 293; 299
Smith A.; 154; 208; 209; 210; 240
Smith D.E.; 153; 299
Smith M.J.; 95; 299
Smith P.; 35; 53; 54; 131; 132; 294
Smith T.E.; 14; 183; 299
Smullyan R.; 263; 270; 299
Sokal A.; 6; 299
Solomon A.I.; 285
Solovay R.N.; 292
Sonis M.; 87; 95; 159; 244; 245; 290; 293; 299; 300
Sonnenschein H.; 285
Spano M.L.; 290
Sprott J.C.; 116; 300
St. Paul; 268
Stacey R.; 224; 300
Stanley H.E.; 46; 151; 248; 295
Staricco L.; 96; 192; 216; 228; 244; 287; 300
Stein D.L.; 286
Stengers I.; 30; 216; 297
Stephenson J.W.; 177; 300
Stewart I.; 31; 192; 199; 211; 258; 262; 271; 289; 300

Stopher P.R.; 287
Strakhov V.N.; 9; 292
Strogatz S.H.; 167; 171; 172; 298; 300
Suber P.; 281; 300
Swift J.B.; 302
Swinney H.L.; 32; 148; 300; 302

Tabor M.; 172; 177; 300
Tadei R.; 287; 295
Takens F.; 32; 103; 111; 172; 298
Taletes; 261
Tanner J.T.; 84; 300
Tarski A.; 264; 300
Tayler P.; 286; 297
Taylor B.; 34; 44
Taylor C.; 298
Taylor G.I.; 32; 300
Thiel D.; 224; 239; 300
Thom R.; 221; 300
Thompson C.J.; 288
Tikhonov A.N.; 9; 10; 286; 292; 300
Tipler F.J.; 199; 286
Torricelli E.; 252
Trappl R.; 293
Tricomi F.G.; 164; 301
Tufillaro N.B.; 171; 172; 301
Turing A.M.; 266; 267; 269; 301

Ulam S.; 18; 291

Vaio F.; 116; 216; 244; 271; 285; 287; 290; 297; 298; 300
Van Dalen D.; 269; 301
Van den Berg I.; 350; 373
Van der Mark J.; 79; 301
Van der Pol B.L.; 79; 301
Van der Zaan A.; 87; 286
Van Dijk D.; 248; 291
Van Heijenoort J.; 302
Van Nes P.; 87; 286
Van Sigt W.P.; 268; 301
Vasano J.A.; 302
Venables A.J.; 183; 292

Verhulst P.-F.; 5; 21; 33; 80; 154; 155; 157; 158; 160; 301
Viète F.; 252
Volterra V.; 6; 72; 73; 76; 80; 81; 82; 84; 85; 87; 88; 89; 90; 92; 94; 96; 97; 101; 152; 186; 236; 241; 301
Von Haeseler F.; 164; 301
Von Neumann J.; 179
Voogd H.; 259; 296
Vulpiani A.; 112; 116; 146; 172; 194; 198; 301

Waldrop W.M.; 192; 216; 301
Walras L.; 3; 14; 46; 301
Wang Y.; 177; 300
Weick K.; 224; 301
Weidlich W.; 3; 243; 244; 301
Weierstrass K.; 110; 124; 253
West B.J.; 79; 292
Weyl H.; 254; 262; 268; 281; 301
Wigner E.P.; 196; 301
Wilder R.L.; 262; 301
Wilke C.; 295
Wilmott P.; 247; 301
Wilson A.G.; 5; 159; 183; 287; 295; 301
Wolf A.; 141; 144; 301; 302
Wolff R.P.H.; 350; 356
Wolfram S.; 37; 302
Wrigley N.; 259; 296

Xia Zh.; 278; 298

Yorke J.A.; 15; 53; 112; 120; 148; 171; 178; 285; 291; 292; 293; 295

Zadeh L.A.; 260; 302
Zanarini G.; 172; 188; 299; 302
Zeldovich Ya.B.; 54; 164; 302
Zeno of Elea; 253
Zermelo E.; 254; 302
Zhang W.-B.; 3; 241; 242; 243; 245; 302
Zurek W.H.; 228; 302

283036